Advances in Intelligent Systems and Computing

Volume 282

Series editor

Janusz Kacprzyk, Polish Academy of Sciences, Warsaw, Poland
e-mail: kacprzyk@ibspan.waw.pl

For further volumes:
http://www.springer.com/series/11156

About this Series

The series "Advances in Intelligent Systems and Computing" contains publications on theory, applications, and design methods of Intelligent Systems and Intelligent Computing. Virtually all disciplines such as engineering, natural sciences, computer and information science, ICT, economics, business, e-commerce, environment, healthcare, life science are covered. The list of topics spans all the areas of modern intelligent systems and computing.

The publications within "Advances in Intelligent Systems and Computing" are primarily textbooks and proceedings of important conferences, symposia and congresses. They cover significant recent developments in the field, both of a foundational and applicable character. An important characteristic feature of the series is the short publication time and world-wide distribution. This permits a rapid and broad dissemination of research results.

Advisory Board

Chairman

Nikhil R. Pal, Indian Statistical Institute, Kolkata, India
e-mail: nikhil@isical.ac.in

Members

Rafael Bello, Universidad Central "Marta Abreu" de Las Villas, Santa Clara, Cuba
e-mail: rbellop@uclv.edu.cu

Emilio S. Corchado, University of Salamanca, Salamanca, Spain
e-mail: escorchado@usal.es

Hani Hagras, University of Essex, Colchester, UK
e-mail: hani@essex.ac.uk

László T. Kóczy, Széchenyi István University, Győr, Hungary
e-mail: koczy@sze.hu

Vladik Kreinovich, University of Texas at El Paso, El Paso, USA
e-mail: vladik@utep.edu

Chin-Teng Lin, National Chiao Tung University, Hsinchu, Taiwan
e-mail: ctlin@mail.nctu.edu.tw

Jie Lu, University of Technology, Sydney, Australia
e-mail: Jie.Lu@uts.edu.au

Patricia Melin, Tijuana Institute of Technology, Tijuana, Mexico
e-mail: epmelin@hafsamx.org

Nadia Nedjah, State University of Rio de Janeiro, Rio de Janeiro, Brazil
e-mail: nadia@eng.uerj.br

Ngoc Thanh Nguyen, Wroclaw University of Technology, Wroclaw, Poland
e-mail: Ngoc-Thanh.Nguyen@pwr.edu.pl

Jun Wang, The Chinese University of Hong Kong, Shatin, Hong Kong
e-mail: jwang@mae.cuhk.edu.hk

Tien Van Do · Hoai An Le Thi
Ngoc Thanh Nguyen
Editors

Advanced Computational Methods for Knowledge Engineering

Proceedings of the 2nd International
Conference on Computer Science,
Applied Mathematics and Applications
(ICCSAMA 2014)

 Springer

Editors
Tien Van Do
Department of Networked Systems
    and Services
Budapest University of Technology
    and Economics
Budapest
Hungary

Ngoc Thanh Nguyen
Institute of Informatics
Wrocław University of Technology
Wrocław
Poland

Hoai An Le Thi
LITA, UFR MIM
Metz
France

ISSN 2194-5357   ISSN 2194-5365   (electronic)
ISBN 978-3-319-06568-7   ISBN 978-3-319-06569-4   (eBook)
DOI 10.1007/978-3-319-06569-4
Springer Cham Heidelberg New York Dordrecht London

Library of Congress Control Number: 2014936764

Printed on acid-free paper

Springer is part of Springer Science+Business Media (www.springer.com)

Preface

This volume contains papers presented at the 2nd *International Conference on Computer Science, Applied Mathematics and Applications* (ICCSAMA 2014) held on 8-9 May, 2014 in Budapest, Hungary. The conference is co-organized by Analysis, Design and Development of ICT systems (AddICT) Laboratory, Budapest University of Technology and Economics, Hungary, Division of Knowledge Management Systems, Wroclaw University of Technology, Poland, and Laboratory of Theoretical & Applied Computer Science, Lorraine University, France in cooperation with IEEE SMC Technical Committee on Computational Collective Intelligence.

The aim of ICCSAMA 2014 is to bring together leading academic scientists, researchers and scholars to discuss and share their newest results in the fields of Computer Science, Applied Mathematics and their applications. After the peer review process, 30 papers by authors from Algeria, Austria, Finland, France, Germany, Hungary, India, Israel, Japan, Republic of Korea, Poland, Slovakia, United Kingdom and Vietnam have been selected for including in this proceedings. The presentations of 30 have been partitioned into 7 sessions: *Advanced Optimization Methods and Their Applications, Queueing Models and Performance Evaluation, Software Development and Testing, Computational Methods for Mobile and Wireless Networks, Computational Methods for Knowledge Engineering, Logic Based Methods for Decision Making and Data Mining, and Nonlinear Systems and Applications.*

The clear message of the proceedings is that the potentials of computational methods for knowledge engineering and optimization algorithms are to be exploited, and this is an opportunity and a challenge for researchers. It is observed that the ICCSAMA 2013 and 2014 clearly generated a significant amount of interaction between members of both communities on Computer Science and Applied Mathematics. The intensive discussions have seeded future exciting development at the interface between computational methods, optimization and engineering.

The works included in this proceedings can be useful for researchers, Ph.D. and graduate students in Optimization Theory and Knowledge Engineering fields. It is the hope of the editors that readers can find many inspiring ideas and use them to their research. Many such challenges are suggested by particular approaches and models presented in the proceedings.

We would like to thank all authors, who contributed to the success of the conference and to this book. Special thanks go to the members of the Steering and Program Committees for their contributions to keeping the high quality of the selected papers. Cordial thanks are due to the Organizing Committee members for their efforts and the organizational work.

Finally, we cordially thank Springer for supports and publishing this volume.

We hope that ICCSAMA 2014 significantly contributes to the fulfilment of the academic excellence and leads to greater success of ICCSAMA events in the future.

May 2014   Tien Van Do
                                                               Hoai An Le Thi
                                                          Ngoc Thanh Nguyen

ICCSAMA 2013 Organization

General Chair

Tien Van Do   Budapest University of Technology and
Economics, Hungary

General Co-Chairs

Le Thi Hoai An   Lorraine University, France
Nguyen Ngoc Thanh   Wroclaw University of Technology, Poland

Program Chairs

Pham Dinh Tao   INSA Rouen, France
Nguyen Hung Son   Warsaw University, Poland
Tien Van Do   Budapest University of Technology and
Economics, Hungary
Duc Truong Pham   University of Birmingham, UK

Doctoral Track Chair

Nguyen Anh Linh   Warsaw University, Poland

Organizing Committee

Nam H. Do   Budapest University of Technology and
Economics, Hungary
Vu Thai Binh   Budapest University of Technology and
Economics, Hungary

Steering Committee

Le Thi Hoai An	Lorraine University, France (Co-chair)
Nguyen Ngoc Thanh	Wroclaw University of Technology, Poland (Co-chair)
Pham Dinh Tao	INSA Rouen, France
Nguyen Van Thoai	Trier University, Germany
Pham Duc Truong	University of Birmingham, UK
Nguyen Hung Son	Warsaw University, Poland
Alain Bui	Université de Versailles-St-Quentin-en-Yvelines, France
Nguyen Anh Linh	Warsaw University, Poland
Tien Van Do	Budapest University of Technology and Economics, Hungary
Tran Dinh Viet	Slovak Academy of Sciences, Slovakia

Program Committee

Bui Alain	Université de Versailles-St-Quentin-en-Yvelines, France
Nguyen Thanh Binh	International Institute for Applied Systems Analysis (IIASA), Austria
Ram Chakka	RGMCET, RGM Group of Institutions, Nandyal, India
Tien Van Do	Budapest University of Technology and Economics, Hungary
Nam H. Do	Budapest University of Technology and Economics, Hungary
Ha Quang Thuy	Vietnam National University, Vietnam
László Jereb	University of West Hungary
László Lengyel	Budapest University of Technology and Economics, Hungary
Le Chi Hieu	University of Greenwich, UK
Le Nguyen-Thinh	Humboldt Universität zu Berlin, Germany
Le Thi Hoai An	Lorraine University, France
Luong Marie	Université Paris 13, France
Ngo Van Sang	University of Rouen, France
Nguyen Anh Linh	Warsaw University, Poland
Nguyen Benjamin	University of Versailles Saint-Quentin-en-Yvelines, France
Nguyen Duc Cuong	International University VNU-HCM, Vietnam
Nguyen Hung Son	Warsaw University, Poland
Nguyen Ngoc Thanh	Wroclaw University of Technology, Poland

Nguyen Van Thoai	Trier University, Germany.
Nguyen Viet Hung	Laboratory of Computer Sciences Paris 6, France
Nguyen-Verger Mai K.	Cergy-Pontoise University, France
Pham Cong Duc	University of Pau and Pays de l'Adour, France
Thong Vinh Ta	INRIA, CITI/INSA-Lyon, France
Pham Dinh Tao	INSA Rouen, France
Pham Duc Truong	University of Birmingham, UK
Phan Duong Hieu	Université Paris 8, France
Tran Dinh Viet	Slovak Academy of Sciences, Slovakia
Truong Trong Tuong	Cergy-Pontoise University, France

Contents

Part III: Software Development and Testing

Part IV: Computational Methods for Knowledge Engineering

Part V: Logic Based Methods for Decision Making and Data Mining

Part VI: Nonlinear Systems and Applications

Part VII: Computational Methods for Mobile and Wireless Networks

Part I

Advanced Optimization Methods and Their Applications

A Collaborative Metaheuristic Optimization Scheme: Methodological Issues

Mohammed Yagouni[1] and Hoai An Le Thi[2]

[1] Laromad, USTHB, Alger, Algérie
mohammed.yagouni@univ-lorraine.fr
[2] Theoretical and Applied Computer Science Lab LITA EA 3097
Université de Lorraine, Ile du Saulcy, 57 045 Metz, France
hoai-an.le-thi@univ-lorraine.fr
http://lita.sciences.univ-metz.fr/ lethi/

Abstract. A so called MetaStorming scheme is proposed to solve hard Combinatorial Optimization Problems (COPs). It is an innovative parallel-distributed collaborative approach based on metaheuristics. The idea is inspired from brainstorming, an efficient meeting mode for collectively solving company's problems. Different metaheuristic algorithms are used parallely for collectively solving COPs. These algorithms collaborate by exchanging the best current solution obtained after each running cycle via an MPI (Message Passing Interface) library. Several collaborative ways can be investigated in the generic scheme. As an illustrative example, we show how the MetaStorming works on an instance of the well known Traveling Salesman Problem (TSP).

Keywords: MetaStorming, Brainstorming, Heuristics, Metaheuristics, collaboration, Distributed Programming, Parallel Programming.

1 Introduction

Combinatorial Optimization problems (COPs) are widespread in the real world and are usually NP-Hard [19]. Metaheuristics are practical techniques which can provide, in a reasonable amount of time, good quality solutions for COPs [23]. Furthermore, as pointed out in [15], [16], [17], intensification and diversification properties are desirable and crucial for the performance of this kind of methods. The intensification is used whenever the exploration process reaches on promising areas, while the diversification is performed within sterile areas of the search space. These properties are unevenly present in meta-heuristic techniques. Various strategies have been proposed in the literature to reinforce intensification and/or diversification and to manage the trade-off between them. The main drawback of all these strategies usually lies in the increase of the search time without any substantial impact on both quality solutions and method robustness. Metaheuristics, as most mathematical methods used in various fields, have greatly benefited from the development of parallelism and parallel programming. Thank to these techniques, larger size instances of hard optimization problems can be solved in a reasonable search time. Many models of parallel meta-heuristics have

T.V. Do et al. (eds.), *Advanced Computational Methods for Knowledge Engineering,*
Advances in Intelligent Systems and Computing 282,
DOI: 10.1007/978-3-319-06569-4_1, © Springer International Publishing Switzerland 2014

been developed on parallel architectures. For a survey, the reader may refer, e.g., to [1], [4], [14], [16] and the references therein. More recently, further generalizations on hybrid and parallel meta-heuristics have been suggested [13], [15], [31], [32]. Today, the steady evolution, the variety and the availability of the computing resources allow the meta-heuristics users to imagine, design and implement various strategies and techniques in order to enhance the effectiveness and the robustness of their approaches.

The present work deals with a new distributed and cooperative approach for solving COPs. This approach is based on a collaboration between several meta-heuristics (or heuristics). The idea is inspired from the brainstorming. Such a technique has been already demonstrated to be an efficient meeting mode for collectively solving company's problems. The proposed approach consists of implementing different heuristics and meta-heuristics on multi-computers platform (launched on multi-processors) which solve the same instance of a optimization problem. All the implemented heuristics and metaheuristics collaborate in order to provide a better solution. This collaboration is built on the basis of information exchange, following a message-passing model, where the MPI (Message Passing Interface) library is used. The implemented methods are among metaheuristics, namely, Simulated Annealing (SA), Variable Neighborhood Search (VNS), Genetic Algorithms (GA) and other heuristics such as the Nearest Neighbor (NN), K-Opt, Max-Regret (MaxR) [33].

The remain part of the paper is organized as follows. The Brainstorming technique is presented in Section 2, and Section 3 is devoted to its adaptation to the solution of COPs. In Section 4, we show, with an illustrative example, how our approach works on an instance of the well-known Traveling Salesman Problem. Finally, Section 5 concludes the paper.

2 Brainstorming

The word brainstorming and its technique were coined by Alex Osborn Faickney in 1935 and presented for the first time in 1948 in his book, "Your Creative Power" [24]. It was initially thought as a method of holding group meetings to find adequate solutions to advertising problems (spot making, label creating...) [24], [28]. Its success and effectiveness [3], [8], [9] are today recognized and rightly given a lasting and outstanding place in animation techniques used in management teams companies. Thus this technique is taught today in Business and Management schools [2].

2.1 Definitions and Basic Concepts

The original words used by the inventor of brainstorming are *the storming of a problem in a commando fashion* [8], [24]. Since then many different definitions have been proposed in the literature. The method has been adopted in various areas of creation and collective decisions, and adapted as a flexible tool to successfully solve organization, administration as well as management problems of large companies in the world [3], [8], [9]. The formal brainstorming process is the exchange of ideas in the context that encourages individuals to exchange as many ideas as possible without given any importance to their relevance. The assumption is that through the free (uninhibited) exchange of diversified ideas, more interesting ideas will come out.

According to [8], [28], Brainstorming is defined as a meeting group for solving problems. In [3], the authors give another definition seen as a technique to provide solutions to a problem overlapping through a working group. On another hand, Brainstorming, as pointed out in [2], is a technique to solve collective problems appearing in Business. All these definitions share the idea that the resolution of a problem is done in a collective way, where a group of individuals cooperate and combine their efforts to find the best solution to the problem. To ensure the effectiveness of the method, the rules, as imposed by Osborn should be strictly followed [3], [24].

2.2   Brainstorming Rules

The rules as established by Osborn concern first the basic principles, next the team composition, and last the chronology of a brainstorming session [28]. Details are given in the following subsections.

Brainstorming Basic Principles. The important factor behind success and excellent results obtained by the brainstorming are mainly the ban on ideas criticism during the meeting. As imposed by Osborn, the basics to be observed are as follows :

- Principle of the tabula rasa = leave out prejudices, taboos, ...
- Each participant has to give as many ideas as possible;
- Any criticism of participant's ideas, by others is prohibited;
- Any criticism or justification of a personally suggested idea is prohibited;
- All ideas must be openly expressed (even the erroneous ones as they can lead to a more convincing and relevant one);
- Each participant may make use of others ideas, thus triggering off news and more convincing ones.

Composition and Team Organization. A moderator leads a brainstorming session. He should not be the supervisor of any member of the group (most of time an external consultant). The group must be composed of at least three participants of a heterogeneous competence according to the studied problem. During the brainstorming session, a "recording device" of all the mentioned ideas is necessary.

For more details, the reader may refer to the seminal books published by Clark [9] and Goguelin et al [3].

Main Stages of a Brainstorming Session. The conduct of a brainstorming session develops in three main stages:

Stage 1 - *Brainstorming Launching*: the moderator recalls the brainstorming operating rules (previously defined laws) and then presents the problem under question form.

Stage 2 - *Production of ideas*: the participants present their own ideas in turn, freely and without any predefined order. The expressed ideas are recorded in the memory device. At this stage, participant's idea may be directed, created, completed, or combined with another already expressed idea, or about to be expressed by another participant.

Stage 3 - *Exploitation of produced ideas*: all the ideas recorded during the previous stage, are revised in order to put together, to restate those considered as unclear, and to eliminate those, which are infeasible. Afterward, it would be possible to value those ideas and sort them out, and then choose the best solution. Several evaluation processes can be adopted at this stage (to call upon a vote, marking the ideas) according to the problem and its context.

The last brainstorming session stage suggests us to use this idea for solving COPs.

3 Adaptation of the Brainstorming Technique to Combinatorial Optimization Problems

The strong analogy existing between the conduct of a brainstorming session and the optimization process has facilitated the adaptation and implementation of this approach for COPs. The implementation of this system is based on an accurate brainstorming conduct rules. We realized this adaptation in two complementary levels:

Level 1: *Composition of the System*

- Moderator
- Participants (assistants)
- Recording device

Level 2: *Conduct of the brainstorming session*

- Each participant is asked to emit his ideas;
- All ideas should be expressed;
- Each participant is encouraged to copy and combine already expressed ideas.

Table 1 summarizes the correspondences between elements in brainstorming and optimization process.

Table 1. Analogy between brainstorming and optimization process

Brainstorming	Optimization
Problem solving by brainstorming	An instance of a COP
Idea in a meeting	Solution (feasible, partial, complete)
Evaluation of an idea	Evaluation of a solution *(objective function)*
Moderator	Sorting Algorithm
Participants (Assistants)	Resolution methods
Recording device	Adaptative Memory
Combination and association of ideas	Combination of solutions

3.1   MetaStorming: Brainstorming of Metaheuristics for COPs

The brainstorming approach proposed for COPs is a general framework for collaboration and cooperation between methods [1] in order to solve collectively a given COP. The basic principle is based on the real brainstorming. The collaboration aspects are achieved through asynchronous exchanges of information (solutions). These solutions are "filed" and made available to the participating methods via a device called "Moderator-Memory (MM)". MM is a mechanism which allows the storage and the exchanges of information between the different methods.

The implemented methods, which consist mainly of heuristics and metaheuristics such as Simulated Annealing (SA), Genetic Algorithms (GA), Variable Neighborhood Search (VNS), Nearest Neighbor (NN), Maximum Regret (Max_R) and K-Opt of LinKernighan (LK), collaborate by exchanging information via an MPI library (Message Passing Interface) [18], [29], as illustrated by Figure 1.

3.2   Two Distributed Implementations

The implementation of MetaStorming can be realized using a parallel architecture with distributed processors . This platform can be modeled by a complete graph (Figure 2), where the number of vertices represents the number of independent processors. Each of these processors implements a particular heuristic or metaheuristic method. Edges in this graph represent the links of communications between any pair of processors.

The choice of "Participating methods" to be implemented is important. To exploit the efficiency of all methods, this choice should be driven by the diversity of their nature, including the various types of approximate methods, such as *constructive* and *improvement heuristics*; *Single-Solution based metaheuristics* and *Population-Based metaheuristics*. To evaluate the efficiency of the proposed

---

[1] Meta-Storming: we use mainly heuristics and metaheuristics methods

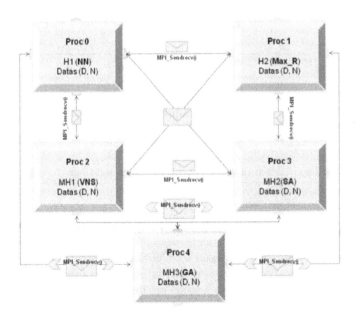

Fig. 1. Metaheuristics collaboration with exchanging message: a distributed implementation

approach, the well-know and benchmark problem of COPs, TSP (Travelling Salesman Problem), should be tested.

We propose here two ways to implement the system. The first consists of a *concurrent distributed implementation*, where independent searches are executed by various methods (Figure 3). All the searches are executed independently and the best solution is chosen at the end. Such a system is not really a MetaStorming approach since there is no cooperation.

The second is a *distributed collaborative implementation* (Figure 4) which corresponds to our MetaStorming approach presented above. Here the collaboration is implemented by information exchanges between the various process.

The aspects of collaboration and cooperation between the different methods, according to the brainstorming rules, are achieved through asynchronous exchange of information (solutions) using the routines of the MPI (Message Passing Interface) library. Solutions are shared via the "Moderator-Memory (MM)" in charge of "recording" and "broadcast" of all proposed solutions by the various participants. In such a implementation scheme, a process sends data to MM and takes the external information from MM.

This method achieves two main objectives. The first one concerns the optimization of the communication by avoiding, all the complex operations related to the diffusion and mechanisms selection of process senders, receivers, and the

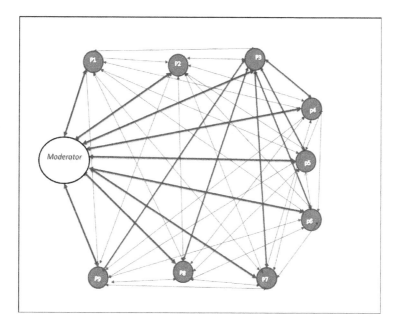

Fig. 2. Modeling of MetaStorming on complete processors network

control of senders's/receivers's informations. The second is to ensure a faithful implementation of the general brainstorming pattern, where, a participant idea will be immediately recorded without judgment nor justification (basic principle of brainstorming) and each participant consults all ideas emitted by others on the same support.

Bouthillier et al. [23] and Crainic et al. [15], specified that the cooperation between methods can be totally characterized by specifying the information which must be shared, the moment when the communications occur, and the end of these communications (say, the moment where every process received the information). In our MetaStorming approach, feasible solutions of various participating methods, - SA, GA, VNS, Max_R, NN, LK-, are deposited in MM and constitute the information to share. Each of the process, according to the corresponding implemented metaheuristic, takes a solution or a set of solutions, in order to integrate it (them), either as a solution of initialization or reset (RS, VNS), or as solution(s) of diversification (GA, LK).

4   Illustrative Example: A MetaStorming Scheme for Solving Traveling Salesman Problem (TSP)

As an illustrative example, we show below how the MetaStorming scheme works on an instance of the well known TSP which is stated as follows: a salesperson must visit n cities, passing through each city only once, beginning from one of

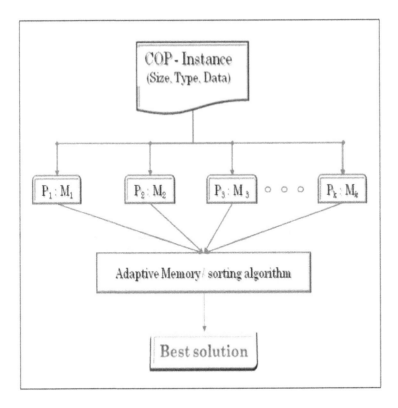

Fig. 3. Concurrent Distributed Implementation

the city that is considered as a base or starting city and returns to it. The cost
of the transportation among the cities is given. The problem is to find the order
of visiting the cities in such a way that the total cost is the minimum. Hence
an instance of TSP is defined by n, the number of cities and $D = (d_{ij})$, the
distance matrix between the cities. In our MetaStorming scheme, the participating algorithms implemented for TSP are Simulated Annealing (SA), Genetic
Algorithms (GA), Variable Neighborhood Search (VNS). By the limitation of
the length of the paper, we omit here a detailed description of these algorithms.
The MetaStorming is composed of three phases. In the first phase we find an initial solution (resp. an initial population) of SA and VNS by using, respectively,
Nearest Neighbor (NN) and Maximum Regret (MaxR) (resp. of GA by using
NN and MaxR). Starting from any city, the basic of the heuristic NN is to build
a solution by integrating, at each step, the city (which is not yet visited) closest
to the last visited city. MaxR is a heuristic that builds tours with maximum
regret (which is computed via the regret of each edge in the network). After this
initialization phase, n tours are obtained ($n/2$ by NN and $n/2$ by MaxR).

    The main phase of MetaStorming is the second which is a collaborative and
exchange information iterative procedure. At each running cycle we apply paral-

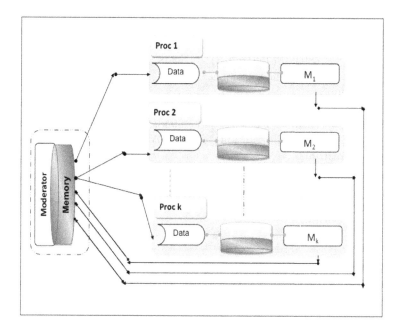

Fig. 4. Distributed Collaborative Implementation

lely all algorithms to get an good solution (in SA and VNS) or a good population of solutions (in GA), and then by exchanging information we take the best solution (or the best population in GA) to restart each algorithm in the next cycle (for GA, the best solution is added in the best current population). In the third phase the stopping criterion holds and the best solution is identified.

Below is an example with $n = 130$. The parameters are chosen as follows. In SA: the initial temperature is set to 1000000, the number of iterations per temperature level is $2n$, the factor of decreasing the temperature is 0.85. For VNS: the number of neighborhoods is $Kmax = 4$, the transformations used are 2-Opt and 3-Opt. For GA: the size of each population is n, the selection is elitist, the cross-over from two points of cuts and the probability to cross-over is 0.5 and the one to mutation is 0.01.

Let us denote, respectively, $T\#(SA)$, $T\#(VNS)$, $T\#(GA)$ the best solution given by SA, VNS, and GA. Also, denote by, respectively, $Z(T)$ and $R(T)$ the total cost and the total regret of the solution T.

Phase 1: Initialization.

Applying NN and MaxR we get n initial solutions that constitute the initial population of GA, with $Z(T) \in [11241, 31372]$ in NN and $Z(T) \in [28340, 26000]$ and $R(T) \in [9633, 9700]$ in MaxR. The initial solution of SA (resp. VNS) has the cost $Z(T) = 11241$ (resp. the cost $Z(T) = 26000$ and $R(T) = 9700$).

Phase 2 : collaboration scheme

1st iteration :

$Z(T\#(SA)) = 8893; Z(T\#(VNS)) = 8808; Z(T\#(GA)) = 8079$. By exchanging information, the solution $Z(T\#(GA)) = 8079$ is used for restarting each algorithm.

2nd iteration :

$Z(T\#(SA)) = 7855; Z(T\#(VNS)) = 6948; Z(T\#(GA)) = 7008$. By exchanging information, the solution ; $Z(T\#(VNS)) = 6948$ is used for restarting each algorithm.

3th iteration :

$Z(T\#(SA)) = 6852$; $Z(T\#(VNS)) = 6546; Z(T\#(GA)) = 6623$. By exchanging information, the solution $Z(T\#(VNS)) = 6546$ is used for restarting each algorithm.

. . . .

5th iteration :

$Z(T\#(SA)) = 6269$; $Z(T\#(VNS)) = 6110$; $Z(T\#(GA)) = 6157$.
By exchanging information, the solution $Z(T\#(VNS)) = 6110$ is used for restarting each algorithm.

6th iteration

$Z(T\#(SA)) = 6110; Z(T\#(VNS)) = 6110; Z(T\#(GA)) = 6110$.

Phase 3 : terminate, the optimal solution is obtained with the cost 6110.

5   Conclusion

We have presented methodological issues for a new distributed collaborative and cooperative approach for solving combinatorial optimization problems. The idea of collaboration between metaheuristics is inspired from the brainstorming technique. We show how to adapt the main rules of brainstorming to solve COPs. In the next step we will implement this generic MetaStorming scheme on several benchmark COPs to study the performance of our algorithms in comparing with standard methods. Works for the TSP is in progress. Further, we will develop the same idea for continuous nonconvex optimization and multi-objective programming.

References

1. Alba, E. (ed.): Parallel metaheuristics: A new class of Algorithms. Wiley-Interscience, John Wiley et Sons, Hoboken, New Jersey (2008)
2. Bachelet, R.: Comment animer/organiser un brainstorming? Ecole centrale de Lille (2008)
3. Bize, P.R., Goguelin, P., Carpentier, R.: Le penser efficace, la Problémation Société dÕédition de lÕ Enseignement Supérieur (1967)
4. Blesa, M.J., Blum, C., Raidl, G., Roli, A., Sampels, M. (eds.): HM 2010. LNCS, vol. 6373. Springer, Heidelberg (2010)

5. Blesa, M.J., Blum, C., Raidl, G., Roli, A., Sampels, M. (eds.): HM 2010. LNCS, vol. 6373. Springer, Heidelberg (2010)
6. Blum, C., Roli, A.: Metaheuristics in combinatorial optimization: Overview and conceptual comparison. ACM Computing Surveys 35(3), 268–308 (2003)
7. Calvo, Á.-L., Cortés, A., Giménez, D., Pozuelo, C.: Using metaheuristics in a parallel computing course. In: Bubak, M., van Albada, G.D., Dongarra, J., Sloot, P.M.A. (eds.) ICCS 2008, Part II. LNCS, vol. 5102, pp. 659–668. Springer, Heidelberg (2008)
8. Clark, C.: Brainstorming: The dynamic new way to create successful ideas. Garden City, NY (1958)
9. Clark, C.: Brainstorming. Edition Dunod, Paris (1971)
10. Consoli, S., Dowman, K.D.: Combinatorial Optimization and Meta-heuristics. ORR-U Brun 47 (2006)
11. Cotta, C., Talbi, E.G., Alba, E.: Parallel hybrid metaheuristics. In: Parallel Metaheuristics, A New Class of Algorithms, pp. 347–370. John Wiley (2005)
12. Crainic, T.G., Gendreau, M., Hansen, P., Mladenovic, N.: Cooperative parallel VNS for the p-median. Journal of Heuristics 10(3), 293–314 (2004)
13. Crainic, T.G., Gendreau, M., Rousseau, L.M.: Special issue on recent advances in metaheuristics. J. Heuristics 16(3), 235–237 (2010)
14. Crainic, T.G., Toulouse, M.: Explicit and emergent cooperation schemes for search algorithms. In: Maniezzo, V., Battiti, R., Watson, J.-P. (eds.) LION 2007 II. LNCS, vol. 5313, pp. 95–109. Springer, Heidelberg (2008)
15. Crainic, T.G., Toulouse, M.: Parallel metaheuristics. In Cirrelt, 2009-22 (2009)
16. Dreo, J., Petrowski, A., Siarry, P., Taillard, P.: Meta-heuristics for Hard Optimization. Springer, Berlin (2006)
17. Hansen, P., Mladenović, N.: Variable Neighborhood Search: principles and Applications: Gerad Canada (2001)
18. High Performance Computing Center Stuttgart: MPI: Message Passing Interface Standard, Version 2.2. University of Stuttgart, Germany (2009)
19. Garey, M., Johnson, D.: Computers and Intractability: A Guide to the Theory of NP-Completeness. W.H. Freeman, San Francisco (1979)
20. Kirkpatrick, S., Gelatt, C., Vecchi, M.: Optimization by Simulated Annealing. Science 220(4598), 671–680 (1983)
21. Lavault, C.: Evaluation des algorithmes distribués, Analyse, complexité, méthodes. Edition Hermès, Paris (1995)
22. Lazarova, M., Borovska, P.: Comparaison of Parallel Meta-heuristics for Solving the TSP. In: CompSysTech 2008 (2008)
23. Le Bouthillier, A., Crainic, T.G.: A cooperative parallel metaheuristic for the vehicle routing problem with time windows. Computers & OR 32, 1685–1708 (2005)
24. Osborn, A.F.: Your Creative Power: How to Use Imagination to brighten life, to get ahead, ch. XXXIII, How To Organize a Squad To Create Ideas, pp. 265–274. Charles Scribner's Sons, New York (1948)
25. Osborn, A.F.: How to Think Up. McGraw-Hill Book Co., New York (1942)
26. Osborn, A.F.: Applied Imagination: Principles and Procedures of Creative Problem Solving. Charles Scribner's Sons, New York (1963)
27. Osborn, A., How, F.: to Think Up. New York London McGraw-Hill Book Co. (1942)
28. Osborn, A., F.: L'imagination constructive : Comment tirer partie de ses idées. Principes et processus de la Pensé créative et du Brainstorming. Dunod, Paris France (1971)

29. Pacheco, P.S.: A User's Guide MPI. DMSF USA (2009)
30. Raidl, G.: A Unified View on Hybrid metaheuristics. VU. Austria (2005)
31. Talbi, E.G.: metaheuristics: from design to implementation. John Wiley and Sons, Inc., Hoboken (2009)
32. Talbi, E.G.: Hybrid Metaheuristics. Springer London, Limited (2012)
33. Yagouni, M., Le Thi, A.H., Ait Haddadène, H.: Solving hard optimizationproblems with metaheuristics hybridization, case of MaxR-SA-VNS for TSP (subbmited)

DC Programming and DCA
for General DC Programs

Hoai An Le Thi[1], Van Ngai Huynh[2], and Tao Pham Dinh[3]

[1] Laboratory of Theorical and Applied computer Science LITA EA 3097,
University of Lorraine, Ile du Saulcy-Metz 57045, France
[2] Department of Mathematics, University of Quynhon,
170 An Duong Vuong, Quy Nhon, Vietnam
[3] Laboratory of Mathematics, INSA - Rouen, University of Normandie
76801 Saint Etienne du Rouvray, France
hoai-an.le-thi@univ-lorraine.fr,
ngaivn@yahoo.com,
pham@insa-rouen.fr

Abstract. We present a natural extension of DC programming and DCA for modeling and solving general DC programs with DC constraints. Two resulting approaches consist in reformulating those programs as standard DC programs in order to use standard DCAs for their solutions. The first one is based on penalty techniques in DC programming, while the second linearizes concave functions in DC constraints to build convex inner approximations of the feasible set. They are proved to converge to KKT points of general DC programs under usual constraints qualifications. Both designed algorithms can be viewed as a sequence of standard DCAs with updated penalty (resp. relaxation) parameters.

Keywords: DC programming, DCA, DC constraints, subdifferential, nonsmooth penalty function, constraint qualification.

1 Introduction

DC (Difference of Convex functions) Programming and DCA (DC Algorithms), which constitute the backbone of nonconvex programming and global optimization, are introduced in 1985 by Pham Dinh Tao in the preliminary state, and extensively developed by Le Thi Hoai An and Pham Dinh Tao since 1994 to become now classic and increasingly popular ([2,3,5,8,10,8,1] and references quoted therein). *Their original key idea* relies on the structure DC of objective function and constraint functions in nonconvex programs which are explored and exploited in a deep and suitable way. *The resulting DCA introduces the nice and elegant concept of approximating a nonconvex (DC) program by a sequence of convex ones*: each iteration of DCA requires solution of a convex program.

Their popularity resides in their rich and deep and mathematical foundations, and the versality, flexibility, robustness, inexpensiveness and efficiency of DCA's compared to existing methods, their adaptation to specific structures of

T.V. Do et al. (eds.), *Advanced Computational Methods for Knowledge Engineering,*   15
Advances in Intelligent Systems and Computing 282,
DOI: 10.1007/978-3-319-06569-4_2, © Springer International Publishing Switzerland 2014

addressed problems and their ability to solve real-world large-scale nonconvex programs. Recent developments in convex programming are mainly devoted to reformulation techniques and scalable algorithms in order to handle large-scale problems. Obviously, they allow for enhancement of DC programming and DCA in high dimensional nonconvex programming.

A standard DC program is of the form (with the usual convention $(+\infty) - (+\infty) = +\infty$)

$$\alpha = \inf\{f(x) := g(x) - h(x) \ : \ x \in \mathbb{R}^n\} \quad (P_{dc})$$

where $g, h \in \Gamma_0(\mathbb{R}^n)$, the convex cone of all the lower semicontinuous proper (*i.e.*, not identically equal to $+\infty$) convex functions defined on $\mathbb{R}^n$ and taking values in $\mathbb{R} \cup \{+\infty\}$. Such a function f is called a DC function, and $g - h$, a DC decomposition of f while the convex functions g and h are DC components of f. Note that the finiteness of the optimal value of α implies that dom $g \subset$ dom h where the effective domain of a real-extended value $\theta : \mathbb{R}^n \to \mathbb{R} \cup \{+\infty\}$ is dom $\theta := \{x \in \mathbb{R}^n : \theta(x) < +\infty\}$. A standard DC program with a convex constraint C (a nonempty closed convex set in $\mathbb{R}^n$)

$$\alpha = \inf\{f(x) := g(x) - h(x) \ : \ x \in C\} \tag{1}$$

can be expressed in the form of (P_{dc}) by adding the indicator function χ_C of C ($\chi_C(x) = 0$ if $x \in C$, $+\infty$ otherwise) to the function g. The vector space of DC functions, $DC(\mathbb{R}^n) = \Gamma_0(\mathbb{R}^n) - \Gamma_0(\mathbb{R}^n)$, forms a wide class encompassing most real-life objective functions and is closed with respect to usual operations in optimization. DC programming constitutes so an extension of convex programming, sufficiently large to cover most nonconvex programs (([2,3,5,8,10,8,1] and references quoted therein), but not too in order to leverage the powerful arsenal of the latter.

DC duality associates a primal DC program with its dual, which is also a DC program

$$\alpha = \inf\{h^*(y) - g^*(y) : y \in \mathbb{R}^n\} \quad (D_{dc})$$

by using the fact that every function $\varphi \in \Gamma_0(\mathbb{R}^n)$ is characterized as a pointwise supremum of a collection of affine functions, say

$$\varphi(x) = \sup\{\langle x, y \rangle - \varphi^*(y) : y \in \mathbb{R}^n\}, \ \forall x \in \mathbb{R}^n,$$

where φ^* defined by

$$\varphi^*(y) := \sup\{\langle x, y \rangle - \varphi(x) \ : \ x \in \mathbb{R}^n\}, \ \forall x \in \mathbb{R}^n$$

is the conjugate of φ. There is a perfect symmetry between (P_{dc}) and its dual (D_{dc}): the dual of (D_{dc}) is exactly (P_{dc}).

The subdifferential of φ at x_0, denoted by $\partial\varphi(x_0)$, is defined by

$$\partial\varphi(x_0) := \{y_0 \in \mathbb{R}^n : \varphi(x) \geq \varphi(x_0) + \langle x - x_0, y_0 \rangle, \forall x \in \mathbb{R}^n\}$$

The subdifferential $\partial\varphi(x_0)$ - which is a closed convex set - is an extension of the derivative notion for convex functions. An element $y_0 \in \partial\varphi(x_0)$ is called subgradient of φ in x_0. The convex function φ is differentiable at x_0 iff $\partial\varphi(x_0)$ is reduced to a singleton and one has $\partial\varphi(x_0) = \{\nabla\varphi(x_0)\}$. For dom $\partial\varphi := \{x \in \mathbb{R}^n : \partial\varphi(x) \neq \emptyset\}$ there holds (ri C stands for the relative interior of the convex set $C \subset \mathbb{R}^n$)

$$ri \ (\text{dom} \ \varphi) \subset \text{dom} \ \partial\varphi \subset \text{dom} \ \varphi$$

The function φ is called $\rho-$convex, (for some $\rho \geq 0$) on a convex set C, if for all $x, y \in C$, $\lambda \in [0, 1]$ one has

$$\varphi(\lambda x + (1 - \lambda)y) \leq \lambda\varphi(x) + (1 - \lambda)\varphi(y) - \frac{\rho}{2}\lambda(1 - \lambda)\|x - y\|^2.$$

The supremum of all $\rho \geq 0$ such that the above inequality is verified is denoted by $\rho(\varphi, C)$.It is called the strong convexity modulus of φ on C if $\rho(\varphi, C) > 0$. If $C = \mathbb{R}^n$, $\rho(\varphi, C)$ is simply written as $\rho(\varphi)$. The effective domain of a real-extended value $\theta : \mathbb{R}^n \to \mathbb{R} \cup \{+\infty\}$ is dom $\theta := \{x \in \mathbb{R}^n : \theta(x) < +\infty\}$. The function φ is said to be polyhedral convex if it is the sum of the indicator of a nonempty polyhedral convex set and the pointwise supremum of a finite collection of affine functions.

Polyhedral DC program is a DC program in which at least one of the functions g and h is polyhedral convex. Polyhedral DC programming, which plays a central role in nonconvex optimization and global optimization and is the foundation of DC programming and DCA, has interesting properties (from both a theoretical and an algorithmic point of view) on local optimality conditions and the finiteness of DCA's convergence.

DC programming investigates the structure of $DC(\mathbb{R}^n)$, DC duality and local and global optimality conditions for DC programs. The complexity of DC programs clearly lies in the distinction between local and global solution and, consequently; the lack of verifiable global optimality conditions.

We have developed necessary local optimality conditions for the primal DC program (P_{dc}), by symmetry those relating to dual DC program (D_{dc}) are trivially deduced

$$\partial h(x^*) \cap \partial g(x^*) \neq \emptyset \tag{2}$$

(such a point x^* is called critical point of $g - h$ or (2) a generalized Karusk-Kuhn-Tucker (KKT) condition for (P_{dc})), and

$$\emptyset \neq \partial h(x^*) \subset \partial g(x^*). \tag{3}$$

The condition (3) is also sufficient (for local optimality) in many important classes of DC programs. In particular it is sufficient for the next cases quite often encountered in practice:

- In polyhedral DC programs with h being a polyhedral convex function. In this case, if h is differentiable at a critical point x^*, then x^* is actually a local minimizer for (P_{dc}). Since a convex function is differentiable everywhere except for a set of measure zero, one can say that a critical point x^* is almost always a local minimizer for (P_{dc}).

- In case the function f is locally convex at x^*. Note that, if h is polyhedral convex, then $f = g - h$ is locally convex everywhere h is differentiable.

The transportation of global solutions between (P_{dc}) and (D_{dc}) is expressed by:

$$[\bigcup_{y^* \in \mathcal{D}} \partial g^*(y^*)] \subset \mathcal{P} \, , \, [\bigcup_{x^* \in \mathcal{P}} \partial h(x^*)] \subset \mathcal{D} \qquad (4)$$

where $\mathcal{P}$ and $\mathcal{D}$ denote the solution sets of (P_{dc}) and (D_{dc}) respectively. The first (second) inclusion becomes equality if the function h (resp. g^*) is subdifferentiable on $\mathcal{P}$ (resp. $\mathcal{D}$). They show that solving a DC program implies solving its dual. Note also that, under technical conditions, this transportation also holds for local solutions of (P_{dc}) and (D_{dc}).([2,3,5,8,10,8,1] and references quoted therein).

Based on local optimality conditions and duality in DC programming, the DCA consists in constructing of two sequences $\{x^k\}$ and $\{y^k\}$ of trial solutions of the primal and dual programs respectively, such that the sequences $\{g(x^k) - h(x^k)\}$ and $\{h^*(y^k) - g^*(y^k)\}$ are decreasing, and $\{x^k\}$ (resp. $\{y^k\}$) converges to a primal feasible solution x^* (resp. a dual feasible solution y^*) satisfying local optimality conditions and

$$x^* \in \partial g^*(y^*), \quad y^* \in \partial h(x^*). \qquad (5)$$

The sequences $\{x^k\}$ and $\{y^k\}$ are determined in the way that x^{k+1} (resp. y^{k+1}) is a solution to the convex program (P_k) (resp. (D_{k+1})) defined by ($x^0 \in$ dom ∂h being a given initial point and $y^0 \in \partial h(x^0)$ being chosen)

$$(P_k) \quad \inf\{g(x) - [h(x^k) + \langle x - x^k, y^k \rangle] : x \in \mathbb{R}^n\}, \qquad (6)$$

$$(D_{k+1}) \quad \inf\{h^*(y) - [g^*(y^k) + \langle y - y^k, x^{k+1} \rangle] : y \in \mathbb{R}^n\}. \qquad (7)$$

The DCA has the quite simple interpretation: at the k-th iteration, one replaces in the primal DC program (P_{dc}) the second component h by its affine minorization $h^{(k)}(x) := h(x^k) + \langle x - x^k, y^k \rangle$ defined by a subgradient y^k of h at x^k to give birth to the primal convex program (P_k), the solution of which is nothing but $\partial g^*(y^k)$. Dually, a solution x^{k+1} of (P_k) is then used to define the dual convex program (D_{k+1}) obtained from (D_{dc}) by replacing the second DC component g^* with its affine minorization $(g^*)^{(k)}(y) := g^*(y^k) + \langle y - y^k, x^{k+1} \rangle$ defined by the subgradient x^{k+1} of g^* at y^k : the solution set of (D_{k+1}) is exactly $\partial h(x^{k+1})$. The process is repeated until convergence. DCA performs a double linearization with the help of the subgradients of h and g^* and the DCA then yields the next scheme: (starting from given $x^0 \in$ dom ∂h)

$$y^k \in \partial h(x^k); \quad x^{k+1} \in \partial g^*(y^k), \quad \forall k \geq 0. \qquad (8)$$

DCA's convergence properties:

DCA is a descent method without line search, but with global convergence, which enjoys the following properties: (C and D are two convex sets in $\mathbb{R}^n$, containing the sequences $\{x^k\}$ and $\{y^k\}$ respectively).

i) The sequences $\{g(x^k) - h(x^k)\}$ and $\{h^*(y^k) - g^*(y^k)\}$ are decreasing and

- • $g(x^{k+1}) - h(x^{k+1}) = g(x^k) - h(x^k)$ iff $y^k \in \partial g(x^k) \cap \partial h(x^k)$, $y^k \in \partial g(x^{k+1}) \cap \partial h(x^{k+1})$ and $[\rho(g, C) + \rho(h, C)] \|x^{k+1} - x^k\| = 0$. Moreover if g or h are strictly convex on C then $x^k = x^{k+1}$.
  In such a case DCA terminates at the kth iteration (finite convergence of DCA)

- • $h^*(y^{k+1}) - g^*(y^{k+1}) = h^*(y^k) - g^*(y^k)$ iff $x^{k+1} \in \partial g^*(y^k) \cap \partial h^*(y^k)$, $x^{k+1} \in \partial g^*(y^{k+1}) \cap \partial h^*(y^{k+1})$ and $[\rho(g^*, D) + \rho(h^*, D)] \|y^{k+1} - y^k\| = 0$. Moreover if g^* or h^* are strictly convex on D, then $y^{k+1} = y^k$.
  In such a case DCA terminates at the kth iteration (finite convergence of DCA).

ii) If $\rho(g, C) + \rho(h, C) > 0$ (resp. $\rho(g^*, D) + \rho(h^*, D) > 0$)) then the series $\{\|x^{k+1} - x^k\|^2$ (resp. $\{\|y^{k+1} - y^k\|^2\}$ converges.

iii) If the optimal value α of problem (P_{dc}) is finite and the infinite sequences $\{x^k\}$ and $\{y^k\}$ are bounded then

every limit point $\widetilde{x}$ (resp. $\widetilde{y}$) of the sequence $\{x^k\}$ (resp. $\{y^k\}$) is a critical point of $g - h$ (resp. $h^* - g^*$).

iv) DCA has a linear convergence for general DC programs.

v) DCA has a finite convergence for polyhedral DC programs.

vi) Convergence of the whole sequences $\{x^k\}$ and $\{y^k\}$ in DCA with its convergence rate for DC programming with subanalytic data [15].

DCA's distinctive feature relies upon the fact that DCA deals with the convex DC components g and h but not with the DC function f itself. DCA is one of the rare algorithms for nonconvex nonsmooth programming. Moreover, a DC function f has infinitely many DC decompositions which have crucial implications for the qualities (convergence speed, robustness, efficiency, globality of computed solutions,...) of DCA. For a given DC program, the choice of optimal DC decompositions is still open. Of course, this depends strongly on the very specific structure of the problem being considered. In order to tackle the large-scale setting, one tries in practice to choose g and h such that sequences $\{x^k\}$ and $\{y^k\}$ can be easily calculated, i.e., either they are in an explicit form or their computations are inexpensive. Very often in practice, the solution of (D_k) to compute the sequence $\{y^k\}$ is explicit because the calculation of a subgradient of h is explicitly obtained by using the usual rules for calculating subdifferential of convex functions. But the solution of the convex program (P_k), if not explicit, should be achieved by efficient algorithms well-adapted to its special structure, in order to handle the large-scale setting.

This very simple scheme (8) hides the extrapolating character of DCA. Indeed, we show that, at the limit, the primal (dual) solution x^* (resp. y^*) computed by

DCA is also a global solution of the DC program obtained from (P_{dc}) (resp. (D_{dc})) by replacing the function h (resp. g^*) with the supremum $\sup_{k \geq 1} h_k$ (resp. $\sup_{k \geq 1}(g^*)_k$ of all the affine minorizations h_k (resp. $(g^*)_k$) of h (resp. g^*) generated by DCA. These DC programs are closer to (P_{dc}) and (D_{dc}) than (P_k) and (D_k) respectively, because the function $\sup_{k \geq 1} h_k$ (resp. $\sup_{k \geq 1}(g^*)_k$) better approximates the function h (resp. g^*) than h_k (resp. $(g^*)_k$) . Moreover if $\sup_{k \geq 1} h_k$ (resp.$\sup_{k \geq 1}(g^*)_k$) coincides with h (resp. g^*) at an optimal solution of (P_{dc}) (resp. (D_{dc})), then x^* and y^* are also primal and dual optimal solutions respectively. These original and distinctive features explain in part the effective convergence of suitably customized DCA, with a reasonable choice of a starting point, towards global optimal solutions of DC programs. In practice, DCA quite often converges to global optimal solutions. The globality of DCA may be assessed either when the optimal values are known a priori, or through global optimization techniques, the most popular among them remains Branch-and-Bound (BB), [4,3,7,13,1]. Last but not least, note that with appropriate DC decompositions and suitably equivalent DC reformulations, DCA permits to recover most of standard methods in convex and nonconvex programming. Moreover, DCA is a global algorithm (i.e. providing global solutions) when applied to convex programs recast as DC programs.

In the spirit of unifying existing theoretical and algorithmic tools in nonconvex programming and global optimization and providing a deeper insight - which will help researchers and practitioners better model and efficiently solve nonconvex programming problems - of a very general and powerful theory based on the DC structure and the related convex approximation, it is convenient to point out that the four methods EM (Expectation-Maximization) by Dempster-Laird-Rubin [53], SLA (Succesive Linear Approximation) by Bradley-Mangasarian [51], ISTA (Iterative Shrinkage-Thresholding Algorithms) by Chambolle, DeVore, Lee, and Lucier [52], and CCCP (Convex-Concave Procedure) by Yuille-Rangarajan [73], better known, in a certain period, to data miners not aware of the state of the art in optimization, are special cases of DCA. Since then, this fact has been acknowledged by leading experts in the field in their publications.

DCA is a descent method without line search, (greatly appreciated in the large-scale setting), with global convergence (i.e., from an arbitrary starting point), which is successfully applied to lots of nonconvex optimization problems in many fields of Applied Sciences: Transport Logistics, Telecommunications, Genomics, Finance, Data Mining-Machine Learning,Cryptology, Computational Biology, Computational Chemistry, Combinatorial Optimization, Mechanics, Image Processing, Robotics & Computer Vision, Petrochemicals, Optimal Control and Automatic, Inverse Problems and Ill-posed Problems, Multiobjective Programming, Game Theory, Variational Inequalities Problems (VIP), Mathematical Programming with Equilibrium Constraints (MPEC), to cite but a few(see [1] and references therein).

It is certain that developments of nonconvex programming and global optimization via DC Programming and DCA for modeling and solving real-world nonconvex optimization problems (in many branches of applied sciences) will intensify yet in the years to come and for long, because Nonconvex Programming and Global Optimization are endless. Therefore a deep mastery of these theoretical and algorithmic tools would, no doubt, be a major asset in achieving the goal-oriented research with potential transfer of the state of the art technology in nonconvex programming and global optimization: it will improve the efficiency and the scalability of DC programming and DCA and consolidates their key role in these fields.

Our present work is concerned with the extension of DC programming and DCA to solve general DC programs with DC constraints (as opposed to standard ones (P_{dc}) with only convex constraints)

$$(P_{gdc}) \qquad \min \quad f_0(x) \qquad \text{subject to} \qquad \begin{cases} f_i(x) \leq 0, & i = 1, ..., m \\ x \in C, \end{cases}$$

where $C \subseteq \mathbb{R}^n$ is a nonempty closed convex set; $f, f_i : \mathbb{R}^n \to \mathbb{R}$ $(i = 0, 1, ..., s)$ are DC functions. This class of nonconvex programs is the most general in DC Programming and, a fortiori, more difficult to treat than that (standard) DC programs (P_{dc}) because of the nonconvexity of the constraints. It is not new and has been addressed in [2]. Its renewed interests is due to the fact that this class appears, increasingly, in many models of nonconvex variational approaches. That is the reason why we present in this paper the extended DCAs for solving (P_{gdc}) that we introduced in the research report [9] not yet published to date. There are two approaches for modeling and solving (P_{gdc}). Both consist in iteratively approximating (P_{gdc}) by a sequence of convex programs, according to the philosophy of DC programming and DCA. The first one is based on penalty techniques in DC programming while the second approach linearizes concave parts in DC constraints to build convex inner approximations of the feasible set. These algorithms can be viewed as a sequence of standard DCAs with updated penalty (resp. relaxation) parameters.

Let us recall some notions from Convex Analysis and Nonsmooth Analysis, which will be needed thereafter (see [56], [68], [70]). In the sequel, the space $\mathbb{R}^n$ is equipped with the canonical inner product $\langle \cdot \rangle$. Its dual space is identified with $\mathbb{R}^n$ itself. $\mathcal{S}(\mathbb{R}^n)$ denotes the set of lower semicontinuous functions $f : \mathbb{R}^n \to \mathbb{R} \cup \{+\infty\}$, while $\Gamma_0(\mathbb{R}^n)$ stands for the set of proper lower semicontinuous convex functions $f : \mathbb{R}^n \to \mathbb{R} \cup \{+\infty\}$. Recall that a function $f : \mathbb{R}^n \to \mathbb{R} \cup \{+\infty\}$ is said to be a DC function if there exists $g, h \in \Gamma_0(\mathbb{R}^n)$ such that $f = g - h$, where, the convention $+\infty - (+\infty) = +\infty$ is used. The open ball with the center $x \in \mathbb{R}^n$ and radius $\varepsilon > 0$ is denoted by $B(x, \varepsilon)$; while the unit ball (i.e., the ball with the center at the origin and unit radius) is denoted by B.

Let $f : \mathbb{R}^n \to \mathbb{R} \cup \{+\infty\}$ be a locally Lipschitz function at a given $x \in \mathbb{R}^n$. The *Clarke directional derivative* and the *Clarke subdifferential* of f at x is given by the following formulas.

$$f^\uparrow(x, v) := \limsup_{(t,y) \to (0^+, x)} \frac{f(y + tv) - f(y)}{t}.$$

$$\partial f(x) := \left\{ x^* \in \mathbb{R}^n : \langle x^*, v \rangle \leq f^\uparrow(x, v) \, \forall v \in \mathbb{R}^n \right\}.$$

If f is continuously differentiable at x then $\partial^{\uparrow} f(x) = \nabla f(x)$ (the Fréchet derivative of f at x). When f is a convex function, then $\partial f(x)$ coincides with the subdifferential in the sense of Convex Analysis, i.e.,

$$\partial f(x) = \{y \in \mathbb{R}^n : \ \langle y, d \rangle \le f(x + d) - f(x), \ \ \forall d \in \mathbb{R}^n\}.$$

We list the following calculus rules for the Clarke subdifferential which are needed thereafter.

1. For given two locally Lipschitz functions f, g at a given $x \in \mathbb{R}^n$, one has

$$\partial(-f)(x) = -\partial f(x); \quad \partial(f + g)(x) \subseteq \partial f(x) + \partial g(x), \tag{9}$$

and, the equality in the latter inclusion holds if f is continuously differentiable at x.

2. For given functions $f_i : \mathbb{R}^n \to \mathbb{R}$ $(i = 1, ..., m)$, let $f : \mathbb{R}^n \to \mathbb{R}$ be the max-type function defined by

$$f(x) := \{f_i(x) : \ i = 1, ..., m\}, \ \ x \in \mathbb{R}^n.$$

Then one has

$$\partial f(x) \subseteq \left\{ \sum_{i=1}^{m} \lambda_i \partial f_i(x) : \ \ \lambda_i \ge 0, \ \sum_{i=1}^{m} \lambda_i = 1; \ \lambda_i = 0 \text{ if } f_i(x) < f(x) \right\}, \tag{10}$$

and, the equality holds if $f_i : \mathbb{R}^n \to \mathbb{R}$ $(i = 1, ..., m)$ are continuously differentiable at $x \in \mathbb{R}^n$.

Let C be a nonempty closed subset of $\mathbb{R}^n$. The indicator function χ_C of C is defined by $\chi_C(x) = 0$ if $x \in C$, $+\infty$ otherwise. For a closed subset C of $\mathbb{R}^n$, the normal cone of C at $x \in C$, denoted $N(C, x)$, is defined by

$$N(C, x) := \partial \chi_C(x) = \{u \in \mathbb{R}^n : \ \langle u, y - x \rangle \le 0 \ \forall y \in C\}.$$

Before investigating the extension of DC programming and DCA for solving general DC programs with DC constraints, let us summarize the standard DCA for solving standard DC programs outlined above.

DC Algorithm (DCA)

1. Let $x^1 \in \text{dom } \partial h$. Set $k = 1$ and let ϵ be a sufficiently small positive number.
2. Compute $y^k \in \partial h(x^k)$.
3. Compute $x^{k+1} \in \partial g^*(y^k)$; i.e., x^{k+1} is a solution of the convex program

$$\min\{g(x) - \langle x, y^k \rangle : \ x \in \mathbb{R}^n\} \tag{11}$$

4. If stopping criterion is met, then stop and we have x^{k+1} is the computed solution, otherwise, set $k = k + 1$ and go to Step 2.

Output x^k and $f(x^k)$ as the best known solution and objective function value.

2   DC Algorithm Using l_∞-penalty Function with Updated Parameter: DCA1

General DC programs with DC constraints are of the form

$$(P_{gdc}) \qquad \min \quad f_0(x) \qquad \text{subject to} \quad \begin{cases} f_i(x) \le 0, \quad i = 1, ..., m \\ x \in C, \end{cases} \qquad (12)$$

where $C \subseteq \mathbb{R}^n$ is a nonempty closed convex set $f_i : \mathbb{R}^n \to \mathbb{R} \cap \{+\infty\}$, $i = 0, 1, ..., m$ are DC functions. Denote by $F = \{x \in C : f_i(x) \le 0, \ i = 1, ..., m\}$ the feasible set of (P_{gdc})-(12), and

$$p(x) := \max\{f_1(x), ..., f_m(x)\}; \quad I(x) := \{i \in \{1, ..., m\} : f_i(x) = p(x)\};$$
$$p^+(x) = \max\{p(x), 0\}.$$

We consider the following penalty problems

$$\begin{cases} \text{minimize} \quad \varphi_k(x) := f_0(x) + \beta_k p^+(x) \\ \text{subject to} \quad x \in C, \end{cases} \qquad (13)$$

where, β_k are penalty parameters. Exact penalty relative to DC function p^+ is said to be exact in (13) if there exists $\beta \ge 0$ such that for every $\beta_k > \beta$ (P_{gdc}) and (13) are equivalent in the sense that they have the same optimal value and the same solution set.

Remark. p^+ is a DC function on C whenever f_i $(i = 1, ..., m)$ are DC functions on C. Note that if $f_i := g_i - h_i$, $i = 1, ..., m$ with g_i, h_i are finite convex functions on C, then p is a DC function on C with the standard DC decomposition (see, e.g., [2])

$$p(x) = g(x) - h(x),$$

here,

$$g(x) := \max_{i=1,...,m} \left\{ g_i(x) + \sum_{j=1, j \ne i}^{m} h_j(x) \right\}; \quad h(x) := \sum_{j=1}^{m} h_j(x).$$

It follows that p^+ is a DC function with the following DC decomposition

$$p^+(x) = \max\{g(x), h(x)\} - h(x),$$

i.e.,

$$p^+(x) = \max \left\{ \max_{i=1,...,m} \left\{ g_i(x) + \sum_{j=1, j \ne i}^{m} h_j(x) \right\}, \sum_{j=1}^{m} h_j(x) \right\} - \sum_{j=1}^{m} h_j(x).$$

Let DC decompositions of f_0 and p^+ be given by

$$f_0(x) = g_0(x) - h_0(x); \qquad (14)$$

$$p^+(x) = p_1(x) - p_2(x), \tag{15}$$

where g_0, h_0 p_1, p_2 are convex functions defined on the whole space. Then, we have the following DC decomposition for φ_k

$$\varphi_k(x) = g_k(x) - h_k(x), \quad x \in \mathbb{R}^n \tag{16}$$

where,

$$g_k(x) := g_0(x) + \beta_k p_1(x); \quad h_k(x) := h_0(x) + \beta_k p_2(x). \tag{17}$$

We make the following assumptions that will be used in the sequel.

Assumption 1. $f'_i s$ $(i = 0, ..., m)$ are locally Lipschitz functions at every point of C.

Assumption 2. Either g_k or h_k is differentiable on C, and $\rho(g_0) + \rho(h_0) + \rho(p_1) + \rho(p_2) > 0$.

2.1   DC Algorithm with Updated Penalty Parameter

DCA1

Initialization: Take an initial point $x^1 \in C$; $\delta > 0$; an initial penalty parameter $\beta_1 > 0$ and set $k := 1$.

1. Compute $y^k \in \partial h_k(x_k)$.
2. Compute $x^{k+1} \in \partial(g_k + \chi_C)^*(y^k)$, i.e., x^{k+1} is a solution of the convex program

$$\min\{g_k(x) - \langle x, y^k \rangle : x \in C\}. \tag{18}$$

3. **Stopping test.**
   Stop if $x^{k+1} = x^k$ and $p(x^k) \leq 0$.
4. **Penalty parameter update.**
   Compute $r_k := \min\{p(x^k), p(x^{k+1})\}$ and set

$$\beta_{k+1} = \begin{cases} \beta_k & \text{if either} \quad \beta_k \geq \|x^{k+1} - x^k\|^{-1} \quad \text{or} \quad r_k \leq 0, \\ \beta_k + \delta \text{ if} \quad \beta_k < \|x^{k+1} - x^k\|^{-1} \quad \text{and} \quad r_k > 0, \end{cases}$$

5. Set $k := k + 1$ and go to Step 1.

Note that the rule of penalty parameter is to ensure that if the sequence $\{\beta_k\}$ is unbounded then $\|x^{k+1} - x^k\| \to 0$ and $r_k > 0$. As will be shown later, under the Mangasarian-Fromowitz type constraint qualification, this case cannot occur, hence $\{\beta_k\}$ must be bounded. DCA1 is particularly important when exact penalty doesn't hold in (13) or when exact penalty occurs but upper bounds for the penalty parameter β are computationally intractable.

2.2   Global Convergence

Recall that a point $x^* \in F$ (the feasible set of (P_{gdc})-(12)) is a Karush-Kuhn-Tucker (KKT) point for the problem (P_{gdc})-(12) if there exist nonnegative scalars λ_i, $i = 1, ..., s$ such that

$$\begin{cases} 0 \in \partial f_0(x^*) + \sum_{i=1}^{m} \lambda_i \partial f_i(x^*) + N(C, x), \\ \lambda_i f_i(x^*) = 0, \quad i = 1, ..., m. \end{cases} \tag{19}$$

We say that the *extended Mangasarian-Fromowitz constraint qualification* (EM-FCQ) is satisfied at $x^* \in F$ with $I(x^*) \neq \emptyset$ if

$(MFCQ)$ there is a vector $d \in \mathrm{cone}(C - \{x^*\})$ (the *cone hull of* $C - \{x^*\}$) such that $f_i^\uparrow(x^*, d) < 0$ for all $i \in I(x^*)$.

When $f_i's$ are continuously differentiable, then $f_i^\uparrow(x^*, d) = \langle \nabla f(x^*), d \rangle$. Therefore, (EMFCQ) becomes the well-known Mangasarian-Fromowitz constraint qualification. It is well known that if the (extended) Mangasarian-Fromowitz constraint qualification is satisfied at a local minimizer x^* of (P_{gdc})-(12) then the KKT first order necessary condition (19) holds (see [61], [63]). In the global convergence theorem, we make use the following assumption

Assumption 3. The (extended) Mangasarian-Fromowitz constraint qualification (EMFCQ) is satisfied at any $x \in \mathbb{R}^n$ with $p(x) \geq 0$.

When f_i, $i = 1, ..., s$ are all convex functions, then it is obvious that this assumption is satisfied under the *Slater regular condition*, i.e., there exists $x \in C$ such that

$$f_i(x) < 0 \quad \text{for all} \quad i = 1, ..., m.$$

Theorem 1. *Suppose that $C \subseteq \mathbb{R}^n$ is a nonempty closed convex set and f_i, $i = 1, ..., m$ are DC functions on C. Suppose further that Assumptions 1-3 are verified.*

Let $\delta > 0$, $\beta_1 > 0$ be given. Let $\{x^k\}$ be a sequence generated by DCA1. Then DCA1 either stops, after finitely many iterations, at a KKT point x^k for problem (P_{gdc})-(12) or generates an infinite sequence $\{x^k\}$ of iterates such that $\lim_{k \to \infty} \|x^{k+1} - x^k\| = 0$ and every limit point x^∞ of the sequence $\{x^k\}$ is a KKT point of problem (P_{gdc})-(12).

Proof. Suppose that at each iterate step $x^k \in \mathbb{R}^n$, the DC decompositions

$$f_0(x) = g_0(x) - h_0(x); \tag{20}$$

$$p^+(x) = p_1(x) - p_2(x), \tag{21}$$

satisfy Assumption 1 and 2. Then, consider the following DC decomposition for $\varphi_k(x)$

$$\varphi_k(x) = g_k(x) - h_k(x) \tag{22}$$

with

$$g_k(x) := g_k(x) + \beta_k p_1(x); \tag{23}$$

$$h_k(x) := h_0(x) + \beta_k p_2(x). \tag{24}$$

By Assumption 2, one has

$$\rho_k := \rho(g_k) + \rho(h_k) \geq \rho(g_0) + \rho(h_0) + \beta_k \left(\rho(p_1) + \rho(p_2) \right) > 0.$$

For every $k = 0, 1, ...$; since $y^k \in \partial h_k(x^k)$, then

$$\langle y^k, x^{k+1} - x^k \rangle \leq h_k(x^{k+1}) - h_k(x^k) - \frac{\rho(h_k)}{2} \|x^k - x^{k+1}\|^2. \tag{25}$$

On the other hand, since $x^{k+1} \in \partial(g + \chi_C)^*(y^k)$, then

$$\langle y^k, x^k - x^{k+1} \rangle \leq g_k(x^k) - g_k(x^{k+1}) - \frac{\rho(g_k)}{2} \|x^k - x^{k+1}\|^2. \tag{26}$$

By inequalities (25) and (26), one obtains

$$\varphi_k(x^k) - \varphi_k(x^{k+1}) \geq \frac{\rho_k}{2} \|x^k - x^{k+1}\|^2. \tag{27}$$

Suppose that DCA1 stops, after finitely many iterations, at x^k, i.e., $x^{k+1} = x^k$ and $p(x^k) \leq 0$. Thus, x^k is a feasible point of the original problem (P_{gdc})-(12). Hence,

$$y^k \in \partial h_k(x^k) \cap \partial(g_k + \chi_C)(x^k) = \partial g_k(x^k) + N(C, x^k).$$

Consequently, by assumption 1, either g_k or h_k is continuously differentiable at x^k, one has

$$0 \in \partial \varphi_k(x^k) + N(C, x^k).$$

According to the suddifferential calculus, there exist $\lambda_i \geq 0$ $(i = 1, ..., m)$ such that $\lambda_i = 0$ if $i \notin I(x^k)$ and

$$0 \in \partial f_0(x^k) + \sum_{i \in I(x^k)} \lambda_i \partial f_i(x^k) + N(C, x^k).$$

That is, x^k is a KKT point of (P_{gdc})-(12).

Now, suppose that $\{x^k\}$ is an infinite sequence generated by DCA1. Let x^∞ be a limit point of the sequence $\{x^k\}$, say, $\lim_{i \to \infty} x^{k_i} = x^\infty$ for some subsequence $\{x^{k_i}\}$ of $\{x^k\}$.

We prove that there exists an index k_0 such that $\beta_k = \beta_{k_0}$ for all $k \geq k_0$. Assume the contrary, i.e, $\lim_{k \to \infty} \beta_k = +\infty$. Then, there exist infinitely many indices i such that

$$\beta_{k_i} < \|x^{k_i+1} - x^{k_i}\|^{-1} \quad \text{as well as} \quad p(x^{k_i}) > 0; p(x^{k_i+1}) > 0.$$

By considering a subsequence if necessarily, without loss of generality, we can assume that

$$\lim_{i \to \infty} \|x^{k_i+1} - x^{k_i}\| = 0 \quad \text{and} \quad p(x^{k_i}) > 0; p(x^{k_i+1}) > 0 \quad \text{for all } i.$$

Since

$$x^{k_i+1} \in \partial (g_k + \chi_C)^* (y^{k_i}),$$

then

$$y^{k_i} \in \partial g_k(x^{k_i+1}) + N(C, x^{k_i+1}).$$

Therefore,

$$0 \in \partial g_k(x^{k_i+1}) - \partial h_k(x^k) + N(C, x^{k_i+1}). \tag{28}$$

If g_k is differentiable then

$$0 \in \nabla g_k(x^{k_i+1}) - \nabla g_k(x^{k_i}) + \partial \varphi_k(x^{k_i}) + N(C, x^{k_i+1}), \tag{29}$$

otherwise,

$$0 \in \partial \varphi_k(x^{k_i+1}) + \nabla h_k(x^{k_i+1}) - \nabla h_k(x^{k_i}) + N(C, x^{k_i+1}). \tag{30}$$

Since $p(x^{k_i}) > 0$ and $p(x^{k_i+1}) > 0$, then, from the upper semicontinuity of the Clarke subdifferential operator, by dividing formula (28) by β_{k_i} and letting $i \to \infty$ one obtains

$$0 \in \partial p(x^\infty) + N(C, x^\infty). \tag{31}$$

Note that

$$\partial p(x^\infty) \subseteq \left\{ \sum_{i \in I(x^\infty)} \lambda_i \partial f_i(x^\infty) : \lambda_i \geq 0, \sum_{i \in I(x^\infty)} \lambda_i = 1 \right\}. \tag{32}$$

Since $r_{k_i} = \min\{p(x^k), p(x^{k_i+1})\} > 0$, then $p(x^\infty) \geq 0$, thus relation (44) contradicts Assumption 3. Hence, there exists an index k_0 such that

$$\beta_k = \beta_{k_0} \quad \text{for all} \quad k \geq k_0.$$

In virtue of inequality (27), $\lim_{k \to \infty} \|x^{k+1} - x^k\| = 0$. Hence, $r_k \leq 0$ for all k sufficiently large, which follows $p(x^\infty) \leq 0$, i.e., x^∞ is a feasible point of the original problem. By letting $i \to \infty$ in relation (28), we derive that

$$0 \in \partial f_0(x^\infty) + \partial p^+(x^\infty) + N(C, x^\infty). \tag{33}$$

By noting that

$$\partial p^+(x^\infty) = \begin{cases} \{\lambda \partial p(x^\infty) : \lambda \in [0,1]\} & \text{if } p(x^\infty) = 0 \\ \{0\} & \text{if } p(x^\infty) < 0; \end{cases}$$

relations (32), (33) show that x^∞ is a KKT point of problem (P_{gdc})-(12), which completes the proof.

Instead of the penalty function p^+, we could also use the following one

$$\psi(x) := \sum_{i=1}^{m} f_i^+(x), \quad x \in \mathbb{R}^n.$$

It is easy to see that the convergence result also holds for this penalty function.

3   DC Algorithm Using Slack Variables with Updated Relaxation Parameter: DCA2

As in the preceding section, let f_i $(i = 0, ..., m)$ be DC functions with the following DC decompositions

$$f_i(x) := g_i(x) - h_i(x), \quad x \in \mathbb{R}^n, \ i = 0, ..., m. \tag{34}$$

By using the main idea of the DCA that linearizes the concave part of the DC structure, we can derive a sequential convex programming method based on solving the convex subproblems of the following form:

$$\min \ g_0(x) - \langle y_0^k, x \rangle \quad \text{s.t.} \quad \begin{cases} g_i(x) - h_i(x^k) - \langle y_i^k, x - x^k \rangle \leq 0, & i = 1, ..., m; \\ x \in C, \end{cases} \tag{35}$$

where, $x^k \in \mathbb{R}^n$ is the current iterate, $y_i^k \in \partial h_i(x^k)$ for $i = 0, ..., m$.

This linearization introduces an inner convex approximation of the feasible set of (P_{gdc})-(12). However, this convex approximation is quite often poor and can lead to infeasibility of convex subproblem (35). We propose a relaxation technique to deal with the feasibility of subproblems. Instead of (35), we consider the subproblem

$$\min \ g_0(x) - \langle y_0^k, x \rangle + \beta_k t \quad \text{s.t.} \quad \begin{cases} g_i(x) - h_i(x^k) - \langle y_i^k, x - x^k \rangle \leq t, & i = 1, ..., m; \ t \geq 0 \\ x \in C, \end{cases} \tag{36}$$

where $\beta_k > 0$ is a penalty parameter. Obviously, (36) is a convex problem that is always feasible. Furthermore, the Slater constraint qualification is satisfied for the constraints of (36), thus the Karush-Kuhn-Tucker (KKT) optimality condition holds for some solution (x^{k+1}, t^{k+1}) of (36): there exist some $\lambda_i^{k+1} \in \mathbb{R}$, $i = 1, ..., m$, and $\mu^{k+1} \in \mathbb{R}$ such that

$$\begin{cases} 0 \in \partial g_0(x^{k+1}) - y_0^k + \sum_{i=1}^{m} \lambda_i^{k+1} \left(\partial g_i(x^{k+1}) - y_i^k \right) + N(C, x^{k+1}), \\ \beta_k - \sum_{i=1}^{m} \lambda_i^{k+1} - \mu^{k+1} = 0, \\ g_i(x^{k+1}) - h_i(x^k) - \langle y_i^k, x^{k+1} - x^k \rangle \leq t^{k+1}, \ \lambda_i^{k+1} \geq 0 \ \ i = 1, ..., m, \ x^{k+1} \in C, \\ \lambda_i^{k+1} \left(g_i(x^{k+1}) - h_i(x^k) - \langle y_i^k, x^{k+1} - x^k \rangle - t^{k+1} \right) = 0, \ i = 1, ..., m, \\ t^{k+1} \geq 0, \ \ \mu^{k+1} \geq 0, \ \ t^{k+1} \mu^{k+1} = 0. \end{cases} \tag{37}$$

Our relaxation DC algorithm is stated as follows.

DCA2
Initialization: Take an initial point $x^1 \in C$; $\delta_1, \delta_2 > 0$; an initial penalty parameter $\beta_1 > 0$ and set $k := 1$.

1. Compute $y_i^k \in \partial h_i(x^k)$, $i = 0, ..., m$.
2. Compute (x^{k+1}, t^{k+1}) as the solution of (36), and the associated Lagrange multipliers $(\lambda^{k+1}, \mu^{k+1})$.
3. **Stopping test.**
   Stop if $x^{k+1} = x^k$ and $t^{k+1} = 0$.

4. **Penalty parameter update.**
Compute $r_k := \min\{\|x^{k+1} - x^k\|^{-1}, \|\lambda^{k+1}\|_1 + \delta_1\}$, where $\|\lambda^{k+1}\|_1 = \sum_{i=1}^m |\lambda_i^{k+1}|$, and set

$$\beta_{k+1} = \begin{cases} \beta_k & \text{if} \quad \beta_k \geq r_k, \\ \beta_k + \delta_2 & \text{if} \quad \beta_k < r_k. \end{cases}$$

5. Set $k := k+1$ and go to Step 1.

Note that the updated penalty parameter rule is inspired by Solodov in [71], that is to ensure that the unboundedness of $\{\beta^k\}$ leads to the unboundedness of $\{\|\lambda^k\|_1\}$ as well as $\|x^{k+1} - x^k\| \to 0$.

As in the preceding section, we note

$$\varphi_k(x) := f_0(x) + \beta_k p^+(x), \quad x \in \mathbb{R}^n.$$

The following lemma is needed to investigate the convergence of DCA2.

Lemma 1. *The sequence* (x^k, t^k) *generated by DCA2 satisfies the following inequality*

$$\varphi_k(x^k) - \varphi_k(x^{k+1}) \geq \frac{\rho}{2} \|x^{k+1} - x^k\|^2, \quad \text{for all } k = 1, 2, \dots \tag{38}$$

where, $\rho := \rho(g_0) + \rho(h_0) + \min\{\rho(g_i) : i = 1, \dots m\}$.

Proof. According to the first inclusion of (37), there exist $z_i^k \in \partial g_i(x^{k+1})$ $(i = 0, \dots, m)$ such that

$$y_0^k + \sum_{i=1}^m \lambda_i^{k+1} y_i^k = z_0^k + \sum_{i=1}^m \lambda_i^{k+1} z_i^k + N(C, x^{k+1}). \tag{39}$$

On the other hand, by the convexity of g_i, h_i $(i = 0, \dots m)$ one has

$$\langle z_0^k - y_0^k, x^k - x^{k+1} \rangle \leq f_0(x^k) - f_0(x^{k+1}) - \frac{\rho(g_0) + \rho(h_0)}{2} \|x^k - x^{k+1}\|^2, \tag{40}$$

and for $i = 1, \dots, m$,

$$\langle z_i^k - y_i^k, x^k - x^{k+1} \rangle \leq f_i(x^k) - (g(x^{k+1}) - h_i(x^k) - \langle y_i^k, x^{k+1} - x^k \rangle) \tag{41}$$
$$- \frac{\rho(g_i)}{2} \|x^k - x^{k+1}\|^2,$$

By (37), it immediately follows that

$$\sum_{i=1}^m \lambda_i^{k+1} \langle z_i^k - y_i^k, x^k - x^{k+1} \rangle \leq (\beta_k - \mu^{k+1})(p^+(x^k) - t^{k+1}) - \frac{\rho_1}{2} \|x^k - x^{k+1}\|^2$$
$$\leq \beta_k(p^+(x^k) - t^{k+1}) - \frac{\rho_1}{2} \|x^k - x^{k+1}\|^2, \tag{42}$$

where, $\rho_1 := \min\{\rho(g_i) : i = 1, \dots m\}$. By noting that

$$t^{k+1} \geq g_i(x^{k+1}) - h_i(x^k) - \langle y_i^k, x^{k+1} - x^k \rangle \geq f_i(x^{k+1}) \quad \text{for all } i = 1, \dots, m,$$

inequality (38) follows directly from relations (39), (40), (42).

We are now ready to prove global convergence of DCA2.

Theorem 2. *Suppose that $C \subseteq \mathbb{R}^n$ is a nonempty closed convex set and f_i, $i = 1, ..., m$ are DC functions on C such that assumptions 1 and 3 are verified. Suppose further that for each $i = 0, ..., m$, either g_i or h_i is differentiable on C and that*

$$\rho := \rho(g_0) + \rho(h_0) + \min\{\rho(g_i) : i = 1, ...m\} > 0.$$

Let $\delta_1, \delta_2 > 0$, $\beta_1 > 0$ be given. Let $\{x^k\}$ be a sequence generated by DCA2. Then DCA2 either stops, after finitely many iterations, at a KKT point x^k for problem (P_{gdc})-(12) or generates an infinite sequence $\{x^k\}$ of iterates such that $\lim_{k\to\infty} \|x^{k+1} - x^k\| = 0$ and every limit point x^∞ of the sequence $\{x^k\}$ is a KKT point of problem (P_{gdc})-(12).

Proof. Suppose that DCA2 stops, after finitely many iterations, at x^k, i.e., $x^{k+1} = x^k$ and $t^{k+1} = 0$. Then, x^k is a feasible point of the original problem (P_{gdc})-(12). Since either g_i or h_i is differentiable, then

$$\partial f_i(x) = \partial g_i(x) - \partial h_i(x), \quad x \in C, \ i = 0, ..., m.$$

By (37), one has

$$\begin{cases} 0 \in \partial f_0(x^k + \sum_{i=1}^m \lambda_i^{k+1} \partial f_i(x^{k+1}) + N(C, x^k), \\ f_i(x^k) \leq 0, \ \lambda_i^{k+1} \geq 0 \quad i = 1, ..., m, \ x^k \in C, \ \lambda_i^{k+1} f_i(x^k) = 0, \ i = 1, ..., m. \end{cases} \tag{43}$$

That is, x^k is a KKT point of (P_{gdc})-(12).

Now, suppose that $\{x^k\}$ is an infinite sequence generated by DCA2. Let x^∞ be a limit point of the sequence $\{x^k\}$, say, $\lim_{j\to\infty} x^{k_j} = x^\infty$ for some subsequence $\{x^{k_j}\}$ of $\{x^k\}$.

We prove that there exists an index k_0 such that $\beta_k = \beta_{k_0}$ for all $k \geq k_0$. Assume the contrary, i.e, $\lim_{k\to\infty} \beta_k = +\infty$. Then, there exist infinitely many indices i such that

$$\beta_{k_j} < \|x^{k_j+1} - x^{k_j}\|^{-1} \quad \text{as well as} \quad \beta_{k_i} < \|\lambda^{k_i+1}\|_1 + \delta_1.$$

By considering a subsequence if necessarily, without loss of generality, we can assume that

$$\lim_{j\to\infty} \|x^{k_j+1} - x^{k_j}\| = 0 \quad \text{and} \quad \lim_{j\to\infty} \|\lambda^{k_j+1}\|_1 = +\infty.$$

If $p(x^\infty) < 0$ then when k_j is sufficiently large, one has

$$g_i(x^{k_j+1} - h_i(x^{k_j}) - \langle y^{k_j}, x^{k_j+1} - x^{k_j} \rangle - t^{k_j+1} < 0, \quad i = 1, ..., m.$$

Therefore, by (37), $\lambda_i^{k_j+1} = 0$ for all $i = 1, ..., m$, which contradicts $\|\lambda^{k_j+1}\|_1 \to +\infty$. Hence, $p(x^\infty) \geq 0$. Without loss of generality, assume further that

$$\lim_{j\to\infty} \frac{\lambda_i^{k_j+1}}{\|\lambda^{k_j+1}\|_1} = \eta_i, \ i = 1, ..., m.$$

Then,

$$\eta_i = 0 \text{ if } i \notin I(x^\infty) \quad \text{and} \quad \sum_{i \in I(x^\infty)} \eta_i = 1.$$

According to the upper semicontinuity of the Clarke subdifferential operator, by dividing first inclusion of (37) by $\|\lambda^{k_j+1}\|_1$ and letting $j \to \infty$, one obtains

$$0 \in \sum_{i \in I(x^\infty)} \eta_i \partial f_i(x^\infty) + N(C, x^\infty). \tag{44}$$

This relation contradicts obviously Assumption 3.

Hence, there exists an index k_0 such that

$$\beta_k = \beta_{k_0} \quad \text{for all} \quad k \geq k_0.$$

In virtue of inequality (38), $\lim_{k \to \infty} \|x^{k+1} - x^k\| = 0$. Hence, $\beta_k \geq \|\lambda^{k+1}\|_1 + \delta_1$ for all k sufficiently large. In view of the second relation of (37), it follows $\mu^{k+1} > 0$, thus, $t^{k+1} = 0$ when k is sufficiently large. Consequently, x^∞ is a feasible point of the original problem. As $\{\|\lambda^{k+1}\|_1\}$ is bounded, we can assume that

$$\lim_{j \to \infty} \lambda_i^{k_j+1} = \lambda_i^\infty, \quad i = 1, ..., m.$$

By letting $j \to \infty$ in relations of (37), we derive that

$$\begin{cases} 0 \in \partial f_0(x^\infty) + \sum_{i=1}^m \lambda_i^\infty \partial f_i(x^\infty) + N(C, x^\infty), \\ f_i(x^\infty) \leq 0, \; \lambda_i^\infty \geq 0 \quad i = 1, ..., m, \; x^\infty \in C, \; \lambda_i^\infty f_i(x^\infty) = 0, \; i = 1, ..., m. \end{cases} \tag{45}$$

It shows that x^∞ is a KKT point of problem (P_{gdc})-(12) and the proof is completed.

Note that, as shown in Theorem 1 and Theorem 2, the penalty parameter β_k is constant when k is sufficiently large. Observing from the proof of these convergence theorems, the sequence $\{\varphi(x^k)\}$ of values of the function $\varphi_k(x) = f_0(x) + \beta_k p^+(x)$ along the sequence $\{x^k\}$ generated by DCA1 and DCA2 is decreasing.

4   Conclusion

We have presented the natural extension of DC programming and DCA for general DC programs with DC constraints. Two approaches consisting of reformulation of those programs as standard DC programs give rise to DCA1 and DCA2. The first one is based on penalty techniques in DC programming while the second linearizes concave functions in DC constraints to build convex inner approximations of the feasible set. Both designed algorithms can be viewed as a sequence of standard DCAs with updated penalty (resp. relaxation) parameters, which marks the passage from standard DCAs to extended DCAs. This fact shows clearly that standard DCA is a special case of DCA1/DCA2 applied to standard DC programs. That is the reason why we will not distinguish between standard and extended DCAs and all these algorithms will be called DCA for short. We hope this work will provide a deep understanding of these theoretical and algorithmic tools for researchers and practitioners to efficiently model and solve their real-word nonconvex programs, especially in the large-scale setting.

References

1. DC Programming and DCA,
   http://lita.sciences.univ-metz.fr/~lethi/DCA.html
2. Pham Dinh, T., Le Thi, H.A.: Convex analysis approach to DC programming: Theory, Algorithms and Applications. Acta Mathematica Vietnamica 22(1), 289–355 (1997)
3. Pham Dinh, T., Le Thi, H.A.: A DC Optimization algorithm for solving the trust region subproblem. SIAM J. Optim. 8(2), 476–505 (1998)
4. Le Thi, H.A., Pham Dinh, T.: Solving a class of linearly constrained indefinite quadratic problems by DC Algorithms. Journal of Global Optimization 11, 253–285 (1997)
5. Le Thi, H.A., Pham Dinh, T.: DC Programming: Theory, Algorithms and Applications. The State of the Art (28 pages). In: Proceedings of the First International Workshop on Global Constrained Optimization and Constraint Satisfaction (Cocos 2002), Valbonne-Sophia Antipolis, France, October 2-4 (2002)
6. Le Thi, H.A., Pham Dinh, T., Le, D.M.: Exact penalty in DC programming. Vietnam Journal of Mathematics 27(2), 169–178 (1999)
7. Le Thi, H.A.: An efficient algorithm for globally minimizing a quadratic function under convex quadratic constraints. Mathematical Programming, Ser. A 87(3), 401–426 (2000)
8. Le Thi, H.A., Pham Dinh, T.: Large scale global molecular optimization from distance matrices by a DC optimization appoach. SIAM J. Optim. 14(1), 77–116 (2003)
9. Le Thi, H.A., Huynh, V.N., Pham Dinh, T.: DC Programming and DCA for solving general DC programs, Research Report, National Institute for Applied Sciences (2004)
10. Le Thi, H.A., Pham Dinh, T.: The DC (Difference of Convex functions) Programming and DCA revisited with DC models of real-world nonconvex optimization problems. Annals of Operations Research 133, 23–48 (2005)
11. Le Thi, H.A., Nguyen, T.P., Pham Dinh, T.: A continuous approach for solving the concave cost supply problem by combining DCA and B&B techniques. European Journal of Operational Research 183, 1001–1012 (2007)
12. Le Thi, H.A., Pham Dinh, T.: A continuous approach for the concave cost supply problem via DC Programming and DCA. Discrete Applied Mathematics 156, 325–338 (2008)
13. Pham Dinh, T., Nguyen, C.N., Le Thi, H.A.: An efficient combination of DCA and B&B using DC/SDP relaxation for globally solving binary quadratic programs. Journal of Global Optimization 48(4), 595–632 (2010)
14. Thiao, M., Pham Dinh, T., Le Thi, H.A.: A DC programming approach for Sparse Eigenvalue Problem. In: ICML 2010, pp. 1063–1070 (2010)
15. Le Thi, H.A., Huynh, V.N., Pham Dinh, T.: Convergence Analysis of DC Algorithms for DC programming with subanalytic data, Research Report, National Institute for Applied Sciences, Rouen, France (2009)

16. Thiao, M., Pham Dinh, T., Le Thi, H.A.: A DC programming approach for Sparse Eigenvalue Problem, Research Report, National Institute for Applied Sciences, Rouen, France (2011)

17. Le Thi, H.A., Pham Dinh, T.: Approximation and Penalization of the ℓ_0-norm in DC Programming, Research Report, National Institute for Applied Sciences, Rouen, France (2010)

18. Le Thi, H.A., Pham Dinh, T.: DC Programming and DCA for solving nonconvex programs involving ℓ_0-norm, Research Report, National Institute for Applied Sciences, Rouen, France (2010)

19. Le Thi, H.A., Moeini, M.: Long-Short Portfolio Optimization Under Cardinality Constraints by Difference of Convex Functions Algorithm. Journal of Optimization Theory & Applications, 27 pages (October 2012), doi:10.1007/s10957-012-0197-0

20. Le Thi, H.A., Pham Dinh, T., Huynh, V.N.: Exact penalty and Error Bounds in DC programming. Journal of Global Optimization, Special Issue in Memory of Reiner Horst, Founder of the Journal 52(3), 509–535 (2012)

21. Le Thi, H.A., Pham Dinh, T.: Exact Penalty in Mixed Integer DC Programming, Research Report, Lorraine University, France (2011)

22. Le Thi, H.A., Pham Dinh, T., Nguyen, D.Y.: Properties of two DC algorithms in quadratic programming. Journal of Global Optimization 49(3), 481–495 (2011)

23. Le Thi, H.A., Pham Dinh, T.: On Solving Linear Complementarity Problems by DC programming and DCA. Journal on Computational Optimization and Applications 50(3), 507–524 (2011)

24. Le Thi, H.A., Madhi, M., Pham Dinh, T., Judice, J.: Solving Eigenvalue Symmetric problem by DC programming and DCA. Journal on Computational Optimization and Applications 51(3), 1097–1117 (2012)

25. Le Thi, H.A., Nguyen, D.M., Pham Dinh, T.: Globally solving a Nonlinear UAV Task Assignment Problem by stochastic and derministic optimization approaches. Optimization Letters 6(2), 315–329 (2012)

26. Le Thi, H.A., Pham Dinh, T., Nguyen, D.Y.: Behavior of DCA sequences for solving the trust-region subproblem, Journal of Global Optimization. Journal of Global Optimization 53(2), 317–329 (2012)

27. Le Thi, H.A., Tran Duc, Q.: Solving Continuous Min Max Problem for Single Period Portfolio Selection with Discrete Constraints by DCA. Optimization 61(8) (2012)

28. Schleich, J., Le Thi, H.A., Bouvry, P.: Solving the Minimum m-Dominating Set problem by a Continuous Optimization Approach based on DC Programming and DCA. Journal of Combinatorial Optimization 24(4), 397–412 (2012)

29. Le Thi, H.A., Vaz, A.I.F., Vicente, L.N.: Optimizing radial basis functions by D.C. programming and its use in direct search for global derivative-free optimization. TOP 20(1), 190–214 (2012)

30. Le Thi, H.A., Le Hoai, M., Pham Dinh, T., Huynh, V.N.: Spherical separation by DC programming and DCA. To appear in Journal of Global Optimization, 17 pages (Online first Feabruary 2012), doi:10.1007/s10898-012-9859-6

31. Muu, L.D., Tran Dinh, Q., Le Thi, H.A., Pham Dinh, T.: A new decomposition algorithm for globally solving mathematical programs with affine equilibrium constraints. Acta Mathematica Vietnamica 37(2), 201–218 (2012)

32. Niu, Y.S., Pham Dinh, T., Le Thi, H.A., Judice, J.: Efficient DC Programming Approaches for Asymmetric Eigenvalue Complementarity Problem. Optimization Methods and Software 28(4), 812–829 (2013)

33. Ta, A.S., Le Thi, H.A., Khadraoui, D., Pham Dinh, T.: Solving Partitioning-Hub Location-Routing Problem using DCA. Journal of Industrial and Management Optimization 8(1), 87–102 (2012)
34. Le Thi, H.A., Pham Dinh, T., Tran Duc, Q.: A DC programming approach for a class of bilevel programming problems and its application in portfolio selection. NACO Numerical Algebra, Control and Optimization 2(1), 167–185 (2012)
35. Cheng, S.O., Le Thi, H.A.: Learning sparse classifiers with Difference of Convex functions Algorithms. Optimization Methods and Software 28(4), 830–854 (2013)
36. Anh, P.N., Le Thi, H.A.: An Armijo-type method for pseudomonotone equilibrium problems and its applications. Journal of Global Optimization 57, 803–820 (2013)
37. Le Thi, H.A., Moeini, M.: Long-Short Portfolio Optimization Under Cardinality Constraints by Difference of Convex Functions Algorithm. Journal of Optimization Theory & Applications, 26 pages (October 2012), doi:10.1007/s10957-012-0197-0
38. Nguyen, D.M., Le Thi, H.A., Pham Dinh, T.: Solving the Multidimensional Assignment Problem by a Cross-Entropy method. Journal of Combinatorial Optimization, 16 pages (Online first October 2012), doi:10.1007/s10878-012-9554-z
39. Le Thi, H.A., Pham Dinh, T., Nguyen, D.M.: A deterministic optimization approach for planning a multisensor multizone search for a target. Computer & Operations Research 41, 231–239 (2014)
40. Anh, P.N., Le Thi, H.A.: The Subgradient Extragradient Method Extended to Equilibrium Problems, Optimization (online first December 2012), doi:10.1080/02331934.2012.745528
41. Le Hoai, M., Le Thi, H.A., Pham Dinh, T., Huynh, V.N.: Block Clustering based on DC programming and DCA. NECO Neural Computation 25(10), 2776–2807 (2013)
42. Le Thi, H.A., Tran Duc, Q.: New and efficient algorithms for transfer prices and inventory holding policies in two-enterprise supply chains. Journal of Global Optimization (in press)
43. Le Thi, H.A., Le Hoai, M., Pham Dinh, T.: New and efficient DCA based algorithms for Minimum Sum-of-Squares Clustering. Pattern Recognition (in press)
44. Le Thi, H.A., Pham Dinh, T., Nguyen, C.N., Le Hoai, M.: DC Programming and DCA for Binary Quadratic Programming in Diversity Data Mining. To appear in Optimization
45. Le Thi, H.A., Tran Duc, Q., Adjallah, K.H.: A Difference of Convex functions Algorithm for Optimal Scheduling and real-time assignment of preventive maintenance jobs on parallel processors. To appear in JIMO Journal of Industrial and Management Optimization
46. An, L.T.H., Quynh, T.D.: Optimizing a multi-stage production/inventory system by DC programming based approaches. Computational Optimization an Applications (in press)
47. An, L.T.H., Tao, P.D., Belghiti, T.: DCA based algorithms for Multiple Sequence Alignment (MSA). Central European Journal of Operations Research (in press)
48. Tao, P.D., An, L.T.H.: Recent advances in DC programming and DCA. To appear in Transactions on Computational Collective Intelligence, 37 pages (2013)
49. An, L.T.H., Tao, P.D.: DC programming in Communication Systems: challenging models and methods. To appear in Vietnam Journal of Computer Science, 21 pages. Springer (invited issue)
50. An, L.T.H., Tao, P.D.: DC programming approaches for Distance Geometry problems. In: Mucherino, A., Lavor, C., Liberti, L., Maculan, N. (eds.) Distance Geometry: Theory, Methods and Applications, vol. XVI, 57, 420 pages. Springer (2013)

51. Bradley, P.S., Mangasarian, O.L.: Feature selection via concave minimization and support vector machines. In: Proceedings of the Fifteenth International Conference on Machine Learning (ICML 1998), pp. 82–90 (1998)
52. Chambolle, A., DeVore, R.A., Lee, N.Y., Lucier, B.J.: Nonlinear wavelet image processing: Variational problems, compression, and noise removal through wavelet shrinkage. IEEE Trans. Image Process. 7, 319–335 (1998)
53. Dempster, A.P., Laird, N.M., Rubin, D.B.: Maximum likelihood from incomplete data via the EM algorithm. J. Roy. Stat. Soc. B 39, 1–38 (1977)
54. Bertsekas, D.: Nonlinear Programming. Athenta Scientific, Belmont (1995)
55. Bogg, P.T., Tolle, J.W.: Sequential Quadratic Programming. Acta Numerica, 1–51 (1995)
56. Clarke, F.H.: Optimization and Nonsmooth Analysis. Wiley, New York (1983)
57. Tuy, H.: Convex Analysis and Global Optimization. Kluwer Academic (2000)
58. Fletcher, R., Leyfer, S.: Nonlinear programming without a penalty function. Math. Program. 91, 239–270 (2002)
59. An, L.T.H., Tao, P.D.: Large Scale Molecular Optimization From Distance Matrices by a D.C. Optimization Approach. SIAM Journal on Optimization 14(1), 77–116 (2003)
60. An, L.T.H., Tao, P.D.: The DC (Difference of Convex functions) Programming and DCA revisited with DC models of real world nonconvex optimization problems. Annals of Operations Research 133, 23–46 (2005)
61. Mangasarian, O.L.: Nonlinear Programming, McGraw-Hill, New York (1969)
62. Lawrence, C.T., Tits, A.: A computationally efficient feasible sequential quadratic programming algorithm. SIAM J. Optim. 11, 1092–1118 (2001)
63. Mangasarian, O.L., Fromovitz, S.: The Fritz John necessay optimality conditions in the presence of equality constraints. J. Math. Anal. Appl. 17, 34–47 (1967)
64. Mordukhovich, B.S.: Variational Analysis and Generalized Differentiation, Vol. 1. Springer, Heidelberg (2006)
65. Nocedal, J., Wright, S.J.: Numerical Optimization, Springer, Berlin (2006)
66. Pang, J.-S.: Exact penalty functions for mathematical programs with linear complementary constraints. Optimization 42, 1–8 (1997)
67. Polak, E.: Optimization. Springer. New York (1997)
68. Rockafellar, R.T.: Convex Analysis. Princeton University Press (1970)
69. Rockafellar, R.T.: Penalty methods and augmanted Lagrangians nonlinear programming. In: Conti, R., Ruberti, A. (eds.) 5th Conference on Optimization Techniques Part I. LNCS, vol. 3, pp. 418–425. Springer, Heidelberg (1973)
70. Rockafellar, R.T., Wets, J.-B.: Variational Analysis. Springer, Heidelberg (1998)
71. Solodov, M.V.: On the sequential quadratically constrained quadratic programming methods. Mathematics of Oper. Research 29, 64–79 (2004)
72. Zaslavski, A.J.: A sufficient condition for exact penalty constrained optimization. SIAM J. Optim. 16, 250–262 (2005)
73. Yuille, A.L., Rangarajan, A.: The concave-convex procedure. Neural Computation 15(4), 915–936 (2003)

DC Programming Approaches for BMI and QMI Feasibility Problems

Yi-Shuai Niu[1] and Tao Pham Dinh[2]

[1] University of Paris 6, France
niuyishuai@hotmail.com
[2] National Institute for Applied Sciences - Rouen, France
pham@insa-rouen.fr

Abstract. We propose some new DC (difference of convex functions) programming approaches for solving the Bilinear Matrix Inequality (BMI) Feasibility Problems and the Quadratic Matrix Inequality (QMI) Feasibility Problems. They are both important NP-hard problems in the field of robust control and system theory. The inherent difficulty lies in the nonconvex set of feasible solutions. In this paper, we will firstly reformulate these problems as a DC program (minimization of a concave function over a convex set). Then efficient approaches based on the DC Algorithm (DCA) are proposed for the numerical solution. A semidefinite program (SDP) is required to be solved during each iteration of our algorithm. Moreover, a hybrid method combining DCA with an adaptive Branch and Bound is established for guaranteeing the feasibility of the BMI and QMI. A concept of partial solution of SDP via DCA is proposed to improve the convergence of our algorithm when handling more large-scale cases. Numerical simulations of the proposed approaches and comparison with PENBMI are also reported.

Keywords: BMI/QMI, DC program, DCA, Branch and Bound, SDP.

1 Introduction

The optimization problem with *Bilinear Matrix Inequality* (BMI) constraints is considered as the central problem in the filed of robust control. A wide range of difficult control synthesis problems, such as the fixed order $\mathcal{H}^\infty$ control, the μ/K_m-synthesis, the decentralized control, and the robust gain-scheduling can be reduced to problems involving BMIs [8, 9]. Given the symmetric matrices $F_{ij} \in \mathcal{S}^k (i = 0, \ldots, n; j = 0, \ldots, m)$, where $\mathcal{S}^k$ denotes the space of real symmetric $k \times k$ matrices. We call the *Bilinear Matrix Inequality* (BMI) an inequality for which the biaffine combination of the given matrices is a negative (or positive) semidefinite (or definite) matrix, mathematically defined as:

$$F(x, y) := F_{00} + \sum_{i=1}^{n} x_i F_{i0} + \sum_{j=1}^{m} y_j F_{0j} + \sum_{i=1}^{n} \sum_{j=1}^{m} x_i y_j F_{ij} \preceq 0 \qquad (1)$$

T.V. Do et al. (eds.), *Advanced Computational Methods for Knowledge Engineering*, 37
Advances in Intelligent Systems and Computing 282,
DOI: 10.1007/978-3-319-06569-4_3, © Springer International Publishing Switzerland 2014

where $F(x, y) \in \mathcal{S}^k$, $x \in \mathbb{R}^n$ and $y \in \mathbb{R}^m$. The inequality $A \preceq 0$ means the matrix A is negative semidefinite. The BMI could be considered as the generalization of bilinear inequality constraint (including quadratic inequalities) since once F_{ij}'s are diagonals, it becomes a set of bilinear inequality constraints. Therefore, any bilinear and quadratic constraint can be represented as a BMI [39]. Furthermore, it is well-known that BMI has a biconvexity structure since the BMI constraint becomes convex *Linear Matrix Inequality* (LMI) when x or y is fixed. Since LMI is equivalent to *Semi-Definite Program* (SDP), then BMI could also be considered as a generalization of SDP. Therefore, we can see that the research on BMI is not only very important in the application of robust control, but also very important on theoretical aspect.

We will consider the *BMI Feasibility Problem* (BMIFP) whose objective is to find a feasible solution $(x, y) \in \mathbb{R}^n \times \mathbb{R}^m$ of the matrix inequality (1). In practice, many problems in robust control could be formulated as an optimization problem with a linear/quadratic objective function within BMI constraints, and the variables x and y are often bounded, i.e. (x, y) is bounded in a hyper-rectangle $\mathcal{H} := \{(x, y) \in \mathbb{R}^n \times \mathbb{R}^m : \underline{x} \leq x \leq \bar{x}, \underline{y} \leq y \leq \bar{y}\}$, where $\underline{x}, \bar{x} \in \mathbb{R}^n$ and $\underline{y}, \bar{y} \in \mathbb{R}^m$ denote the upper and the lower bounds for the variables x and y. Solving BMIFP is quite difficult and classified as a NP-hard problem [42].

Some previous researches on BMI will be briefly described here to give a view about what has been already done. Of course, it won't cover all works in this topic. The first paper that formally introduced the BMI in the control theory is probably due to Safonov et al. [38] in 1994. They have shown that the BMIFP is equivalent to checking whether the diameter of a certain convex set is greater than two. And this is equivalent to a maximization of a convex function via a convex set which is proved to be a NP-hard problem. Later, Goh et al. [8] presented the first implemented algorithm based on convex relaxation and branch-and-bound (B&B) scheme for finding a global solution of a particular case of BMIFP. The same authors also proposed algorithms to approximate local optima. Other authors such as VanAntwerp [45], Fujioka [5] improved the lower bound of the B&B algorithm through a better convex relaxation. Kojima et al. [7] proposed a Branch and Cut algorithm and improved the B&B algorithm of Goh [8] by applying a better convex relaxation of the BMI Eigenvalue Problem (BMIEP). Kawanish et al. [14] and Takano et al. [41] presented some results for reducing the feasible region of the problem using informations of local optimal solutions. Tuan et al. [43] had pointed out that BMIFP belongs to the class of *DC (difference of convex functions)* optimization problems. Some DC reformulation of BMIFP and algorithms based on branch and bound approach have been investigated in that paper. Later, a Lagrangian dual global optimization algorithm is proposed by the same author in [44]. A sequence of concave minimization problems and DC programs were employed by Liu and Papavassilopoulos [23] on the same problem of Tuan. Mesbahi and Papavassilopoulos [26] have studied the theoretical aspects of the BMI and established equivalent formulations of cone programming. More recently, the generalized Benders decomposition approach has been extended to BMIFP by Floudas, Beran et al. [2, 4].

In this paper, we will investigate a DC Programming formulation of BMIFP and propose a solution method based on an efficient DC algorithm (DCA). A hybrid method combining the DCA with a Branch-and-Bound scheme is also proposed to solve the general BMIFP when DCA could not found a solution (e.g. the infeasible BMI). As we've known in [43], the BMI can be reformulated as DC programming problem (concave minimization problem over a convex set). Our proposed method DCA is efficient to solve such DC programs. More specifically, our DCA is an iterative method consists of solving a SDP subproblem in each iteration. The fast convergence rate of DCA raises interest in combining DCA with a global optimization framework Branch-and-Bound in order to efficiently solve this concave minimization formulation. Particularly, if we can find some good initial point to start DCA, then our algorithm will give very good upper bound (quite often global one) with mineur iterations. Furthermore, it's important to notice that solving BMIFP does not need globally solving the DC program, since the particular features of the BMIFP and its DC reformulation structure make it possible to stop our algorithm before obtaining a global optimal solution (even not to be a local solution in some cases). This could be frequently encountered in feasible BMIFPs. Numerical experiments compared with the commercial BMI solver (PENBMI) [15] will be also reported.

2   QMIFP Reformulation

Firstly, we present an equivalent reformulation of BMIFP that is referred to as *Quadratic Matrix Inequality Feasibility Problem*(QMIFP). Given the symmetric matrices $Q_i \in \mathcal{S}^k (i = 0, \ldots, N)$ and $Q_{ij} \in \mathcal{S}^k (i, j = 1, \ldots, N)$. QMIFP is to find a point x in a hyper-rectangle $\mathcal{H} := \{x \in \mathbb{R}^N : \underline{x} \leq x \leq \bar{x}\}$ which satisfies the following Quadratic Matrix Inequality:

$$Q(x) := Q_0 + \sum_{i=1}^{N} x_i Q_i + \sum_{i,j=1}^{N} x_i x_j Q_{ij} \preceq 0. \tag{2}$$

Any BMIFP (1) can be easily casted into a QMIFP (2) according to:

Proposition 1. *Let $N := n + m$. The BMIFP (1) is equivalent to the QMIFP (2) with*

$$x_i := \begin{cases} x_i & i=1,\ldots,n \\ y_{(i-m)} & i=n+1,\ldots,N. \end{cases}$$

$$Q_0 := F_{00}; Q_i := \begin{cases} F_{i0} & i=1,\ldots,n; \\ F_{0(i-n)} & i=n+1,\ldots,N. \end{cases}$$

$$Q_{ij} := \begin{cases} F_{i(j-n)} & i=1,\ldots,n; j=n+1,\ldots,N \\ F_{(i-m)j} & i=m+1,\ldots,N; j=1,\ldots,m \\ 0 & otherwise. \end{cases}$$

$$\underline{x}_i := \begin{cases} \underline{x}_i & i=1,\dots,n \\ \underline{y}_{(i-m)} & i=n+1,\dots,N \end{cases} \text{ and } \bar{x}_i := \begin{cases} \bar{x}_i & i=1,\dots,n \\ \bar{y}_{(i-m)} & i=n+1,\dots,N. \end{cases}$$

The proof of the proposition 1 is obvious. This result demonstrates that QMIFP and BMIFP are equivalent in feasibility, and their solution sets. Obviously, solving such a QMIFP is also a NP-hard problem, but QMIFP is easy to be reformulated as a DC program which could be solved efficiently with our proposed methods. In next section, we will focus on the DC reformulation of QMIFP.

3   DC Reformulation of QMIFP

We know that QMIFP (2) is equivalent to a concave minimization problem over compact convex set consists of LMIs and bounds in the variables (see [43]). It is derived as follows.

Firstly, we have the following proposition:

Proposition 2. *Let $W = (w_{ij}) = xx^T$ (where x^T denotes the transpose of the vector x). Considering the matrix $Q(x)$ of the QMIFP (2) in which we can replace all bilinear terms $x_i x_j$ with new variables w_{ij} to get:*

$$Q_L(x,W) := Q_0 + \sum_{i=1}^{N} x_i Q_i + \sum_{i,j=1}^{N} w_{ij} Q_{ij} \preceq 0 \tag{3}$$

where

$$(x,W) \in \{(x,W) \in \mathbb{R}^N \times S^N : x \in \mathcal{H}, W = xx^T\}.$$

The formulation (3) is equivalent to the QMIFP (2) in the sense that they are both feasible or infeasible. More specifically, x^* is a feasible point of (2) if and only if $(x^*, x^* x^{*T})$ is a feasible point of (3).

In the formulation (2), the difficulty lies in the nonconvex QMI constraint, while in the formulation (3), the quadratic matrix function $Q(x)$ becomes an affine matrix function $Q_L(x,W)$, so that the nonconvex QMI constraint is relaxed to a convex LMI constraint. However, a new nonlinear nonconvex equality constraint $W = xx^T$ is introduced in the later one which destroyed the convexity of (3). The following lemma gives us an equivalent formulation to this equality.

Lemma 1 (see [43]). *Let $W \in S_+^N$ defined as $W := xx^T$. The constraint $W = xx^T$ is equivalent to*

$$\begin{bmatrix} W & x \\ x^T & 1 \end{bmatrix} \succeq 0 \tag{4}$$

$$Trace(W - xx^T) = Trace(W) - \sum_{i=1}^{N} x_i^2 \leq 0 \tag{5}$$

This lemma represents the nonlinear equality $W = xx^T$ as a LMI (4) with a reverse convex constraint (5). Therefore, we get from Proposition (2) and Lemma (1) an equivalent formulation of QMIFP as:

$$Q_0 + \sum_{i=1}^{N} x_i Q_i + \sum_{i,j=1}^{N} w_{ij} Q_{ij} \preceq 0; \begin{bmatrix} W & x \\ x^T & 1 \end{bmatrix} \succeq 0; Trace(W) - \sum_{i=1}^{N} x_i^2 \leq 0; x \in \mathcal{H}.$$

$$(6)$$

The system (6) is the intersection of the convex set $\mathcal{C} := \{(x, W) : Q_0 + \sum_{i=1}^{N} x_i Q_i + \sum_{i,j=1}^{N} w_{ij} Q_{ij} \preceq 0, \begin{bmatrix} W & x \\ x^T & 1 \end{bmatrix} \succeq 0, x \in \mathcal{H}\}$ and the nonconvex set (reverse convex set) $\mathcal{RC} := \{(x, W) : Trace(W) - \sum_{i=1}^{N} x_i^2 \leq 0\}$.

Theorem 1. *The QMIFP (6) is feasible if and only if there exists a point (x, W) in $\mathcal{C} \neq \emptyset$ and $Trace(W) - \sum_{i=1}^{N} x_i^2 = 0$. Otherwise, the QMIFP is infeasible if $\mathcal{C} = \emptyset$ or $Trace(W) - \sum_{i=1}^{N} x_i^2 > 0$ for any point (x, W) in $\mathcal{C} \neq \emptyset$.*

Proof. If $\mathcal{C} = \emptyset$, then the system (6) is obviously infeasible. Otherwise, suppose $(x, W) \in \mathcal{C} \neq \emptyset$, applying Schur complements to $\begin{bmatrix} W & x \\ x^T & 1 \end{bmatrix} \succeq 0$, we have $W - xx^T \succeq 0$ which means $Trace(W - xx^T) = Trace(W) - \sum_{i=1}^{N} x_i^2 \geq 0$. This inequality is held for every point in $\mathcal{C}$. Combining this inequality with the nonconvex set $\mathcal{RC}$, we must have $Trace(W) - \sum_{i=1}^{N} x_i^2 = 0$ for any feasible point of the system (6). Clearly, if $Trace(W) - \sum_{i=1}^{N} x_i^2 > 0, \forall (x, W) \in \mathcal{C}$, then (6) is infeasible. $\square$

Based on the Theorem 1, we get that (6) is feasible if and only if the following concave minimization problem is true.

$$\begin{aligned} 0 = \min \ & f(x, W) = Trace(W) - \sum_{i=1}^{N} x_i^2 \\ s.t. \ & Q_L(x, W) := Q_0 + \sum_{i=1}^{N} x_i Q_i + \sum_{i,j=1}^{N} w_{ij} Q_{ij} \preceq 0 \\ & \begin{bmatrix} W & x \\ x^T & 1 \end{bmatrix} \succeq 0 \\ & (x, W) \in \mathcal{H} \times S^N. \end{aligned}$$

$$(7)$$

As a conclusion, BMIFP (1) is feasible if and only if QMIFP (2) is feasible, the QMIFP (2) is equivalent to (6), and the formulation (6) is feasible if and only if 0 is the optimal value of the concave minimization problem (7). Therefore, we can solve the problem (7) to determine whether the BMIFP (1) is feasible or not, and if (x, W) is feasible to (7) such that $f(x, W) \leq 0$, then x must be feasible to QMIFP (2).

Concerning on the bounds of the variables (x, W) in (6), we have that W is bounded if x is bounded due to the definition of W as $w_{ij} = x_i x_j$. Thus w_{ij} has upper and lower bounds as:

$$\underline{w}_{ij} \leq w_{ij} \leq \bar{w}_{ij} (i, j = 1, \ldots, N)$$

$$(8)$$

where

$$\underline{w}_{ij} = \min\{\underline{x}_i \underline{x}_j, \underline{x}_i \bar{x}_j, \bar{x}_i \underline{x}_j, \bar{x}_i \bar{x}_j\}; \bar{w}_{ij} = \max\{\underline{x}_i \underline{x}_j, \underline{x}_i \bar{x}_j, \bar{x}_i \underline{x}_j, \bar{x}_i \bar{x}_j\}.$$

Therefore, we can introduce the bounds $(x, W) \in \mathcal{H}_W := \{(x, W) \in \mathbb{R}^N \times S^N : \underline{x} \leq x \leq \bar{x}, \underline{w}_{ij} \leq w_{ij} \leq \bar{w}_{ij}, i, j = 1, \ldots, N\}$ into the problem (7) to reduce the searching space.

Note that several LMIs can be expressed as one single LMI according to:

Lemma 2. *Suppose that we have p LMIs as*

$$F^j(x) := F_0^j + \sum_{i=1}^N x_i F_i^j \preceq 0, (j = 1, \ldots, p). \tag{9}$$

They are equivalent to the following LMI:

$$\hat{F}(x) := \hat{F}_0 + \sum_{i=1}^N x_i \hat{F}_i \preceq 0 \tag{10}$$

where

$$\hat{F}_i = \begin{bmatrix} F_i^1 & & 0 \\ & \ddots & \\ 0 & & F_i^p \end{bmatrix}, (i = 0, \ldots, N).$$

Proof. The formula (10) is exactly

$$\hat{F}(x) := \begin{bmatrix} F^1(x) & & 0 \\ & \ddots & \\ 0 & & F^p(x) \end{bmatrix} \preceq 0. \tag{11}$$

If x^* is feasible to (9), i.e. $F^j(x^*) \preceq 0, (j = 1, \ldots, p)$, then it is obviously feasible to the formula (11). Inversely, if x^* is feasible to (11), since all principal minors of a negative semidefinite matrix $\hat{F}(x^*)$ must be negative semidefinite, then $F^j(x^*) \preceq 0, (j = 1, \ldots, p)$. □

In the problem (7), the LMI $\begin{bmatrix} W & x \\ x^T & 1 \end{bmatrix} \succeq 0$ can be represented as the standard form as:

$$F(x, W) := F_0 + \sum_{i=1}^N x_i F_i + \sum_{i,j=1}^N w_{ij} F_{ij} \preceq 0 \tag{12}$$

where

$$F_0 := \begin{bmatrix} 0 & 0 \\ 0 & -1 \end{bmatrix}; F_i := \begin{bmatrix} 0 & -e_i \\ -e_i^T & 0 \end{bmatrix} (i = 1, \ldots, N); F_{ij} := \begin{bmatrix} M_{ij} & 0 \\ 0 & 0 \end{bmatrix} (i, j = 1, \ldots, N).$$

F_i's and F_{ij}'s are $(N+1) \times (N+1)$ symmetric matrices. The $N \times N$ symmetric matrix M_{ij} has only nonzero elements at (i,j) and (j,i) with value $-\frac{1}{2}$ for $i \neq j$; If $i = j$ then M_{ii} becomes a diagonal matrix whose only nonzero element is the i-th element in the diagonal which will take -1 (i.e. $M_{ii} := diag(e_i)$, where $e_i \in \mathbb{R}^N$ is the i-th canonical basis vector of $\mathbb{R}^N$). Note that all matrices F_i's and F_{ij}'s are sparses (only at most 2 nonzero elements for each). The sparse structure of LMI (12) can be handled more efficiently than the dense structure in numerical computation of SDP.

Therefore, the two LMIs in (7) are rewritten as one single LMI as

$$\hat{Q}(x, W) := \hat{Q}_0 + \sum_{i=1}^{N} x_i \hat{Q}_i + \sum_{i,j=1}^{N} w_{ij} \hat{Q}_{ij} \preceq 0 \tag{13}$$

where

$$\hat{Q}_0 := \begin{bmatrix} Q_i & 0 \\ 0 & F_i \end{bmatrix}, (i = 0, \ldots, N); \hat{Q}_{ij} := \begin{bmatrix} Q_{ij} & 0 \\ 0 & Fij \end{bmatrix} (i, j = 1, \ldots, N).$$

Q_i, Q_{ij} belong to S^k, and F_i, F_{ij} belong to S^{N+1}. Hence, $\hat{Q}_i$ and $\hat{Q}_{ij}$ belong to S^{k+N+1} with sparse structure.

Finally, from (8) and (13), the problem (7) is equivalent to:

$$\begin{aligned} 0 = \min \ & f(x, W) = Trace(W) - \sum_{i=1}^{N} x_i^2 \\ s.t. \ & (x, W) \in \mathcal{C}.\hat{Q}(x, W) := \hat{Q}_0 + \sum_{i=1}^{N} x_i \hat{Q}_i + \sum_{i,j=1}^{N} w_{ij} \hat{Q}_{ij} \preceq 0 \\ & (x, W) \in \mathcal{H}_W. \end{aligned} \tag{14}$$

where $\mathcal{C} := \{(x, W) \in \mathcal{H}_W : \hat{Q}(x, W) := \hat{Q}_0 + \sum_{i=1}^{N} x_i \hat{Q}_i + \sum_{i,j=1}^{N} w_{ij} \hat{Q}_{ij} \preceq 0\}$. We have to minimize a quadratic concave function within a compact convex set $\mathcal{C}$ involving LMI. Clearly, it is a DC programming problem since concave function is a special DC function. The DC programming formulation of (14) is given by

$$\min \ f(x, W) = g(x, W) - h(x, W). \tag{15}$$

where $g(x, W) := \chi_{\mathcal{C}}(x, W)$ is a convex function ($\chi_{\mathcal{C}}(x, W)$ denotes the indicator function as $\chi_{\mathcal{C}}(x, W) := 0$ if $(x, W) \in \mathcal{C}$, $+\infty$ otherwise), and $h(x, W) := \sum_{i=1}^{N}(x_i^2 - w_{ii})$ is a quadratic convex function.

4   DC Programming and DCA

In order to give readers a brief view of DC programming and DCA, we will outline some essential theoretical results in this section. DC programming and DCA (DC Algorithm) were first introduced by Pham Dinh Tao in their preliminary form in 1985. They have been extensively developed since 1994 by Le Thi Hoai An and Pham Dinh Tao, see e.g. ([10, 11, 13, 16–22, 28–36]) and also the webpage http://lita.sciences.univ-metz.fr/~lethi/index.php/dca.html.   DCA has been successfully applied to many large-scale (smooth/nonsmooth) nonconvex programs in different fields of applied sciences for which they often give global solutions. The algorithm has proved to be more robust and efficient than many standard methods in nonlinear programming.

4.1   General DC Programming

Let $X = \mathbb{R}^n$ and $\|.\|$ be the Euclidean norm on X and let $\Gamma_0(X)$ denote the convex cone of all lower semi-continuous and proper (i.e. not identically equal to $+\infty$) convex functions defined on X and taking values in $\mathbb{R} \cup \{+\infty\}$. The dual space of X denotes by Y which could be identified with X itself. For $\theta \in \Gamma_0(X)$, the effective domain of θ, denoted dom θ, is defined by

$$\text{dom } \theta := \{x \in X : \theta(x) < +\infty\}. \tag{16}$$

DC programming investigates the structure of the vector space $DC(X) := \Gamma_0(X) - \Gamma_0(X)$, DC duality and optimality conditions for DC programs.

A general DC program is in the form of

$$(P_{dc}) \quad \alpha = \inf\{f(x) := g(x) - h(x) \; : \; x \in X\}, \tag{17}$$

with $g, h \in \Gamma_0 (X)$. Such a function f is called a DC function, and $g - h$, a DC decomposition of f while the convex functions g and h are DC components of f. In DC programming, the convention

$$(+\infty) - (+\infty) := +\infty \tag{18}$$

has been adopted to avoid the ambiguity on the determination of $(+\infty) - (+\infty)$. Such a case does not present any interest and can be discarded. We are actually concerned with the following problem

$$\alpha = \inf\{f(x) := g(x) - h(x) \; : \; x \in \text{dom } h\}, \tag{19}$$

which is equivalent to (P_{dc}) under the convention (18). Remark that if the optimal value α is finite then dom $g \subset$ dom h. It should be noted that a constrained DC program (C being a nonempty closed convex set)

$$\alpha = \inf\{\varphi(x) - \psi(x) \; : \; x \in C\} \tag{20}$$

is equivalent to the unconstrained DC program by adding the indicator function χ_C of C ($\chi_C (x) = 0$ if $x \in C$, $+\infty$ otherwise) to the first DC component φ :

$$\alpha = \inf\{g(x) - h(x) \; : \; x \in X\}, \tag{21}$$

where $g := \varphi + \chi_C$ and $h := \psi$. Let

$$g^*(y) := \sup\{\langle x, y \rangle - g(x) \; : \; x \in X\} \tag{22}$$

be the conjugate function of g. By using the fact that every function $h \in \Gamma_0(X)$ is characterized as a pointwise supremum of a collection of affine functions, say

$$h(x) := \sup\{\langle x, y \rangle - h^*(y) : y \in Y\}, \tag{23}$$

we have

$$\alpha = \inf\{h^*(y) - g^*(y) : y \in \text{dom } h^*\}, \tag{24}$$

that is written, in virtue of the convention (18):

$$(D_{dc}) \quad \alpha = \inf\{h^*(y) - g^*(y) : y \in Y\}, \tag{25}$$

which is the dual DC program of (P_{dc}). Finally the finiteness of α implies both inclusions dom $g \subset$ dom h and dom $h^* \subset$ dom g^*. We will denote by $\mathcal{P}$ and $\mathcal{D}$ be the solution sets of (P_{dc}) and (D_{dc}) respectively.

Recall that, for $\theta \in \Gamma_0(X)$ and $x_0 \in$ dom θ, $\partial\theta(x_0)$ denotes the subdifferential of θ at x_0, i.e.,

$$\partial\theta(x_0) := \{y \in Y : \theta(x) \ge \theta(x_0) + \langle x - x_0, y \rangle, \ \forall x \in X\} \tag{26}$$

(see [12, 37]). The subdifferential $\partial\theta(x_0)$ is a closed convex set in Y. It generalizes the derivative in the sense that θ is differentiable at x_0 if and only if $\partial\theta(x_0)$ is a singleton, which is exactly $\{\nabla\theta(x_0)\}$.

The domain of $\partial\theta$, denoted dom $\partial\theta$, is defined by dom $\partial\theta := \{x \in$ dom $\theta :$ $\partial\theta(x) \ne \emptyset\}$.

$$ri(\text{dom } \theta) \subset \text{dom } \partial\theta \subset \text{dom } \theta \tag{27}$$

where ri stands for the relative interior.

A function $\phi \in \Gamma_0(X)$ is a polyhedral convex function if it can be expressed as

$$\phi(x) = \sup\{\langle a_i, x \rangle - \gamma_i : i = 1, \ldots, m\} + \chi_K(x) \tag{28}$$

where $a_i \in Y$, $\gamma_i \in \mathbb{R}$ for $i = 1, \ldots, m$ and K is a nonempty polyhedral convex set in X (see [37]). A DC program is called polyhedral if either g or h is a polyhedral convex function. This class of DC programs, which is frequently encountered in real-life optimization problems and was extensively developed in our previous works (e.g. [16] and references therein), enjoys interesting properties (from both theoretical and practical viewpoints) concerning the local optimality and the finite convergence of DCA.

In the lack of verifiable conditions for global optimality of DC programs, we developed instead the following necessary local optimality conditions for DC programs in their primal part, by symmetry their dual part is trivial (see ([33, 35, 36]):

$$\partial h(x^*) \cap \partial g(x^*) \ne \emptyset \tag{29}$$

Such a point x^* is called critical point of $g - h$ or generalized KKT point for (P_{dc}), and

$$\emptyset \ne \partial h(x^*) \subset \partial g(x^*). \tag{30}$$

The condition (30) is also sufficient (for local optimality) in many important classes of DC programs. In particular it is sufficient for the next cases quite often encountered in practice ([33, 35, 36] and references therein):

- In polyhedral DC programs with h being a polyhedral convex function . In this case, if h is differentiable at a critical point x^*, then x^* is actually a local minimizer for (P_{dc}). Since a convex function is differentiable everywhere except for a set of measure zero, one can say that a critical point x^* is almost always a local minimizer for (P_{dc}).

– In case the function f is locally convex at x^*. Note that, if h is polyhedral convex, then $f = g - h$ is locally convex everywhere h is differentiable.

The transportation of global solutions between (P_{dc}) and (D_{dc}) can be expressed by:

$$\bigcup_{y^* \in \mathcal{D}} \partial g^*(y^*) \subset \mathcal{P}, \quad \bigcup_{x^* \in \mathcal{P}} \partial h(x^*) \subset \mathcal{D}. \tag{31}$$

Moreover, the equality holds in the first inclusion of (31) if $\mathcal{P} \subseteq \mathrm{dom}\ \partial h$, in particular if $\mathcal{P} \subseteq \mathrm{ri}(\mathrm{dom}\ h)$ according to (27). Similar property relative to the second inclusion can be stated by duality. On the other hand, under technical conditions this transportation holds also for local solutions of (P_{dc}) and (D_{dc}) (see [33–36]).

4.2 DC Algorithm (DCA)

Based on local optimality conditions and DC duality, the DCA consists in constructing of two sequences $\{x^k\}$ and $\{y^k\}$ of trial solutions of the primal and dual programs respectively, such that the sequences $\{g(x^k) - h(x^k)\}$ and $\{h^*(y^k) - g^*(y^k)\}$ are decreasing, and $\{x^k\}$ (resp. $\{y^k\}$) converges to a primal feasible solution $\tilde{x}$ (resp. a dual feasible solution $\tilde{y}$) satisfying local optimality conditions and

$$\tilde{x} \in \partial g^*(\tilde{y}), \quad \tilde{y} \in \partial h(\tilde{x}). \tag{32}$$

The sequences $\{x^k\}$ and $\{y^k\}$ are determined in the way that x^{k+1} (resp. y^{k+1}) is a solution to the convex program (P_k) (resp. (D_{k+1})) defined by ($x^0 \in \mathrm{dom}\ \partial h$ being a given initial point and $y^0 \in \partial h(x^0)$ being chosen)

$$(P_k) \quad \inf\{g(x) - [h(x^k) + \langle x - x^k, y^k \rangle] : x \in \mathrm{X}\}, \tag{33}$$

$$(D_{k+1}) \quad \inf\{h^*(y) - [g^*(y^k) + \langle y - y^k, x^{k+1} \rangle] : y \in \mathrm{Y}\}. \tag{34}$$

The DCA has the quite simple interpretation: at the k-th iteration, one replaces in the primal DC program (P_{dc}) the second component h by its affine minorization $h^{(k)}(x) := h(x^k) + \langle x - x^k, y^k \rangle$ defined by a subgradient y^k of h at x^k to give birth to the primal convex program (P_k), the solution of which is nothing but $\partial g^*(y^k)$. Dually, a solution x^{k+1} of (P_k) is then used to define the dual convex program (D_{k+1}) obtained from (D_{dc}) by replacing the second DC component g^* with its affine minorization $(g^*)^{(k)}(y) := g^*(y^k) + \langle y - y^k, x^{k+1} \rangle$ defined by the subgradient x^{k+1} of g^* at y^k. The process is repeated until convergence. DCA performs a double linearization with the help of the subgradients of h and g^* and the DCA then yields the next scheme: (starting from given $x^0 \in \mathrm{dom}\ \partial h$)

$$y^k \in \partial h(x^k); \quad x^{k+1} \in \partial g^*(y^k), \ \forall k \geq 0. \tag{35}$$

DCA's distinctive feature relies upon the fact that DCA deals with the convex DC components g and h but not with the DC function f itself. DCA is one of the rare algorithms for nonconvex nonsmooth programming. Moreover, a DC function f has infinitely many DC decompositions which have crucial implications for the qualities (speed of convergence, robustness, efficiency, globality of computed solutions,...) of DCA. For a given DC program, the choice of *optimal* DC decompositions is still open. Of course, this depends strongly on the very specific structure of the problem being considered. In order to tackle the large-scale setting, one tries in practice to choose g and h such that sequences $\{x^k\}$ and $\{y^k\}$ can be easily calculated, *i.e.*, either they are in an explicit form or their computations are inexpensive.

Theorem 2 (Convergence Theorem of DCA). *([33, 35, 36]) DCA is a descent method without line-search which enjoys the following primal properties (the dual ones can be formulated in a similar way):*

1. *The sequences $\{g(x^k) - h(x^k)\}$ and $\{h^*(y^k) - g^*(y^k)\}$ are decreasing and*
   - *$g(x^{k+1}) - h(x^{k+1}) = g(x^k) - h(x^k)$ if and only if $y^k \in \partial g(x^k) \cap \partial h(x^k)$, $y^k \in \partial g(x^{k+1}) \cap \partial h(x^{k+1})$ and $[\rho(g, C) + \rho(h, C)]\|x^{k+1} - x^k\| = 0$. Moreover if g or h are strictly convex on C, then $x^k = x^{k+1}$. In such a case DCA terminates at finitely many iterations.*
     Here C (resp. D) denotes a convex set containing the sequence $\{x^k\}$ (resp. $\{y^k\}$) and $\rho(g, C)$ denotes the modulus of strong convexity of g on C given by:

     $$\rho(g, C) := \sup\{\rho \geq 0 : g - (\rho/2)\|.\|^2 \text{ be convex on } C\}.$$

   - *$h^*(y^{k+1}) - g^*(y^{k+1}) = h^*(y^k) - g^*(y^k)$ if and only if $x^{k+1} \in \partial g^*(y^k) \cap \partial h^*(y^k)$, $x^{k+1} \in \partial g^*(y^{k+1}) \cap \partial h^*(y^{k+1})$ and $[\rho(g^*, D) + \rho(h^*, D)]\|y^{k+1} - y^k\| = 0$. Moreover if g^* or h^* are strictly convex on D, then $y^k = y^{k+1}$. In such a case DCA terminates at finitely many iterations.*

2. *If $\rho(g, C) + \rho(h, C) > 0$ (resp. $\rho(g^*, D) + \rho(h^*, D) > 0$) then the series $\{\|x^{k+1} - x^k\|^2\}$ (resp. $\{\|y^{k+1} - y^k\|^2\}$) converges.*
3. *If the optimal value of P_{dc} is finite and the infinite sequence $\{x^k\}$ and $\{y^k\}$ are bounded then every limit point x^∞ (resp. y^∞) of the sequence $\{x^k\}$ (resp. $\{y^k\}$) is a critical point of $g - h$ (resp. $h^* - g^*$).*
4. *DCA has a linear convergence for general DC programs.*
5. *DCA has a finite convergence for polyhedral DC programs.*

The convergence of DCA is always guarantied whatever an initial point (could be a feasible or even infeasible point of the DC program) is. However, the choice of an initial point is important that could affect the number of iterations and the globality of the obtained solution. Searching a good initial point (leads a convergence to a global optimal solution) is often an open question in applications, since it is highly depending on the specific structure of the DC function and the problem's constraints.

4.3   Geometric Interpretation of DCA

In this subsection, we present a new result on the geometric view of DCA, this result is first introduced by the author Niu in his PhD thesis[29] in 2008. Let's consider in the k-th iteration of DCA, the convex majorization $g(x) - (h(x^k) + \langle x - x^k, y^k \rangle)$ of the DC function $g(x) - h(x)$ is constructed at the point x^k (with $y^k \in \partial g(x^k)$). The following theorem can be proved:

Theorem 3. *Let $f(x) = g(x) - h(x)$ be a DC function on the convex set $\mathcal{C}$. Given a point $x^k \in \mathcal{C}$, let's define the convex function $f^k(x) := g(x) - [h(x^k) + \langle x - x^k, y^k \rangle]$ at the point x^k (with $y^k \in \partial g(x^k)$) and suppose that g is a differentiable function, we have*

1. $f^k(x) \geq f(x), \forall x \in \mathcal{C}$.
2. $f^k(x^k) = f(x^k)$.
3. $\nabla f^k(x^k) = \nabla f(x^k)$.

Proof. Since $h(x)$ is a convex function on $\mathcal{C}$, we have $h(x) \geq h(x^k) + \langle x - x^k, y^k \rangle$, $\forall x \in \mathcal{C}$ which yields $f^k(x) = g(x) - [h(x^k) + \langle x - x^k, y^k \rangle] \geq g(x) - h(x), \forall x \in \mathcal{C}$. Obviously, $f^k(x^k) = g(x^k) - [h(x^k) + \langle x^k - x^k, y^k \rangle] = g(x^k) - h(x^k) = f(x^k)$. And $\nabla f^k(x) = \nabla g(x) - y^k = \nabla g(x) - \nabla h(x^k)$, finally $\nabla f^k(x^k) = \nabla g(x^k) - \nabla h(x^k) = \nabla f(x^k)$. □

Theorem 3 gives us an explicit geometric view between the surfaces of f^k and f. In fact, f^k is coincide with f at the point x^k, and f^k is a convex function and always located above f on $\mathcal{C}$ (i.e., f^k is a convex majorization of f on $\mathcal{C}$). Roughly speaking, we can image the surface of the function f^k as a "bowl" locating above the surface of f and touch with f at the point $(x^k, f(x^k))$. Thus, the principal of DCA can be illustrated in the following figure:

We have an easy understanding how DCA works (find a point x^{k+1} from the point x^k and the convergent tendency). Many important properties of DCA have intuitional explanations on this figure. For instance, we know that DCA will generate a decreasing sequence $\{f(x^k)\}$. This can be observed directly in the figure, since $f \leq f^k$ and the point x^{k+1} is a minimum of the convex function f^k over the convex set $\mathcal{C}$ which yield $f(x^{k+1}) \leq f^k(x^{k+1}) \leq f^k(x^k) = f(x^k)$. The relationships among the values $f(x^{k+1}), f^k(x^{k+1}), f(x^k)$ are directly revealed in the figure. The convergence of DCA can be also pointed out since $\{f(x^k)\}$ is a decreasing sequence and bounded below, i.e. $\{f(x^k)\}$ converges (and probably tends to the global optimal value $f(x^*)$ in this figure). We also observe that starting from some possible positions of x^k, DCA could jump over a minimiser to approach another minimizer. e.g., in the figure, DCA jumps over a local minimizer located between x^k and x^{k+1} and approach the global minimizer x^*. This special feature has very important impact to the solution quality of DCA especially in nonconvex optimization. It would be a key point to make difference on the performance between DCA and many other standard algorithms. It's also easy to understand that the jumping depends on where to jump and where

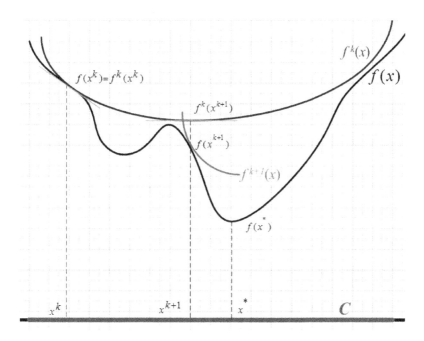

Fig. 1. Geometric interpretation of DCA

is the minimizer of the convex majorization f^k, that is equivalent to say that the performance of DCA depends on *the choice of initial point* and *the DC decomposition of f*. Therefore, these two questions must be carefully considered in practice.

5   DCA for Solving QMIFP

In this section, we focus on how to solve QMIFP via DCA. The problem (2) has been reformulated as a DC program (15) in section 2 as follows:

$$\min \ f(x, W) = g(x, W) - h(x, W). \tag{36}$$

where $h(x, W) := \sum_{i=1}^{N}(x_i^2 - w_{ii})$ is a continuous and differentiable convex quadratic function, and $g(x, W) := \chi_{\mathcal{C}}(x, W)$ is also a convex function since it is the indicator function defined on the convex compact set $\mathcal{C} := \{(x, W) \in \mathcal{H}_W, \hat{Q}(x, W) := \hat{Q}_0 + \sum_{i=1}^{N} x_i \hat{Q}_i + \sum_{i,j=1}^{N} w_{ij} \hat{Q}_{ij} \preceq 0\}$. (here $\mathcal{H}_W := \{(x, W) \in \mathbb{R}^N \times S^N : \underline{x} \leq x \leq \bar{x}, \underline{w}_{ij} \leq w_{ij} \leq \bar{w}_{ij}, i, j = 1, \dots, N\}$).

According to the general framework of DCA, we can generate two sequences $\{(x^k, W^k)\}$ and $\{(y^k, Z^k)\}$ via the following scheme:

$$(x^k, W^k) \to (y^k, Z^k) \in \partial h(x^k, W^k)$$
$$(x^{k+1}, W^{k+1}) \in \partial g^*(y^k, Z^k). \tag{37}$$

Since $h(x, W) := \sum_{i=1}^{N}(x_i^2 - w_{ii})$ is a continuous and differentiable function, then $\partial h(x^k, W^k)$ reduces to a singleton $\{\nabla h(x^k, W^k)\}$ which could be computed explicitly:

$$(y^k, Z^k) \in \partial h(x^k, W^k) \Leftrightarrow \{(y^k, Z^k) = \nabla h(x^k, W^k)\}$$
$$\Leftrightarrow \{y^k = 2x^k; Z^k = -I_N\}. \tag{38}$$

where I_N denotes the $N \times N$ identity matrix.

For computing $(x^{k+1}, W^{k+1}) \in \partial g^*(y^k, Z^k)$, we can solve the following program:

$$(x^{k+1}, W^{k+1}) \in argmin\{Trace(W) - \langle y^k, x \rangle : (x, W) \in \mathcal{C}\}. \tag{39}$$

$Trace(W)$ denotes the trace of the matrix W. In the above problem, Z^k is not been used, i.e., for computing (x^{k+1}, W^{k+1}), we only need x^k (computing Z^k is not necessary). Finally, DCA yields the following iterative scheme:

$$(x^{k+1}, W^{k+1}) \in argmin\{Trace(W) - \langle 2x^k, x \rangle : (x, W) \in \mathcal{C}\}. \tag{40}$$

The problem (40) is to minimize the linear function $Trace(W) - \langle 2x^k, x \rangle$ over the convex set $\mathcal{C}$. It is a *linear LMI program*. Since the dual problem of LMI is a SDP (Semidefinite Program), we can solve this problem via many efficient SDP solvers such as SDPA, SeDuMi, SDPT, Yalmip ...

For determining stopping criterions of DCA, the objective function $f(x, W)$ must be non-negative over the constraint of (7) since $\begin{bmatrix} W & x \\ x^T & 1 \end{bmatrix} \succeq 0 \Rightarrow W - xx^T \succeq 0 \Rightarrow f(x, W) = Trace(W) - \sum_{i=1}^{N} x_i^2 = Trace(W - xx^T) \geq 0$. Therefore, BMIFP/QMIFP is feasible if and only if there exists a feasible point $(\tilde{x}, \tilde{W}) \in \mathcal{C}$ such that $f(\tilde{x}, \tilde{W}) = 0$.

This feature can be used as first stopping criterion of our algorithm as:

Stopping Criterion 1: *DCA will stop at kth iteration if $f(x^k, W^k) \leq \epsilon_1$ (where ϵ_1 is a given small positive number).*

On the other hand, It is also possible that the sequence $\{f(x^k, W^k)\}$ converges to a number different from zero (For instance, when BMIFP is infeasible). In this case, we can stop DCA by its general stopping condition as:

Stopping Criterion 2: *DCA can stop at k-th iteration if $\frac{|f(x^k, W^k) - f(x^{k-1}, W^{k-1})|}{|f(x^k, W^k)|} \leq \epsilon_2$ (where ϵ_2 is a given small positive number).*

The full algorithm DCA for solving QMIFP is described as follows:

DCA1
Initialization:
Choose an initial point $\underline{x} \leq x^0 \leq \bar{x}$, compute $W^0 = x^0 x^{0T}$.
Let ϵ_1, ϵ_2 be sufficiently small positive numbers.
Let iteration number $k = 0$.

Iteration: $k = 0, 1, 2, \ldots$
Solve SDP program (40) to compute (x^{k+1}, W^{k+1}) from (x^k, W^k).

Stopping criterions:
If $f(x^{k+1}, W^{k+1}) \leq \epsilon_1$ then
Stop. return "QMIFP is feasible and (x^{k+1}, W^{k+1}) is a feasible solution".
Else if $|f(x^{k+1}, W^{k+1}) - f(x^k, W^k)| \leq \epsilon_2 |f(x^k, W^k)|$ then
Stop. return "No feasible solution is found".
Else, let $k = k + 1$ and repeat the step **Iteration**.

Theorem 4 (Convergence Theorem of DCA1). *DCA1 can generate a sequence $\{(x^k, W^k)\}$ such that the sequence $\{f(x^k, W^k)\}$ is decreasing and bounded below (i.e. convergent). The sequence $\{(x^k, W^k)\}$ will converge either to a global optimal solution of the DC formulation (36) or to a general KKT (Karush-Kuhn-Tucker) point of (36).*

Proof. The proof of this theorem is obvious which is based on the general convergence theorem of DCA [33, 35, 36]. The Stopping Criterion 1 is a global optimality condition for the feasible QMIFP. Without this condition, DCA1 can also converge to a general KKT point of (36) due to the general convergence of DCA.   □

6   For Solving a Large-Scale DC Program

In the algorithm DCA1, each iteration requires solving a SDP subproblem. However, it's well known that the computational time for solving large-scale SDP will be expensive. We have also tested several SDP solvers: SDPA, SeDuMi, CSDP and DSDP. To our knowledge, they are currently the most popular high performance SDP solvers. We found in our tests that for solving a non-sparse SDP with 500 variables and 300 linear constraints requires about $15-20$ seconds. The SDP subproblem in DCA1 has $m + n + \frac{(m+n)(m+n+1)}{2}$ variables and one LMI constraint of order $[3(m+n)+k+1] \times [3(m+n)+k+1]$. If m, n and k are large numbers, SDP subproblem will be hard to be numerically solved. In order to reduce the computational time for each iteration of DCA, we propose a strategy which could be referees to as *"partial solution"* or *"sub-solution"*. More precisely, we can stop SDP solution process when some conditions are satisfied. For instance, we can stop SDP solution procedure after a fixed number of iterations.

The reason is that, DCA don't really require solving a SDP in each iteration, the objective is to improve the value of the objective function, i.e., find a better feasible point (x^{k+1}, W^{k+1}) from the current feasible point (x^k, W^k) such that $f(x^{k+1}, W^{k+1}) < f(x^k, W^k)$. Therefore, we don't need to find a minimizer of SDP subproblem for each iteration of DCA, it is sufficient to find a better feasible solution than the current one. Therefore, we can stop the SDP solution process when a better feasible point is found. The proposed strategy could be helpful when tackling large-scale problems. Thus, a modified DCA is proposed as follows:

DCA2

Initialization:
Choose an initial point $\underline{x} \le x^0 \le \bar{x}$, compute $W^0 = x^0 x^{0T}$.
Let ϵ_1, ϵ_2 be sufficiently small positive numbers.
Let iteration number $k = 0$.

Iteration: $k = 0, 1, 2, ...$
Execute SDP solution procedure to (40) and check in each iteration if we have found a better feasible solution (x^{k+1}, W^{k+1}) satisfying $f(x^{k+1}, W^{k+1}) < f(x^k, W^k)$.

Stopping criterions:
If $f(x^{k+1}, W^{k+1}) \le \epsilon_1$ then
Stop. return "QMIFP is feasible and (x^{k+1}, W^{k+1}) is a feasible solution".
Else if $|f(x^{k+1}, W^{k+1}) - f(x^k, W^k)| \le \epsilon_2 |f(x^k, W^k)|$ then
Stop. return "No feasible solution is found".
Else, let $k = k + 1$ and repeat **Iteration**.

Theorem 5 (Convergence Theorem of DCA2). *DCA2 can generate a sequence $\{(x^k, W^k)\}$ such that the sequence $\{f(x^k, W^k)\}$ is decreasing and bounded below (i.e. convergent). The sequence $\{(x^k, W^k)\}$ will converge either to a global optimal solution of the DC formulation (36) or to a general KKT point of (36).*

Proof. The proof of this theorem is almost the same as the convergence theorem of DCA1. Globally solving a SDP subproblem is not necessary at each iteration of DCA, so finding a better feasible point instead of a minimizer of SDP will not change the convergence of DCA.

Remark 1. The sub-solution strategy will affect the total number of iterations of DCA and the value of the computed solution. In general, the number of iterations will increase when applying this strategy. However, since the computational time for each iteration is extensively reduced, thus the total computational time could be also reduced.

7   Hybrid Algorithm DCA-B&B for Solving QMIFP

We must emphasize that DCA is a local optimization method for nonconvex dc program. For a given DC formulation, the quality of the computed solution depends on the choice of initial point. In DCA1 and DCA2, if the first stopping condition is not satisfied at the convergence of DCA, we can not decide the feasibility of the problem QMIFP since the computed solution may be not a global minimizer of the DC program (36). In order to determine whether the problem QMI/BMI is feasible or not, we must recur to some global optimization techniques for further research. In this section, we propose a hybrid method that combining DCA with a branch-and-bound (B&B) scheme.

The Branch-and-Bound (B&B) method used in this paper is an alterative procedure in which the hyper-rectangle of $x-space$, $\mathcal{H} := \{x \in \mathbb{R}^N : \underline{x} \leq x \leq \bar{x}\}$, will be partitioned into small partitions (branching procedure), and for each partition M, we will estimate a lower bound $\beta(M)$ of $f(x, W)$ over $\mathcal{C} \cap M$, and determining a upper bound $\alpha(M)$ via DCA (bounding procedure).

The principle of the proposed B&B is described as follows: Those partitions M with $\beta(M) > 0$ cannot provide any optimal solution, and then we can discard them from further consideration. On the other hand, the partition with smallest $\beta(M)$ could be considered as the most promising partition, which will be subdivided in the next step into more refined subsets for further investigation. For those partitions M with $\beta(M) \leq 0$, local search on M via DCA will be carried out in order to reduce the upper bound. For any partition M on which DCA obtains an upper bound $\alpha(M) = 0$, we can then terminate immediately the B&B procedure and the corresponding upper bound solution is the required feasible solution of QMIFP. Finally, if all partitions are discarded and no feasible solution has been found by DCA, then QMIFP is infeasible.

Let's get into details. Given a partition $M := [p, q] = \{x \in \mathbb{R}^N : p \leq x \leq q\} \subseteq \mathcal{H}$. The optimization problem on M (often called node problem) is defined as:

$$\begin{aligned} \min \; & f(x, W) = Trace(W) - \sum_{i=1}^{N} x_i^2 \\ s.t. \; & (x, W) \in \mathcal{C} \cap M. \end{aligned} \qquad (41)$$

The convex underestimation of f on M can be easily constructed as $F_M(x, W) := Trace(W) + \langle c, x \rangle + d$ (where $c = -(p + q)$, $d = \langle p, q \rangle$) satisfying $F_M(x, W) \leq f(x, W), \forall (x, W) \in M$. In fact, F_M is the convex envelope of f on M. We can then define the lower bound problem on the partition M as:

$$\beta(M) = \min\{F_M(x, W) : (x, W) \in \mathcal{C} \cap M\}. \qquad (42)$$

The upper bound on M can be achieved by applying DCA to the problem (41) which yields the following iterative scheme:

$$(x^{k+1}, W^{k+1}) \in argmin\{Trace(W) - \langle 2x^k, x \rangle : (x, W) \in \mathcal{C} \cap M\}. \qquad (43)$$

Each iteration point $(x^{k+1}, W^{k+1}) \in \mathcal{C} \cap M$ provides an upper bound $\alpha(M) := f(x^{k+1}, W^{k+1})$, and the sequence $\{f(x^{k+1}, W^{k+1})\}$ is decreasing and convergent. Therefore, the upper bound $\alpha(M)$ is decreasing and convergent too.

The DCA-B&B method involves three basic operations:

1. **Branching:** For each iteration of B&B, a partition $M = [p, q]$ is selected and subdivided into two subsets via some hyperplan defined as $x_{i_M} = \lambda$ (i_M and $\lambda \in (p_{i_M}, q_{i_M})$ are chosen via some specified rules). When $i_M = argmax\{q_j - p_j : j = 1, \dots, N\}$ and $\lambda = (p_{i_M} + q_{i_M})/2$, this rule of subdivision is called the *standard bisection*. In our paper, instead of using standard bisection, we would like to use *an adaptive subdivision rule* with consideration of the information so far obtained during the exploration on the partition M.

   Adaptive subdivision rule: Given a partition $M = [p, q]$, and apply DCA to (41) for obtaining a computed solution (x^*, W^*). Suppose it is not a feasible solution to QMIFP. We will compute i_M and λ as follows:

   - $(i^*, j^*) \in argmax\{|w_{ij}^* - x_i^* x_j^*| : \forall i, j = 1, \dots, N\}$.
   - $i_M = argmax\{|p_{i^*} - q_{i^*}|, |p_{j^*} - q_{j^*}|\}$.
   - $\lambda = (p_{i_M} + q_{i_M})/2$.

   Then, the partition set M will be divided into two subsets according to the $x_{i_M} = \lambda$.
   The objective of this adaptive subdivision rule is to potentially reduce the gaps between w_{ij} and $x_i x_j$ for all $i, j = 1, \dots, N$, the reason is that any feasible point (x, W) of QMIFP must satisfy the equality $W = xx^T$, i.e. the gaps are equals to 0.

2. **Bounding:** For each partition M, The lower bound $\beta(M)$ obtains by solving the problem (42) and the upper bound $\alpha(M)$ is computed via the scheme (43). It's possible to stop DCA-B&B if $\alpha(M) = 0$.

3. **Discarding:** A partition M will be discarded if one of the conditions is true:
   - The lower bound problem (42) defined on M is infeasible.
   - $\beta(M) > 0$.

For starting DCA on a partition M, we also need an initial point. A good initial point will lead to a fast convergence and a global optimal solution on M. When the compact set $\mathcal{C} \cap M$ is not empty, the existence of such a good initial point is ensured. We propose using the lower bound solution of (42) as an initial point of DCA. The full DCA-B&B algorithm is illustrated as follows:

DCA and B&B for QMIFP
Initialization:
Start with $M_0 = [\underline{x}, \bar{x}]$.
Solving the SDP problem (42) to find an initial point (x^0, W^0) and $\beta(M_0)$.
Define ϵ_1 as a sufficiently small positive number.
Let iteration number $k = 0$ and $\mathcal{S} = \{M_0\}$.

If $f(x^0, W^0) \leq \epsilon_1$, then STOP
QMIFP is feasible and (x^0, W^0) is a feasible solution.
Else, goto Step1.

Step1:
If $\mathcal{S} = \emptyset$, then STOP, QMIFP is infeasible.
Else, find a partition $M_k = [p, q] \in \mathcal{S}$ with the smallest $\beta(M_k)$.
Let (x^*, W^*) be the lower bound solution corresponding to $\beta(M_k)$.
Start DCA with initial point (x^*, W^*) to (41), its computed solution
is denoted as (x^k, W^k).

If $\alpha(M) := f(x^k, W^k) \leq \epsilon_1$, then STOP
QMIFP is feasible and (x^k, W^k) is a feasible solution.
Else, goto Step2.

Step2:
- $(i^*, j^*) \in argmax\{|w_{ij}^* - x_i^* x_j^*| : i, j = 1, \ldots, N\}$.
- $i_{M_k} = argmax\{|p_{i^*} - q_{i^*}|, |p_{j^*} - q_{j^*}|\}$.
- $\lambda = (p_{i_{M_k}} + q_{i_{M_k}})/2$.
- Divide M_k into two partitions $M_{k,1} := M_k \cap \{p_{i_{M_k}} \leq x_{i_{M_k}} \leq \lambda\}$
and $M_{k,2} := M_k \cap \{\lambda \leq x_{i_{M_k}} \leq q_{i_{M_k}}\}$. goto Step3.

Step3:
Compute respectively $\beta(M_{k,1})$ and $\beta(M_{k,2})$.
If $\beta(M_{k,1}) \leq \epsilon_1$ (resp. $\beta(M_{k,2}) \leq \epsilon_1$), then
Set $\mathcal{S} = \mathcal{S} \cup \{M_{k,1}\}$ (resp. $\mathcal{S} = \mathcal{S} \cup \{M_{k,2}\}$).
Endif

$\mathcal{S} = \mathcal{S} \setminus \{M_k\}$.
Set $k = k + 1$ goto Step1.

---

8   Computational Experiments

In this section, we present some numerical results of computational experiments.
Our algorithm DCA and DCA-B&B are implemented in MATLAB 2008A and
tested on PC (Vista, Intel Duo P8400, CPU 2GHz, 2G RAM). YALMIP [24] is
incorporated in the subroutines to create LMI constraints and to invoke SDP
solvers. YALMIP provides an advanced MATLAB interface for using many well-
known optimization solvers. In our experiments, we test these 4 SDP solvers:
SDPA [6] by Kojima, SeDuMi [40] by Sturm, CSDP [3] by Borchers and DSDP
[1] by Benson and Ye. They are all supported by YALMIP. Our computational
results are compared with the commercial software PENBMI version 2.1 [15]
developed by Michal Kočvara and Michael Stingl. PENBMI is an implementation of the exterior penalty and interior barrier methods with the Augmented

Lagrangian method. In our knowledge, it is currently the only commercial solver for BMI problems. It is now a part of the Matlab-based commercial optimisation software TOMLAB.

The test data were randomly generated via Matlab. We use a code "ranbmi.m" given in the package of PENBMI for generating random BMI instances. Q_i and Q_{ij} were symmetric $k \times k$ sparse matrices which are randomly generated with a given density $dens$ (i.e. with approximately $dens \times k \times k$ nonzero entries). The lower and upper bounds of the variables x and y are restricted into the interval $[-10, 10]$.

Some parameters for generating random test data are presented in Table 1.

Table 1. Parameters for generating random test data

parameter	value	description
n	any positive integer	dimension of the variable x
m	any positive integer	dimension of the variable y
k	any positive integer	Q_i and Q_{ij} are matrices of dimension $k \times k$
$dens$	a value in $]0, 1]$	density of sparse matrix for Q_i and Q_{ij}

As we have introduced in section 1, any BMIFP can be easily converted into QMIFP. Therefore, we transforme all BMIFP into QMIFPs. Since a QMI constraint

$$Q_0 + \sum_{i=1}^{N} x_i Q_i + \sum_{i,j=1}^{N} x_i x_j Q_{ij} \preceq 0$$

is defined by the matrices $Q_i, i = 0, \ldots, N$ and $Q_{ij}, i, j = 1, \ldots, N$. For storing these matrices, we only need to store $Q_i, i = 0, \ldots, N$ and $Q_{ij}, i = 1, \ldots, N; j = i, \ldots, N$ since $Q_{ij} = Q_{ji}$. In our program, we have defined a special QMI structure whose fields were summarized in Table 2.

Table 2. QMI structure

field	type	description
$Q0$	matrix	a $k \times k$ matrix for Q_0
QI	list	a list of nonzero matrices for $Q_i, i = 1, \ldots, N$
$IdxQI.N$	integer	the number of total elements in the list QI
$IdxQI.IDX$	list	a list of index numbers for QI
QIJ	list	a list of nonzero matrices for $Q_{ij}, i = 1, \ldots, N; j = i, \ldots, N$
$IdxQIJ.N$	integer	the number of total elements in the list QIJ
$IdxQIJ.IDX$	matrix	the (i, j)-element denotes the index number of Q_{ij} in the list QIJ

Remark 2. For storing all nonzero matrices of $Q_i, i = 1, \ldots, N$, we use the fields: $QI, IdxQI.N$ and $IdxQI.IDX$. The list QI (as cell in Matlab) saves all nonzero matrices of $Q_i, i = 1, \ldots, N$, and $IdxQI.N$ denotes the total number of nonzero matrices. The list $IdxQI.IDX$ saves index i for the nonzero matrices Q_i and

0 for null matrix. For instance, suppose Q_1 and Q_3 are both nonzero matrices, and $Q_2 = 0$. We have $QI = \{Q_1, Q_3\}$, $IdxQI.N = 2$ and $IdxQI.IDX = \{1, 0, 3\}$. Using these three fields (QI, $IdxQI.N$ and $IdxQI.IDX$), we can easily access any matrix Q_i with a given index i form 1 to N. The fields (QIJ, $IdxQIJ.N$ and $IdxQIJ.IDX$) are similarly defined as the fields (QI, $IdxQI.N$ and $IdxQI.IDX$). The only difference is that $IdxQIJ.IDX$ is a 1-D list while $IdxQIJ.IDX$ is a 2-D matrix whose (i, j)-element denotes the index number of Q_{ij} to locate this matrix in the list QIJ. Obviously, $IdxQIJ.IDX$ must be a symmetric matrix since $Q_{ij} = Q_{ji}$ for all i and j from 1 to N. Using this QMI structure, we can define a QMI constraint in a nature way and access easily all matrices Q_i and Q_{ij} in this structure.

The parameters and their preferable values for DCAs and DCA-B&B algorithms are summarized in Table 3.

Table 3. Parameters for DCAs and DCA-B&B Algorithms

parameter	value	description
ϵ_1	10^{-5}	computational zero (for DCA and DCA-B&B)
ϵ_2	10^{-3}	relative error - stopping criterion for DCA

Now, let's present some numerical experiments and their results.
Firstly, we achieve a benchmark test for comparing the performance of DCA with different SDP solvers (SDPA, SeDuMi, CSDP and DSDP). A fast SDP solver is very important since DCA needs solving a SDP subproblem in each iteration. In general, it is known that SDPA is a faster SDP solver for medium and large-scale problems than SeDuMi and other two solvers. See Mittelmann [27] for a comparison report (including SDPA, SeDuMi, CSDP and DSDP) on the SDPLIB test set. In order to confirm the exact performance of these solvers and find the fastest one in combination with DCA, we randomly generate 10 BMIFPs with fixed parameters $n = 2$, $m = 3$, $k = 3$ and $dens = 1$, then solving them via DCA with these SDP solvers. The consuming time were summarized in Table 4.

It's obvious to see that SDPA is the fastest SDP solver (according to the average solution time) incorporated with our DCA. We can rank the SDP solvers in increasing order of the average consuming time as: SDPA, SeDuMi, CSDP and DSDP. The columns *feas* denotes the feasibility of BMIFPs, T means DCA found a feasible solution, while F means no feasible solution was found within 30 iterations. For these problems, they are probably infeasible, since the objective value are far from zero. To guarantee the infeasibility/feasibility, we should use DCA-B&B algorithm. Note that these SDP solvers for solving a given SDP subproblem do not often give same optimal solution, but give the same objective value. It depends on the solution method for SDP as well as the nature of SDP since the optimal solution of a SDP is unique if and only if it is a strictly convex problem or its solution set is a singleton. The different optimal solutions of SDP

Table 4. Performance of DCA with SDPA, SeDuMi, CSDP and DSDP

Problem	SDPA		SeDuMi		CSDP		DSDP	
	time	feas	time	feas	time	feas	time	feas
$P1$	0.148	T	1.203	T	0.960	T	0.736	T
$P2*$	1.158	F	2.434	F	3.887	F	2.120	F
$P3*$	1.169	F	2.464	F	4.154	F	2.180	F
$P4$	0.155	T	0.250	T	0.951	T	0.433	T
$P5$	0.439	T	0.688	T	1.153	T	0.677	T
$P6$	0.192	T	0.260	T	0.546	T	0.291	T
$P7$	0.369	T	0.240	T	0.822	T	1.314	T
$P8*$	1.175	F	0.278	F	4.072	F	1.685	F
$P9$	0.619	T	1.019	T	1.753	T	1.264	T
$P10$	0.233	T	0.313	T	0.796	T	0.324	T
Average time	0.566		0.915		1.909		1.102	

∗ algorithm terminates within 30 iterations at most.

subproblems could affect the final solution of DCA. Even so, we observe in the feasibility results that the impact to the feasibility of BMIFP is not evident. Therefore, we can choose the fastest SDP solver, i.e. SDPA for the further tests.

Table 5 gives some comparison results between DCA1 and PENBMI. The test problems are randomly generated with $dens = 0.1$. Each line of Table 3 consists of numerical results for a set of 20 feasible BMIFPs randomly generated with given dimensions m, n and k. the variables x and y are restricted to the interval $[-10, 10]$ and all matrices Q_i are of order from 3×3 to 100×100. The parameters ϵ_1 and ϵ_2 are fixed as in Table 3. The average iterations (ave.iter) and the average CPU time (ave.time) for DCA1 and PENBMI are presented and compared.

Comparing the numerical results in Table 5, DCA1 needs 2-3 iterations in average for finding a feasible solution, while PENBMI often requires 18-23 iterations. The computed feasible solution found by DCA and PENBMI could be different, and the tackled problem for PENBMI is defined as:

$$\min\{\lambda : F(x,y) \preceq \lambda I, (x,y) \in \mathcal{H}\}. \tag{44}$$

It's easy to prove that BMIFP is feasible if and only if the problem (44) has a feasible solution with $\lambda \leq 0$. In these tests, we compare the computational time of the two methods for determining a feasible solution of BMI. We observe that solving the problems of dimensions $n + m \leq 35$ and $k \leq 100$, DCA is faster than PENBMI. Moreover, an important feature of DCA is the stability in the number of iterations (often between 1-2.8 in average) for these feasible problems.

Solving the set of problems Pset13 - Pset15, DCA seems slightly time consuming than PENBMI. This may be due to the fact that DCA requires solving large-scale SDP subproblems with $m + n + \frac{(m+n)(m+n+1)}{2}$ variables and a LMI constraint of order $[3(m+n)+k+1] \times [3(m+n)+k+1]$. Solving such large-scale SDP subproblems are time consuming. Therefore, DCA2 with sub-solution strategy is proposed and compared with DCA1. We can see in Table 6 that DCA2 can reduce the solution time for some large-scale problems.

Table 5. Comparison between DCA1 and PENBMI for some sets of randomly generated feasible BMIFPs

Problem	Dimension			DCA1		PENBMI	
	n	m	k	ave.iter	ave.time	ave.iter	ave.time
Pset1	1	1	3	1	0.1580	14.6	0.8136
Pset2	2	3	5	1.2	0.1668	15.6	0.8442
Pset3	8	10	10	1.3	1.316	18.6	1.815
Pset4	10	10	40	1.5	1.573	19.2	2.125
Pset5	10	10	60	1.2	2.667	20.2	3.243
Pset6	10	10	70	1.2	3.848	16.8	4.029
Pset7	10	10	100	1.5	7.816	18.2	10.249
Pset8	15	10	30	2.1	2.909	20.5	3.677
Pset9	15	10	60	2.8	5.328	20.5	6.583
Pset10	15	10	80	1.6	9.125	19.5	22.800
Pset11	20	15	20	1.3	5.869	20.5	2.362
Pset12	20	15	60	2.5	25.490	19.8	35.841
Pset13	30	30	10	1.2	27.17	21.5	5.287
Pset14	40	30	6	1.5	30.492	18.5	7.596
Pset15	50	50	10	1.2	50.585	18.2	13.685

Each problem set Psetxx contains 20 problems with given dimensions m,n,k.

For the set of problems Pset12 - Pset15, DCA2 is faster than DCA1. But for small-scale problems (Pset1 - Pset11), DCA1 is faster which is probably due to the fact that DCA1 needs less number of iterations and solving small-scale SDP is very efficient, while DCA2 often requires more iterations than DCA1 that maybe yields more computational time. The number of iterations of DCA2 is also stable between 5.3-8.8 in average. But comparing DCA2 (in table 6) with PENBMI (in table 5), PENBMI is still slightly faster than DCA2 for the set of problems Pset13 - Pset15. In conclusion, we suggest using DCA1 for small-scale and medium-scale problems and DCA2 or PENBMI for relatively large-scales.

We are also interested in some random problems that DCA have not found feasible solution with given initial point, i.e. the objective value is still positive at the convergence of DCA. For these problems, we have tested DCA-B&B and some of them have been determined as feasible problems.

In table 7, we have some instances for which DCA and PENBMI have not found feasible solutions (since the objective values are not equal to zero). Using DCA-B&B, we obtained a feasible solution for BMI after exploring some number of nodes. The number of explored nodes are given in the column "node". In general, the efficiency of B&B based algorithm highly depends on the quality of the lower bound. A tighter and cheaper lower bound will lead to a fast convergence. In our problem, for determining a infeasible BMIFP, the quality of lower bound is also important to get a global optimal solution as soon as possible. However, if BMI is feasible, a special feature of our DCA-B&B is that it can be terminated when the objective value is smaller enough. Therefore, for a feasible BMI, the quality of upper bound is more important for fast convergence than the quality of lower bound. Fortunately, DCA can provide good upper bound.

Table 6. Performance comparison between the DCA and DCA2

Problem	Dimension			DCA1		DCA2	
	n	m	k	ave.iter	ave.time	ave.iter	ave.time
Pset1	1	1	3	1	0.1580	5.8	0.8210
Pset2	2	3	5	1.2	0.1668	5.3	0.662
Pset3	8	10	10	1.3	1.316	5.5	1.938
Pset4	10	10	40	1.5	1.573	6.7	2.982
Pset5	10	10	60	1.2	2.667	5.2	2.332
Pset6	10	10	70	1.2	3.848	5.0	4.398
Pset7	10	10	100	1.5	7.816	6.3	10.210
Pset8	15	10	30	2.1	2.909	7.1	11.897
Pset9	15	10	60	2.8	5.328	8.8	15.199
Pset10	15	10	80	1.6	9.125	5.6	20.928
Pset11	20	15	20	1.3	5.869	5.1	12.360
Pset12	20	15	60	2.5	25.490	6.8	20.868
Pset13	30	30	10	1.2	27.17	5.5	25.836
Pset14	40	30	6	1.5	30.492	6.5	29.630
Pset15	50	50	10	1.2	50.585	5.1	38.331

Each problem set contains 20 problems.

Table 7. Comparison results between DCA, PENBMI and DCA-B&B

Prob	Dimension			DCA			PENBMI			DCA-B&B		
	n	m	k	obj	iter	time	obj	iter	time	obj	node	time
P1	3	5	3	0.035	1	0.178	1.275	19	1.028	1.12E-6	30	6.818
P2	5	8	5	0.088	2	0.339	1.287	20	1.317	3.31E-6	26	28.991
P3	8	10	10	0.092	1	1.512	0.008	20	2.216	5.19E-6	69	268.025
P4	10	10	40	0.011	3	2.025	2.105	19	3.276	8.75E-6	33	236.102
P5	10	10	60	0.059	3	3.311	0.211	20	3.288	9.21E-6	82	356.811

9   Conclusion

We proposed a DC programming approaches for solving the BMI and QMI
Feasibility Problems. We presented the equivalent DC programming formulation
which is solved by our proposed DC algorithms (DCA1, DCA2 and DCA-B&B).
DCA1 requires solving a SDP problem in each iteration. The DCA-B&B can
be used for guaranteeing the feasibility of BMI/QMI. And the partial solution
strategy in DCA2 is particularly suitable for large scale problems. The numerical
results show us good convergence and solution quality of our methods.

References

1. Benson, S.T., Ye, Y.Y.: DSDP: A complete description of the algorithm and a proof
   of convergence can be found in Solving Large-Scale Sparse Semidefinite Programs
   for Combinatorial Optimization. SIAM Journal on Optimization 10(2), 443–461
   (2000), http://www.mcs.anl.gov/hs/software/DSDP/

2. Beran, E.B., Vandenberghe, L., Boyd, S.: A global BMI algorithm based on the generalized Benders decomposition. In: Proceedings of the European Control Conference, Brussels, Belgium (July 1997)
3. Borchers, B.: CSDP: a C library for semidefinite programming, Department of Mathematics, New Mexico Institute of Mining and Technology, Socorro, NM (November 1998), https://projects.coin-or.org/Csdp/
4. Floudas, C.A., Visweswaran, V.: A primal-relaxed dual global optimization approach. Journal of Optimization Theory and Applications 78, 187–225 (1993)
5. Fujioka, H., Hoshijima, K.: Bounds for the BMI eingenvalue problem - a good lower bound and a cheap upper bound. Transactions of the Society of Instrument and Control Engineers 33, 616–621 (1997)
6. Fujisawa, K., Kojima, M., Nakata, K.: SDPA (SemiDefinite Programming Algorithm) - user's manual - version 6.20. Research Report B-359, Department of Mathematical and Computing Sciences, Tokyo Institute of Technology, Tokyo, Japan (January 2005), http://sdpa.indsys.chuo-u.ac.jp/sdpa/download.html (revised May 2005)
7. Fukuda, M., Kojima, M.: Branch-and-Cut Algorithms for the Bilinear Matrix Inequality Eigenvalue Problem. Computational Optimization and Applications 19(1), 79–105 (2001)
8. Goh, K.C., Safonov, M.G., Papavassilopoulos, G.P.: Global optimization for the biaffine matrix inequality problem. Journal of Global Optimization 7, 365–380 (1995)
9. Goh, K.C., Safonov, M.G., Ly, J.H.: Robust synthesis via bilinear matrix inequalities. International Journal of Robust and Nonlinear Control 6, 1079–1095 (1996)
10. Horst, R.: D.C. Optimization: Theory, Methods and Algorithms. In: Horst, R., Pardalos, P.M. (eds.) Handbook of Global Optimization, pp. 149–216. Kluwer Academic Publishers, Dordrecht (1995)
11. Horst, R., Thoai, N.V.: DC Programming: Overview. Journal of Optimization Theory and Applications 103, 1–43 (1999)
12. Hiriart Urruty, J.B., Lemaréchal, C.: Convex Analysis and Minimization Algorithms. Springer, Heidelberg (1993)
13. Horst, R., Pardalos, P.M., Thoai, N.V.: Introduction to Global Optimization, 2nd edn. Kluwer Academic Publishers, Netherlands (2000)
14. Kawanishi, M., Sugie, T., Kanki, H.: BMI global optimization based on branch and bound method taking account of the property of local minima. In: Proceedings of the Conference on Decision and Control, San Diego, CA (December 1997)
15. Kočvara, M., Stingl, M.: PENBMI User's Guide (Version 2.1) (February 16, 2006)
16. Le Thi, H.A., Pham Dinh, T.: Solving a class of linearly constrained indefinite quadratic problems by DC Algorithms. Journal of Global Optimization 11, 253–285 (1997)
17. Le Thi, H.A., Pham Dinh, T., Le Dung, M.: Exact penalty in d.c. programming. Vietnam Journal of Mathematics 27(2), 169–178 (1999)
18. Le Thi, H.A.: An efficient algorithm for globally minimizing a quadratic function under convex quadratic constraints. Mathematical Programming Ser. A. 87(3), 401–426 (2000)
19. Le Thi, H.A., Pham Dinh, T.: A continuous approach for large-scale constrained quadratic zero-one programming (In honor of Professor ELSTER, Founder of the Journal Optimization). Optimization 45(3), 1–28 (2001)
20. Le Thi, H.A., Pham Dinh, T.: Large Scale Molecular Optimization From Distance Matrices by a D.C. Optimization Approach. SIAM Journal on Optimization 4(1), 77–116 (2003)

21. Le Thi, H.A.: Solving large scale molecular distance geometry problems by a smoothing technique via the gaussian transform and d.c. programming. Journal of Global Optimization 27(4), 375–397 (2003)
22. Le Thi, H.A., Pham Dinh, T., François, A.: Combining DCA and Interior Point Techniques for large-scale Nonconvex Quadratic Programming. Optimization Methods & Software 23(4), 609–629 (2008)
23. Liu, S.M., Papavassilopoulos, G.P.: Numerical experience with parallel algorithms for solving the BMI problem. In: 13th Triennial World Congress of IFAC, San Francisco, CA (July 1996)
24. Löfberg, J.: YALMIP: A Toolbox for Modeling and Optimization in MATLAB. In: Proceedings of the CACSD Conference, Taipei, Taiwan (2004), http://control.ee.ethz.ch/~joloef/wiki/pmwiki.php
25. MATLAB R2007a: Documentation and User Guides, http://www.mathworks.com/
26. Mesbahi, M., Papavassilopoulos, G.P.: A cone programming approach to the bilinear matrix inequality problem and its geometry. Mathematical Programming 77, 247–272 (1997)
27. Mittelmann, H.D.: Several SDP-codes on problems from SDPLIB, http://plato.asu.edu/ftp/sdplib.html
28. Niu, Y.S., Pham Dinh, T.: A DC Programming Approach for Mixed-Integer Linear Programs. In: Le Thi, H.A., Bouvry, P., Pham Dinh, T. (eds.) MCO 2008. CCIS, vol. 14, pp. 244–253. Springer, Heidelberg (2008)
29. Niu, Y.S.: DC programming and DCA combinatorial optimization and polynomial optimization via SDP techniques, National Institute of Applied Sciences, Rouen, France (2010)
30. Niu, Y.S., Pham Dinh, T.: An Efficient DC Programming Approach for Portfolio Decision with Higher Moments. Computational Optimization and Applications 50(3), 525–554 (2010)
31. Niu, Y.S., Pham Dinh, T.: Efficient DC programming approaches for mixed-integer quadratic convex programs. In: Proceedings of the International Conference on Industrial Engineering and Systems Management (IESM 2011), Metz, France, pp. 222–231 (2011)
32. Niu, Y.S., Pham Dinh, T., Le Thi, H.A., Judice, J.J.: Efficient DC Programming Approaches for the Asymmetric Eigenvalue Complementarity Problem. Optimization Methods and Software 28(4), 812–829 (2013)
33. Pham Dinh, T., Le Thi, H.A.: Convex analysis approach to D.C. programming: Theory, Algorithms and Applications. Acta Mathematica Vietnamica 22(1), 289–355 (1997)
34. Pham Dinh, T., Le Thi, H.A.: DC optimization algorithms for solving the trust region subproblem. SIAM J. Optimization 8, 476–507 (1998)
35. Pham Dinh, T., Le Thi, H.A.: DC Programming. Theory, Algorithms, Applications: The State of the Art. In: First International Workshop on Global Constrained Optimization and Constraint Satisfaction, Nice, October 2-4 (2002)
36. Pham Dinh, T., Le Thi, H.A.: The DC programming and DCA Revisited with DC Models of Real World Nonconvex Optimization Problems. Annals of Operations Research 133, 23–46 (2005)
37. Rockafellar, R.T.: Convex Analysis. Princeton University Press, N.J. (1970)
38. Safonov, M.G., Goh, K.C., Ly, J.H.: Control system synthesis via bilinear matrix inequalities. In: Proceedings of the American Control Conference, Baltimore, MD (June 1994)
39. Sherali, H.D., Alameddine, A.R.: A new reformulation-linearization technique for bilinear programming problems. Journal of Global Optimization 2, 379–410 (1992)

40. Sturm, J.F.: SeDuMi 1.2: a MATLAB toolbox for optimization over symmetric cones. Department of Quantitative Economics, Maastricht University, Maastricht, The Netherlands (August 1998), http://sedumi.ie.lehigh.edu/
41. Takano, S., Watanabe, T., Yasuda, K.: Branch and bound technique for global solution of BMI. Transactions of the Society of Instrument and Control Engineers 33, 701–708 (1997)
42. Toker, O., Özbay, H.: On the NP-hardness of solving bilinear matrix inequalities and simultaneous stabilization with static output feedback. In: American Control Conference, Seattle, WA (1995)
43. Tuan, H.D., Hosoe, S., Tuy, H.: D.C. optimization approach to robust controls: Feasibility problems. IEEE Transactions on Automatic Control 45, 1903–1909 (2000)
44. Tuan, H.D., Apkarian, P., Nakashima, Y.: A new Lagrangian dual global optimization algorithm for solving bilinear matrix inequalities. International Journal of Robust and Nonlinear Control 10, 561–578 (2000)
45. Van Antwerp, J.G.: Globally optimal robust control for systems with time-varying nonlinear perturbations. Master thesis, University of Illinois at Urbana-Champaign, Urbana, IL (1997)
46. Wolkowicz, H., Saigal, R., Vandenberghe, L.: Handbook of Semidefinite Programming - Theory, Algorithms, and Applications. Kluwer Academic Publishers, USA (2000)

A DC Programming Approach for Sparse Linear Discriminant Analysis

Phan Duy Nhat, Manh Cuong Nguyen, and Hoai An Le Thi

Laboratory of Theoretical and Applied Computer Science EA 3097
University of Lorraine, Ile de Saulcy, 57045 Metz, France
{duy-nhat.phan,manh-cuong.nguyen,hoai-an.le-thi}@univ-lorraine.fr

Abstract. We consider the supervised pattern classification in the high-dimensional setting, in which the number of features is much larger than the number of observations. We present a novel approach to the sparse linear discriminant analysis (LDA) using the zero-norm. The resulting optimization problem is non-convex, discontinuous and very hard to solve. We overcome the discontinuity by using an appropriate continuous approximation to zero-norm such that the resulting problem can be formulated as a DC (Difference of Convex functions) program to which DC programming and DC Algorithms (DCA) can be investigated. The computational results show the efficiency and the superiority of our approach versus the l_1 regularization model on both feature selection and classification.

Keywords: Classification, Sparse Fisher linear discriminant analysis, DC programming, DCA.

1 Introduction

Linear discriminant analysis (LDA) was introduced by R. Fisher [6]. It is widely used in classification problems. This paper concerns Fisher's discriminant problem which can be described as follows.

Let X be an $n \times p$ data matrix with observation on the rows and features on the columns, and suppose that each of the n observations falls into on of Q classes. We assume that the features are centered to have mean 0. Let x_i denote observation or ith row, C_j be the set containing the indices of the observations in jth class and n_j be the cardinality of the set C_j. The standard estimate for the within-class covariance matrix S_w and the between-class covariance matrix S_b are given by

$$S_w = \frac{1}{n} \sum_{j=1}^{Q} \sum_{i \in C_j} (x_i - \mu_j)(x_i - \mu_j)^T \text{ and } S_b = \frac{1}{n} \sum_{j=1}^{Q} n_j \mu_j \mu_j^T,$$

where $\mu_j = \frac{1}{n_j} \sum_{i \in C_j} x_i$ is the mean vector for the jth class. Fisher's discriminant problem seeks a low dimensional projection of the observations such that

T.V. Do et al. (eds.), *Advanced Computational Methods for Knowledge Engineering,*
Advances in Intelligent Systems and Computing 282,
DOI: 10.1007/978-3-319-06569-4_4, © Springer International Publishing Switzerland 2014

the between-class variance is large relative to the within-class variance, i.e. we seek discriminant vectors $w_1, ..., w_{Q-1}$ that successively minimize

$$\min_{w_k \in \mathbb{R}^p} \{-w_k^T S_b w_k : w_k^T S_w w_k \leq 1, w_k^T S_w w_l = 0, l < k\}. \tag{1}$$

In general, there are at most $Q-1$ non-trivial discriminant vectors. A classification rule is obtained by computing $Xw_1, ..., Xw_{Q-1}$ and assigning each observation to its nearest centroid in this transformation space. We can use only the first $s < Q-1$ discriminant vectors in order to perform reduced rank classification.

LDA is straightforward in the case where the number of observations is greater than the number of features. However, in high dimension (when the number of features is much larger than the number of observations) the classical LDA includes two great challenges. The first challenge is the singularity of the within-class covariance matrix of the features and the second one is the difficulty in interpreting the classification rule.

To overcome these difficulties, Friedman [7] and Dudoit et al. [5] suggested using the diagonal estimate $\text{diag}(\sigma_1^2, ..., \sigma_p^2)$, where σ_i^2 is the ith diagonal element of S_w. And other positive definite estimates $\tilde{S}_w$ for S_w are considered by Krzanowski et al. [9] and Xu et al. [24]. The resulting criterion is

$$\min_{w_k \in \mathbb{R}^p} \{-w_k^T S_b w_k : w_k^T \tilde{S}_w w_k \leq 1, w_k^T \tilde{S}_w w_l = 0, l < k\}. \tag{2}$$

Some works involve sparse classifiers have also been considered by Tibshirani et al. [22], Guo et al. [8]. Witten and Tibshirani [23] applied l_1-penalty in the problem (2) in order to obtain sparse discriminant vectors $w_1, ..., w_{Q-1}$ that successively minimize

$$\min_{w_k \in \mathbb{R}^p} \{-w_k^T S_b^k w_k + \lambda_k \|w_k\|_1 : w_k^T \tilde{S}_w w_k \leq 1\}, \tag{3}$$

where

$$S_b^k = \frac{1}{n} X^T Y (Y^T Y)^{-\frac{1}{2}} P_k^\perp (Y^T Y)^{-\frac{1}{2}} Y^T X. \tag{4}$$

Here $Y \in \mathbb{R}^{n \times Q}$ with $Y_{ij} = 1$ if $i \in C_j$ and 0 otherwise, $P_1^\perp = I$ (identity matrix), and $P_k^\perp (k > 1)$ is an orthogonal projection matrix into the orthogonal space of the space generated by $\{(Y^T Y)^{-\frac{1}{2}} Y^T X w_l : l = 1, ..., k-1\}$.

We develop a sparse version of LDA by using l_0-penalty in (2) that leads to the following optimization problem

$$\min_{w_k \in \mathbb{R}^p} \{-w_k^T S_b^k w_k + \lambda_k \|w_k\|_0 : w_k^T \tilde{S}_w w_k \leq 1\}, \tag{5}$$

where l_0-norm of the vector w_k is defined as

$$\|w_k\|_0 = \text{cardinality}\{i : w_{ki} \neq 0\}.$$

Both (3) and (5) are non-convex problems. Moreover, the objective function of (5) is discontinuous. We consider an appropriate continuous approximation to

l_0 such that the resulting problem can be formulated as a DC (Difference of Convex functions) program.

Our method is based on DC programming and DCA (DC Algorithms) introduced by Pham Dinh Tao in their preliminary form in 1985. They have been extensively developed since 1994 by Le Thi Hoai An and Pham Dinh Tao and become now classic and increasingly popular (see e.g.[10,11,19,15,16]). Our motivation is based on the fact that DCA is a fast and scalable approach which has been successfully applied to many large-scale (smooth or non-smooth) non-convex programs in various domains of applied sciences, in particular in data analysis and data mining, for which it provided quite often a global solution and proved to be more robust and efficient than standard methods (see e.g.[10,11,19,15,16] and references therein).

The paper is organized as follows. In Section 2, we present DC programming and DCA for general DC programs, and show how to apply DCA to solve the problem (5). The numerical experiments are reported in Section 3. Finally, Section 4 concludes the paper.

2 DC Programming and DCA for Solving the Problem (5)

2.1 A Brief Presentation of DC Programming and DCA

For a convex function θ, the subdifferential of θ at $x_0 \in \mathrm{dom}\theta := \{x \in \mathbb{R}^n : \theta(x) < +\infty\}$, denoted by $\partial\theta(x_0)$, is defined by

$$\partial\theta(x_0) := \{y \in \mathbb{R}^n : \theta(x) \geq \theta(x_0) + \langle x - x_0, y\rangle, \forall x \in \mathbb{R}^n\},$$

and the conjugate θ^* of θ is

$$\theta^*(y) := \sup\{\langle x, y\rangle - \theta(x) : x \in \mathbb{R}^n\}, \quad y \in \mathbb{R}^n.$$

A general DC program is that of the form:

$$\alpha = \inf\{F(x) := G(x) - H(x) \mid x \in \mathbb{R}^n\} \quad (\mathrm{P}_{dc}),$$

where G, H are lower semi-continuous proper convex functions on $\mathbb{R}^n$. Such a function F is called a DC function, and $G - H$ a DC decomposition of F while G and H are the DC components of F. Note that, the closed convex constraint $x \in C$ can be incorporated in the objective function of (P_{dc}) by using the indicator function on C denoted by χ_C which is defined by $\chi_C(x) = 0$ if $x \in C$, and $+\infty$ otherwise.

A point x^* is called a *critical point* of $G - H$, or a generalized Karush-Kuhn-Tucker point (KKT) of (P_{dc})) if

$$\partial H(x^*) \cap \partial G(x^*) \neq \emptyset. \tag{6}$$

Based on local optimality conditions and duality in DC programming, the DCA consists in constructing two sequences $\{x^k\}$ and $\{y^k\}$ (candidates to be solutions of (P_{dc}) and its dual problem respectively). Each iteration k of DCA approximates the concave part $-H$ by its affine majorization (that corresponds to taking $y^k \in \partial H(x^k)$) and minimizes the resulting convex function (that is equivalent to determining $x^{k+1} \in \partial G^*(y^k)$).

Generic DCA scheme
Initialization: Let $x^0 \in \mathbb{R}^n$ be an initial guess, $0 \leftarrow k$.
Repeat
- Calculate $y^k \in \partial H(x^k)$
- Calculate $x^{k+1} \in \arg\min\{G(x) - \langle x, y^k \rangle : x \in \mathbb{R}^n\}$ (P_k)
- $k + 1 \leftarrow k$
Until convergence of $\{x^k\}$.

Convergences properties of DCA and its theoretical basic can be found in [10,19]. It is worth mentioning that

- DCA is a descent method (*without linesearch*): the sequences $\{G(x^k) - H(x^k)\}$ and $\{H^*(y^k) - G^*(y^k)\}$ are decreasing.
- If $G(x^{k+1}) - H(x^{k+1}) = G(x^k) - H(x^k)$, then x^k is a critical point of $G - H$ and y^k is a critical point of $H^* - G^*$. In such a case, DCA terminates at k-th iteration.
- If the optimal value α of problem (P_{dc}) is finite and the infinite sequences $\{x^k\}$ and $\{y^k\}$ are bounded then every limit point x (resp. y) of the sequences $\{x^k\}$ (resp. $\{x^k\}$) is a critical point of $G - H$ (resp. $H^* - G^*$).
- DCA has a *linear convergence* for general DC programs, and has a finite convergence for polyhedral DC programs.

A deeper insight into DCA has been described in [10]. For instant it is crucial to note the main feature of DCA: DCA is constructed from DC components and their conjugates but not the DC function f itself which has infinitely many DC decompositions, and there are as many DCA as there are DC decompositions. Such decompositions play a critical role in determining the speed of convergence, stability, robustness, and globality of sought solutions. It is important to study various equivalent DC forms of a DC problem. This flexibility of DC programming and DCA is of particular interest from both a theoretical and an algorithmic point of view.

For a complete study of DC programming and DCA the reader is referred to [10,19,20] and the references therein. The solution of a nonconvex program (P_{dc}) by DCA must be composed of two stages: the search of an *appropriate* DC decomposition of f and that of a *good* initial point.

In the last decade, a variety of works in Machine Learning based on DCA have been developed. The efficiency and the scalability of DCA have been proved in a lot of works (see e.g. [12,13,17,18,21] and the list of reference in [14]). These successes of DCA motivated us to investigate it for solving Fisher's discriminant problem.

2.2   DCA for Solving Problem (5)

A good approximation for the l_0-norm proposed in [2] is

$$||w_k||_0 \approx \sum_{i=1}^{p} \eta(w_{ki}),$$

where the function η is defined by

$$\eta(x) = 1 - \varepsilon^{-\alpha|x|}, \alpha > 0, \forall x \in \mathbb{R}.$$

η can be expressed as a DC function

$$\eta(x) = g(x) - h(x),$$

where $g(x) := \alpha|x|$ and $h(x) := \alpha|x| - 1 + \varepsilon^{-\alpha|x|}$. The resulting approximate problem of (5) can be written as follows

$$\min\{-w_k^T S_b^k w_k + \lambda_k \sum_{i=1}^{p} \eta(w_{ki}) : w_k \in \Omega\},$$

where $\Omega = \{w_k \in \mathbb{R}^p : w_k^T \tilde{S}_w w_k \leq 1\}$. This is a non-convex problem and can be reformulated as

$$\min\{G(w_k) - H(w_k) : w_k \in \mathbb{R}^p\}, \tag{7}$$

where

$$G(w_k) := \chi_\Omega(w_k) + \lambda_k \sum_{i=1}^{p} g(w_{ki}),$$

and

$$H(w_k) := w_k^T S_b^k w_k + \lambda_k \sum_{i=1}^{p} h(w_{ki})$$

are clearly convex functions. Therefore (7) is a DC program.

According to the generic DCA cheme, at each iteration l, we have to compute a subgradient v^l of H at w_k^l and then solve the convex program of the form (P_l), namely

$$\min\{G(w_k) - \langle v^l, w_k \rangle : w_k \in \mathbb{R}^p\} \tag{8}$$

$$\Leftrightarrow \min\{\lambda_k \sum_{i=1}^{p} t_i - \langle v^l, w_k \rangle : (w_k, t) \in \Omega_1\}, \tag{9}$$

where

$$\Omega_1 = \{(w_k, t) \in \mathbb{R}^p \times \mathbb{R}^p : w_k \in \Omega, \alpha|w_{ki}| \leq t_i, i = 1, ..., p\}.$$

H is differentiable everywhere, and $v^l = \nabla H(w_k^l)$ is calculated by

$$v_i^l = \begin{cases} 2\langle S_{bi}^k, w_k^l \rangle + \lambda_k \alpha(1 - \varepsilon^{-\alpha w_{ki}^l}) & \text{if } w_{ki}^l \geq 0 \\ 2\langle S_{bi}^k, w_k^l \rangle - \lambda_k \alpha(1 - \varepsilon^{\alpha w_{ki}^l}) & \text{if } w_{ki}^l < 0 \end{cases} \quad i = 1, ..., p, \tag{10}$$

where S_{bi}^k stands for the ith row of S_b^k.

Algorithm 1. DCA for solving (7)

Initialization:
- Let τ be a tolerance sufficient small, set $l = 0$ and choose $(w_k^0, t^0) \in \Omega_1$.
- Compute S_b^k via (4).

Repeat
- Compute $v^l = \nabla H(w_k^l)$ via (10).
- Solve the convex program (9) to obtain (w_k^{l+1}, t^{l+1}).
- Increase the iteration counter: $l \leftarrow l + 1$.

Until $||w_k^{l+1} - w_k^l|| + ||t^{l+1} - t^l|| \leq \tau(||w_k^l|| + ||t^l||)$.

---

3 Numerical Experiments

We use the sparse LDA for supervised classification problems in high dimension, in which the number of features is very large and the number of observations is limited. The sparse LDA transforms the set of labelled data points in the original space into a labelled set in a lower-dimensional space and selects relevant features. The classification rule is obtained by computing $X w_1, ..., X w_s$ and assigning each observation to its nearest centroid in this transformation space, i.e. the predicted class for a test observation x is

$$\text{argmin}_i \sum_{k=1}^{s} ((x - \mu_i)^T w_k)^2 - 2\ln(n_i),$$

where s is the number of discriminant vectors used ($s < Q$).

Three simulation datasets and three real datasets (described in section 3.1) are used to compare our method DCA with the standard model (3) proposed in [23]. This is a non-convex problem which can be solved by DCA. A DC formulation of (3) is

$$\min\{G(w_k) - H(w_k) : w_k \in \mathbb{R}^p\}, \tag{11}$$

where

$$G(w_k) := \lambda_k \|w_k\|_1 + \chi_\Omega(w_k), H(w_k) := w_k^T S_b^k w_k$$

are clearly convex functions. Therefore (11) is a DC program.

According to the generic DCA cheme, at each iteration l, we have to compute $y^l = \nabla H(w_k^l) = 2S_b^k w_k^l$ and then solve the convex program of the form (P_l), namely

$$\min\{\lambda_k \sum_{i=1}^{p} t_i - \langle y^l, w_k \rangle : (w_k, t) \in \Omega_2\}, \tag{12}$$

where $\Omega_2 = \{(w_k, t) \in \mathbb{R}^p \times \mathbb{R}^p : w_k \in \Omega, |w_{ki}| \leq t_i, i = 1, ..., p\}$.

Algorithm 2. DCA for solving (11)

Initialization:
- Let τ be a tolerance sufficient small, set $l = 0$ and choose $(w_k^0, t^0) \in \Omega_2$.
- Compute S^k via (4).

Repeat
- Compute $y^l = \nabla H(w_k^l) = 2S_b^k w_k^l$.
- Solve the convex program (12) to obtain (w_k^{l+1}, t^{l+1}).
- Increase the iteration counter: $l \leftarrow l + 1$.

Until $||w_k^{l+1} - w_k^l|| + ||t^{l+1} - t^l|| \leq \tau(||w_k^l|| + ||t^l||)$.

The algorithms have been coded in VC++ and implemented on a Intel Core^{TM} I7 (2×2.2 Ghz) processor, 4 GB RAM. All convex problems (9) and (12) are solved by the commercial software CPLEX 11.2.

3.1 Datasets

The three simulation datasets are generated as follows. Each of them contains 1100 observations with 500 features which are divided into equally classes. The training set consists of 100 observations and 1000 observations are in the test set.

1. Simulation 1: We generate a four-class classification problem in which the kth class is randomly generated to have the multivariate Gaussian distribution $N(\mu_k, I)$ for $k = 1, 2, 3, 4$ with mean μ_k and covariance I (identity matrix), where $\mu_{1j} = 2$ if $0 \leq j \leq 25$, $\mu_{2j} = -2$ if $26 \leq j \leq 50$, $\mu_{3j} = -4$ if $56 \leq j \leq 75$ and $u_{kj} = 0$ otherwise.
2. Simulation 2: We consider a two-class classification problem. The first class has the distribution $N(0, \Sigma)$ and the second class has the distribution $N(\mu, \Sigma)$, where $\mu_j = 0.6$ if $0 \leq j \leq 200$ and $\mu_j = 0$ otherwise. Σ is the block diagonal matrix with five blocks of dimension 100×100 whose element (j, j') is $0.6^{|j-j'|}$.
3. Simulation 3: We generate a four-class classification problem as follows $i \in C_2$ then $X_{ij} \sim N(2, 1)$ if $j \leq 100$, $i \in C_3$ then $X_{ij} \sim N(-2, 1)$ if $j \leq 100$, $i \in C_4$ then $X_{ij} \sim N(-4, 1)$ if $j \leq 100$ and $X_{ij} \sim N(0, 1)$ otherwise, where $N(\mu, \sigma^2)$ denotes the Gaussian distribution with mean μ and variance σ^2.

The three real datasets have a very large number of features including two problems of gene selection for cancer classification with standard public microarray gene expression datasets (Leukemia, Prostate).

1. The Leukemia dataset represents gene expression patterns of 7129 probes from 6817 human genes. The training dataset consists of 38 bone marrow samples (27 ALL, 11 AML), versus 20 ALL and 14 AML for testing dataset. We can download it from http://www.broad.mit.edu.
2. In the Prostate dataset, the training set contains 52 prostate tumor samples and 50 non-tumor (labelled as "Normal") prostate samples with around 12600 genes. There are 25 tumor and 9 normal samples in the testing set. It can be downloaded from http://www-genome.wi.mit.edu/mpr/prostate.

3. The Arcene dataset is probably the most challenging among all the datasets from the NIPS competition as it is a sparse problem with the smallest examples-to-features ratio.

The informations about datasets is summarized in Table 1.

Table 1. The description of the datasets

Dataset	#feature	#train	#test	#class
Simulation 1 (SM1)	500	100	1000	4
Simulation 2 (SM2)	500	100	1000	2
Simulation 3 (SM3)	500	100	1000	4
Leukemia (LEU)	7129	38	34	2
Arcene (ARC)	10000	100	100	2
Prostate (PRO)	12600	102	34	2

3.2   Experimental Setup

In experiment, we set $\alpha = 5$ and stop tolerance $\tau = 10^{-6}$ for DCA. The starting points of DCA are randomly chosen in $[-1, 1]$. After solving problems (5) or (3) by DCA, we select relevant features as follows. We delete the feature i in which $|w_{ki}| < 10^{-6}$ for all $k = 1, ..., Q-1$. In this experiment, we use the diagonal estimate for $\tilde{S}_w$ ([1]): $\tilde{S}_w \approx \mathrm{diag}(\sigma_1^2, ..., \sigma_p^2)$, where σ_i is the within-class standard deviation for feature i.

The value of λ_k is taken as $\lambda_k = \lambda . \lambda_1^k$, where λ_1^k is the largest eigenvalue of $\tilde{S}_w^{-\frac{1}{2}} S_b^k \tilde{S}_w^{-\frac{1}{2}}$ ([23]). And we chose λ through a 3-fold cross validation procedure on training set from a set of candidates given by

$$\Lambda = \{0.001, 0.003, 0.005, 0.007, 0.009, 0.01, 0.02\}.$$

3.3   Experiment Results

The computational results given by Algorithm 1 (l_0-DCA) and Algorithm 2 (l_1-DCA) are reported in Table 2 and Table 3. We are interested in the efficiency (the sparsity and the accuracy of classifiers) as well as the rapidity of the algorithms.

In all datasets, the classifiers obtained by l_0-DCA are sparser than those obtained by the l_1-DCA. The range of the removed features is $[60.14\%, 99.78\%]$. In particularly for Leukemia dataset, 99.78% features are deleted while the accuracy of classifier is quite good (82.94%).

Moreover, the accuracy of classifiers given by l_0-DCA are greater than 82% (except Arcene dataset). And l_0-DCA is better than l_1-DCA in 4/6 datasets while l_1-DCA is sightly better in otherwise.

The CPU time is given in Table 3 shows that l_0-DCA is faster than l_1-DCA in 3/6 datasets.

Table 2. The average number (average percentage) of selected features, the average percentage of accuracy of classifiers over 10 runs of DCA from random starting points, and the number of discriminant vectors used (DV). The best results are in bold font.

Datasets	Selected features		Accuracy of classifiers		DV	
	l_0-DCA	l_1-DCA	l_0-DCA	l_1-DCA	l_0-DCA	l_1-DCA
SM1	**153.56** (30.71%)	228.40 (45.68%)	**96.73**	89.29	2	3
SM2	**199.30** (39.86%)	342.20 (68.44%)	96.96	**99.99**	1	1
SM3	**100.00** (20.00%)	204.10 (40.82%)	**100.00**	100.00	2	2
LEU	**16.00** (0.22%)	2906.00 (40.76%)	82.94	**83.53**	1	1
ARC	**2533.30** (25.33%)	4128.40 (41.62%)	**69.20**	69.00	1	1
PRO	**4721.10** (37.47%)	6182.40 (49.07%)	**89.41**	89.41	1	1

Table 3. The average CPU time in second. The best results are in bold font.

Datasets	SM1	SM2	SM3	LEU	ARC	PRO
l_0-DCA	1.44	**1.35**	1.21	**12.80**	50.98	**28.17**
l_1-DCA	**1.16**	1.75	**0.87**	48.17	**42.93**	87.43

4 Conclusion

We have developed the sparse LDA based on DC programming and DCA. The computational results show that the proposed algorithm is efficient to Fisher's discriminant problem. Comparative numerical results also show that l_0-DCA approach is more efficient than l_1-DCA approach on sparsity, accuracy as well as rapidity.

In the future we will study other models of LDA such as the optimal scoring problem.

References

1. Bickel, P., Levina, E.: Some theory for Fisher's linear discriminant function, naive Bayes, and some alternatives when there are many more variables than observations. Bernoulli 6, 989–1010 (2004)
2. Bradley, P.S., Magasarian, O.L., Street, W.N.: Feature Selection via mathematical Programming. INFORMS Journal on Computing, 209–217 (1998)
3. Clemmensen, L., Hastie, T., Witten, D., Ersboll, B.: Sparse discriminant analysis. Technometrics 53(4), 406–413 (2011)
4. Duan, K.B., Rajapakse, J.C., Wang, H., Azuaje, F.: Multiple SVM-RFE for Genne Selection in Cancer Classification With Expression Data. IEEE Transactions on Nanobioscience 4, 228–234 (2005)
5. Dudoit, S., Fridlyand, J., Speed, T.: Comparison of discrimination methods for the classification of tumors using gene expression data. J. Amer. Statist. Assoc. 96, 1151–1160 (2001)
6. Fisher, R.A.: The use of multiple measurements in taxonomic problems. Annals of Eugenics 7, 179–188 (1936)

7. Friedman, J.: Regularized discriminant analysis. Journal of the American Statistical Association 84, 165–175 (1989)
8. Guo, Y., Hastie, T., Tibshirani, R.: Regularized linear discriminant analysis and its application in microarrays. Biostatistics 8, 86–100 (2007)
9. Krzanowski, W., Jonathan, P., McCarthy, W., Thomas, M.: Discriminant analysis with singular covariance matrices: methods and applications to spectroscopic data. Journal of the Royal Statistical Society, Series C 44, 101–115 (1995)
10. Le Thi, H.A., Pham Dinh, T.: The DC (difference of convex func-tions) programming and DCA revisited with DC models of real world nonconvex optimization problems. Annals of Operations Research 133, 23–46 (2005)
11. Le Thi, H.A., Pham Dinh, T.: Solving a class of linearly constrained indefinite quadratic problems by DC algorithms. Journal of Global Optimization 11(3), 253–285 (1997)
12. Le Thi, H.A., Le Hoai, M., Pham Dinh, T.: Optimization based DC programming and DCA for Hierarchical Clustering. European Journal of Operational Research 183, 1067–1085 (2007)
13. Le Thi, H.A., Le Hoai, M., Nguyen, N.V., Pham Dinh, T.: A DC Programming approach for Feature Selection in Support Vector Machines learning. Journal of Advances in Data Analysis and Classification 2(3), 259–278 (2008)
14. Le Thi, H.A.: DC Programming and DCA,
   http://lita.sciences.univ-metz.fr/~lethi/DCA.html
15. Le Thi, H.A., Pham Dinh, T.: DC optimization algorithm for solving the trust region subproblem. SIAM Journal of Optimization 8(1), 476–505 (1998)
16. Le Thi, H.A., Huynh, V., Pham Dinh, T.: Exact penalty and error bounds in DC programming. Journal of Global Optimization 52(3), 509–535 (2011)
17. Liu, Y., Shen, X., Doss, H.: Multicategory ψ-Learning and Support Vector Machine: Computational Tools. Journal of Computational and Graphical Statistics 14, 219–236 (2005)
18. Liu, Y., Shen, X.: Multicategory ψ-Learning. Journal of the American Statistical Association 101, 500–509 (2006)
19. Pham Dinh, T., Le Thi, H.A.: Convex analysis approach to D.C. programming: Theory, algorithms and applications. Acta Mathematica Vietnamica 22(1), 289–355 (1997)
20. Pham Dinh, T., Le Thi, H.A.: DC optimization algorithms for solving the trust region subproblem. SIAM J. Opt. 8, 476–505 (1998)
21. Thiao, M., Pham Dinh, T., Le Thi, H.A.: DC programming approach for a class of nonconvex programs involving l0 norm. In: Le Thi, H.A., Bouvry, P., Pham Dinh, T. (eds.) MCO 2014. CCIS, vol. 14, pp. 358–367. Springer, Heidelberg (2008)
22. Tibshirani, R., Hastie, T., Narasimhan, B., Chu, G.: Diagnosis of multiple cancer types by shrunken centroids of gene expression. Proc. Natl. Acad. Sci. 99, 6567–6572 (2002)
23. Witten, Tibshirani: Penalized classification using Fisher's linear discriminant. Journal Royal Statistical Society, 753–772 (2011)
24. Xu, P., Brock, G., Parrish, R.: Modified linear discriminant analysis approaches for classification of high-dimensional microarray data. Computational Statistics and Data Analysis 53, 1674–1687 (2009)

Minimum K-Adjacent Rectangles of Orthogonal Polygons and its Application

Thanh-Hai Nguyen

CEA, LIST, Embedded Real-Time Systems Laboratory,
Point Courrier 172, 91120 Palaiseau Cedex, France
thanhhai.nguyen@cea.fr

Abstract. This paper presents the problem of partitioning a rectangle $\mathfrak{R}$ which contains several non-overlapping orthogonal polygons, into a minimum number of rectangles. By introducing maximally horizontal line segments of largest total length, the number of rectangles intersecting with any vertical scan line over the interior surface of $\mathfrak{R}$ is less than or equal to k, a positive integer. Our methods are based on a construction of the directed acyclic graph $G = (V, E)$ corresponding to the structures of the orthogonal polygons contained in $\mathfrak{R}$. According to this, it is easy to verify whether a horizontal segment can be introduced in the partitioning process. It is demonstrated that an optimal partition exists if only if all path lengths from the source to the sink in G are less than or equal to $k+1$. Using this technique, we propose two integer program formulations with a linear number of constraints to find an optimal partition. Our goal is motivated by a problem involving utilization of memory descriptors applying to a memory protection mechanism for the embedded systems. We discuss our motivation in greater details.

Keywords: Polygon partitioning, computational geometry, linear programming, applied graph theory, memory protection.

1 Introduction

The problem of partitioning *orthogonal polygons* (which are polygons with all edges either horizontal or vertical) into a minimum number of rectangles arises in applications such as manipulation of VLSI artwork data [13], image processing [4], mesh generation for finite element analysis [15], computer graphics [10] and embedded systems. The latter of these fields is the application domain of our work on orthogonal polygon partitioning. We are interested in a memory protection mechanism using descriptors provided by memory management unit or memory protection unit that will be briefly discussed in this paper.

In this work, we consider the problem of partitioning a finite number of nonoverlapping orthogonal polygons entirely contained in a rectangle $\mathfrak{R}$ into a minimum number of rectangular partitions such that the number of these rectangular partitions intersecting with any vertical scan line over the surface of $\mathfrak{R}$ is less than or equal to k, a positive integer. The objective is to find a partition

T.V. Do et al. (eds.), *Advanced Computational Methods for Knowledge Engineering*,
Advances in Intelligent Systems and Computing 282,
DOI: 10.1007/978-3-319-06569-4_5, © Springer International Publishing Switzerland 2014

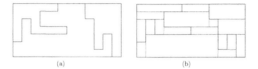

Fig. 1. (a) An input rectangle $\mathfrak{R}$ containing orthogonal polygons, (b) a minimum 4-adjacent rectangular partition of $\mathfrak{R}$ with $k = 4$

whose total length of the partitioning horizontal line segments is maximum. Such a partition is an optimal solution for the problem of Minimum k-Adjacent Rectangles of Orthogonal Polygons, let us denote $k - AROP$ for short (see Fig. 1 for an example of this problem). In this paper, we shall consider $k - AROP$ problem for orthogonal polygons with nondegenerate holes, the holes does not degenerate to points. We shall use both horizontal and vertical segments to find a minimum partition, unlike some other applications [16] that only allow horizontal segments. The problem $k - AROP$ is closely related to the problem of Partitioning Orthogonal Polygons into Rectangles ($POPR$) that has been studied in [9,18,11,4,7]. The development of efficient algorithms for the $POPR$ problem has been a major focus of several researches in computational geometry. A principal step to obtain an optimal partition is to find the maximum number of non-intersecting vertical and horizontal segments between two *concave vertices*. This problem can in turn be solved by finding a maximum matching in a bipartite graph. For more details about orthogonal polygon decompositions, the readers could be referred to [8].

This article is organized as follows: In Section 2, we introduce some basic terminology used in this paper and recall some results of previous researches. In Section 3, we present a construction of the directed acyclic graph corresponding to the structures of the polygons contained in $\mathfrak{R}$. Using this construction, we provide $O(n)$ linear constraints that allow to verify whether a horizontal segment is permissible to introduce in the partitioning process. In Section 4, we propose two integer linear programming formulations : The first one is used to find a maximal non-intersecting *chords* which can be used to partition the rectangle $\mathfrak{R}$ into sub-polygons having no chord. After drawing these non-intersecting chords in $\mathfrak{R}$, the second one is used to find the maximum total length of horizontal segments that will be introduced to obtain a minimal partition. In Section 5, we briefly discuss its application on a memory protection mechanism.

2   Terminologies and Related Works

In this section, we shall recall the terminology already used in [9,18,4,7]. To introduce our approach, we also briefly discuss some of their results.

An *orthogonal polygon* $\mathcal{P}$ on the plan is a polygon whose sides are either vertical or horizontal. A *simple orthogonal* polygon is an orthogonal polygon without holes. Two concave vertices $u_1 = (x_1, y_1)$ and $u_2 = (x_2, y_2)$ which do not belong to the same edge of $\mathcal{P}$ are *co-grid* if they are co-horizontal ($y_1 = y_2$) or co-vertical ($x_1 = x_2$). A *vertical (resp. horizontal) cut* of $\mathcal{P}$ is a line segment entirely contained in $\mathcal{P}$ tracing from a concave vertex until it intersecting with a $\mathcal{P}$ boundary. A *chord* of $\mathcal{P}$ is either vertical or horizontal cut but both extreme points of this line segment are two co-grid vertices. Let $\mathrm{E}^{\mathcal{P}}_{m.n.c}$ denotes a *maximum set of nonintersecting chords* for $\mathcal{P}$. The size of this set is denoted by $|\mathrm{E}^{\mathcal{P}}_{m.n.c}|$. The constraint where the number of rectangular partitions intersecting with any vertical scan line over the interior of $\mathfrak{R}$ is less than or equal to k, is shortly called the *vertical $k-cut$ constraint*. In the remainder of this paper, whenever we refer to polygons and partitioning, we always mean orthogonal polygons and partitioning into rectangles.

If a simple polygon $\mathcal{P}$ contains no chords, then an optimal partition can be easily found by using the following algorithm.

Algorithm 21. *An optimal partition for a simple polygon without chord.*
−For each concave vertex, select one of its incident edges, let e be this edge.
−Extend e until it intersects with an introduced segment or a boundary edge.

In the case where $\mathcal{P}$ contains chords, a theorem described in [9,18,4] shows that the minimum number of partitions is equal to $N - L + 1$ where N is the number of concave vertices and L is equal to $|\mathrm{E}^{\mathcal{P}}_{m.n.c}|$. To find an optimal partition for $\mathcal{P}$, the following algorithm shows the principal construction steps.

Algorithm 22. *An optimal partition for a simple polygon with chords.*
Step 1−Find the set of chords in $\mathcal{P}$.
Step 2−Construct the bipartite graph $G = (V, H, E)$ where each vertex i in V corresponds to a vertical chord v_i and each vertex j in H corresponds to a horizontal chord h_j. Two vertices i and j are connected if only if two chords v_i and h_j are intersected.
Step 3−Find a maximum matching, let M be this matching.
Step 4−Find a maximum independent set based on M, let S be this set (see [19]).
Step 5−Draw L nonintersecting chords corresponding to vertices in S to divide P into L + 1 subpolygons without chords.
Step 6−For each subpolygon without chords, an optimal partition can be found by using Algorithm 21.

In the case of $\mathcal{P}$ containing nondegenerate holes, the method presented in [4] can be applied to transform it into a polygon without holes. The details of this algorithm is briefly described below.

Algorithm 23. *An optimal partition for a polygon containing nondegenerate holes and chords.*
Step 1−From each nondegenerate hole H_i, draw either horizontally or vertically two parallel line segments cutting a boundary of $\mathcal{P}$ that is nearest with H_i, this cut can be seen as an opening hole method that allows to transform $\mathcal{P}$ into a polygon without holes.
Step 2−Apply Algorithm 22 for $\mathcal{P}$ to find an optimal partition.
Step 3−Remove these cuts and reconstitute the boundary of $\mathcal{P}$ and of H_i.

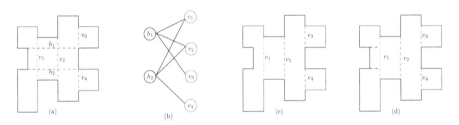

Fig. 2. (a) all vertical and horizontal chords, (b) its corresponding bipartite graph, (c) a representation of maximum number of nonintersecting chords, (d) finding an optimal partition by drawing all horizontal cuts

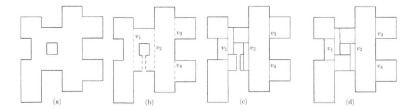

Fig. 3. (a) an input example for polygon with holes, (b) a transformation polygon through an opening hole method, (c) an optimal partition for polygon without holes, (d) an optimal partition with nondegenerate holes

For instance, in Fig. 2(a), the horizontal chords are h_1 and h_2 and the vertical chords are v_1, v_2, v_3, v_4. The corresponding bipartite graph is presented in Fig. 2(b), drawing the chords in $E^{\mathcal{P}}_{m.n.c}$ is presented in Fig. 2(c) and an optimal partition for $\mathcal{P}$ is presented Fig. 2(d) after applying Algorithm 21 with only horizontal cuts. For the case in Fig. 3(a), an opening hole method is shown in Fig. 3(b), an optimal partition by applying Algorithm 22 is shown in Fig. 3(c) and the reconstitution of $\mathcal{P}$ is shown in Fig. 3(d). All the algorithms described above are very important for our approach and they clearly are the starting point of our research. Indeed, if the vertical $k-$ cut constraint, with a big enough value k, is always satisfied, then an optimal partition for $\mathfrak{R}$ is easily found by applying only the techniques described in Algorithm 21 − 23 for each polygon in $\mathfrak{R}$. However, when the value of k is very small, then a horizontal chord or even a horizontal segment introduced during the partitioning process needs to be verified whether it is possible. For this reason, we present our analysis in the next section.

3   Vertical k-Cut Constraints

As we have mentioned above for the vertical $k-$cut constraint, we recall that separately partitioning each polygon contained in $\mathfrak{R}$ does not give a feasible partition. Therefore, we propose here a method allowing to observe and to control globally the partitioning process for them. For this, we present below a construction of a directed acyclic graph that is an incidence graph of $\mathfrak{R}$.

Before showing our ideas, let us introduce some terminology that are useful for our construction and easy for description. We call *vertical partition* of $\mathcal{P}$ a partition formed by drawing all vertical cuts. So, a vertical partition of $\mathfrak{R}$ is union of vertical partitions of the polygons contained in $\mathfrak{R}$. We denote $VP_{\mathfrak{R}}$ this partition (an example for a vertical partition of $\mathfrak{R}$ is shown in Fig. 4). Two rectangles r_i, r_j are called *vertically adjacent* if only if they share one horizontal side. For instance, in Fig. 5(a), the rectangle 3 is vertically adjacent with the rectangles $4, 5$ and 8. In the sequel, we described our construction that proceeds with three principal steps. The first step presents a construction of an incidence $st-$graph G_{st} for $VP_{\mathfrak{R}}$, a graph with a single sink and a single source. The second step gives a mathematical formulation of the vertical $k-$cut constraints by means of the properties of G_{st} and show that the number of $s-t$ paths in G_{st} is invariant by construction. The last step explains how to obtain our integer programs by associating $0/1$ variables for each horizontal cuts.

Step 1– Drawing $VP_{\mathfrak{R}}$ first. After, building- ing a directed acyclic st-graph G_{st} with the principal rules: first add a virtual rectangle s on the top of $\mathfrak{R}$, i.e. a bottom side of s is the top side of $\mathfrak{R}$; after add a virtual rect- angle t on the bottom of $\mathfrak{R}$, i.e. a top side of t is the bottom side of $\mathfrak{R}$.

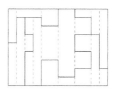

Fig. 4. A vertical partition of $\mathfrak{R}$.

To each rectangular partition in $VP_{\mathfrak{R}}$, associate with a vertex in G_{st}. Two ver- tices i, j of G are connected if only if their incidence rectangular partitions r_i, r_j are adjacent. An outgoing edge of vertex i is created to reach the vertex j if only if the bottom side of r_i and the top side of r_j belong to the same line segment. The graph given by this construction is called the *st-graph* and denoted by G_{st}. For example, two virtual rectangles s, t and an enumerating rectangular partitions of $VP_{\mathfrak{R}}$ are shown in Fig. 5 (a) and its corresponding graph G_{st} to $VP_{\mathfrak{R}}$ is shown in Fig. 5 (b).

Step 2– Find all $s-t$ paths in the $st-$graph. The idea for finding them is straightforward. Indeed, each $s-t$ path is an augmenting path on which the Ford-Fulkerson's algorithm [2] for computing the maximum flow can be applied. So, to each incoming edge of the sink vertex t, we associate a unitary cost 1. For the remaining edges, we associate with a cost equal to the maximum outdegree in the $st-$graph. Hence, all $s-t$ paths can be easily found by means of this algorithm. Let $F_{s,t}$ be the set of these paths and let $(s,t)_i$, respectively $l(s,t)_i$ denote the ith $s-t$ path and its length. According to this construction, the vertical $k-$cut constraint can be mathematically expressed as follows:

$$l(s,t)_i \leq k + 1, \forall (s,t)_i \in F_{s,t} \tag{1}$$

Before describing the $3th$ step, we shall outline ideas to explain how to use the constraints (1) in order to construct our integer program models. We are going to prove Proposition 1, which is the main idea of this section. It states that if

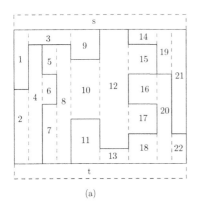

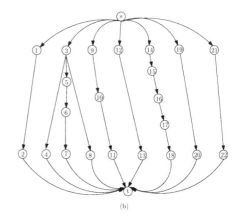

(a)

(b)

Fig. 5. (a) an enumerating rectangular partitions, (b) a construction of the st-graph

any horizontal cut is drawn in $\mathfrak{R}$, then each rectangle in $VP_{\mathfrak{R}}$ intersecting with this cut is divided into two rectangular partitions which are vertically adjacent by this cut, after re-drawing $VP_{\mathfrak{R}}$ and providing a new incidence graph G_{st}, the number of $s - t$ paths in G_{st} is always equal to a constant number.

Let u be a concave vertex in $\mathfrak{R}$. Let N_v be the number of pairwise vertically noncollinear edges of the polygons contained in $\mathfrak{R}$. The boundary of $\mathfrak{R}$ is not taken into account. For example, in Fig. 6, a set of pairwise vertically noncollinear edges are $\{c, a, d\}$ or $\{c, b, d\}$, then $N_v = 3$.

Fig. 6. An example for a set of pairwise vertically noncollinear edges in $\mathfrak{R}$: $\{c, a, d\}$ or $\{c, b, d\}$

Proposition 1. (a) *The number of $s - t$ paths in G_{st} is equal to $N_v + 1$ even if a new G_{st} is defined because of drawing some vertical or horizontal cuts from u during the partitioning process.*
(b) *If a horizontal cut is permissible to draw from u, then there exists at least one $s - t$ path whose length is exactly incremented by one edge.*

Proof. The statement (a) can be done by recurrence. Suppose that $\mathfrak{R}$ has N_v, the number of vertically noncollinear edges and G_{st} has $N_v + 1$ the number of $s - t$ paths. Let e_v the leftmost vertical edge entirely contained in $\mathfrak{R}$. For example, the leftmost vertical edge in Fig. 6 is the edge c. When extending this edge until touching top and bottom boundaries of $\mathfrak{R}$, it is easy to remark that e_v vertically separates $\mathfrak{R}$ into two parts. One part is left to e_v and the other is right to e_v. Note that the left part of e_v corresponds to the leftmost $s - t$ path in the $st-$graph. Let us denote $left_{(s-t)}$ this path. Let $\mathfrak{R}'$ be the rectangle corresponding to the right part of e_v. Therefore, we obtain that $\mathfrak{R}'$ contains

$N_v - 1$ vertically noncollinear edges whereas the st−graph corresponding to $\mathfrak{R}'$ is the one without the path $left_{(s-t)}$, the number of $s - t$ paths is equal to N_v. Thus, the statement (a) recurrently follows.

If a horizontal cut from u, denoted by u_h, is drawn in the partitioning process, then it is trivial that there are always two rectangular partitions in the optimal partition sharing partially or entirely u_h. Thus, there exists at least a rectangle in $VP_{\mathfrak{R}}$ that is sliced into two parts by u_h. Let r_0 denote such rectangle. As each rectangle in $VP_{\mathfrak{R}}$ corresponds to a vertex in the st−graph, when updating G_{st} because of drawing u_h, the incidence vertex of r_0 is replaced by two vertices which correspond to two rectangular partitions. Moreover, only one edge is created to connect these two vertices because they are vertically adjacent by u_h. Thus, the statement (b) follows.

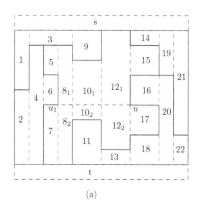

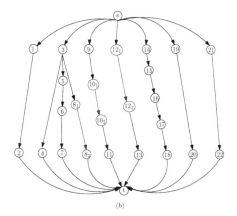

(a)

(b)

Fig. 7. (a) intersection between the rectangular partitions 8, 10, 12 of $VP_{\mathfrak{R}}$ with the horizontal chord uu_1, (b) splitting vertices in G_{st} because of this introduced segment

Proposition 1 implies that if a horizontal cut is traced from any concave vertex, then the number of rectangles in $\mathfrak{R}$ intersecting with any vertical line is only grown by one rectangle. If this number is less than or equal to k, then such horizontal cut can be introduced.

Step 3− Construct $O(n)$ linear constraints with n is the number of concave vertices in $\mathfrak{R}$ by means of the following rules: (a) Draw all horizontal cuts in $VP_{\mathfrak{R}}$. (b) For each path $(s,t)_i$ in G_{st}, let V_i be a set of vertices in this path. For each vertex v in V_i, identify all horizontal cuts intersecting with the corresponding rectangular partition of v. Let $Cut_{(s,t)_i}$ be the index set of these horizontal cuts. (c) To each horizontal cut h_i, we associate a binary variable x_i: $x_i = 1$ if such horizontal cut is permissible, $x_i = 0$ otherwise.

According to Proposition 1 and the constraints (1), we deduce that:

$$\sum_{i \in Cut_{(s,t)_j}} x_i + l(s,t)_j \leq k + 1, \forall (s,t)_j \in F_{s,t} \tag{2}$$

$$x_i \in \{0,1\}$$

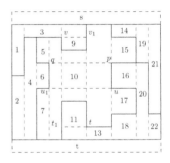

For instance, in Fig. 8, let us consider the $s - t$ path coming through the vertices 9, 10, 11 and its length clearly is equal to 4. Suppose that the binary variables associated with the horizontal cuts $\nu\nu_1$, pq, uv, $\delta\delta_1$ respectively are x_1, x_2, x_3, x_4. Hence, we have a constraint:

$$\sum_{i=1..4} x_i + 4 \leq k + 1 \qquad (3)$$

Fig. 8. An example for illustration of an inequality (2)

In the remainder of this paper we describe application of these constraints to find $\mathrm{E}^{\mathfrak{R}}_{m.n.c}$, a maximum set of nonintersecting chords for $\mathfrak{R}$, *with respects to* (w.r.t) the vertical k−cut constraints.

4   Integer Programming Model

4.1   Integer Linear Program for Finding a Maximum Number of Nonintersecting Chords

First of all, using the same idea that is described in Algorithm 22, we construct the bipartite graphs $G = (V, H, E)$ corresponding to each polygon in $\mathfrak{R}$. To each vertex $i \in V$, we associate a binary variable y_i : $y_i = 1$ if the vertical chord v_i corresponding to the vertex i can be drawn, $y_i = 0$ otherwise. For each horizontal chord i, we associated a binary variable x_i: $x_i = 1$ if the horizontal chord h_i corresponding to the vertex i is permissible, $x_i = 0$ otherwise. For finding a maximum independent vertex set of G, we give here an integer programming formulation (IP):

$$\max \sum_{i \in V} x_i + \sum_{i \in H} y_i$$

$$\sum_{ij \in E} y_j + x_i = 1, \qquad \forall i \in H$$

$$\sum_{ij \in E} x_j + y_i = 1, \qquad \forall i \in V \qquad (4)$$

$$x_i \geq 0, y_j \geq 0 \qquad \forall i \in V, j \in H$$

$$x_i, y_j \;\; integer$$

When dropping the integrality constraints, we then obtain the linear programming relaxation (LP) of the above integer program (4). But whether an optimal integer solution can be found through this LP. For answering this request, we can refer to [5]. This LP is always guaranteed to produce the same optimum cost value as the corresponding integer program (4) even though the LP solution may not be integer but there always exists at least one optimal integer solution. For taking into account the vertical k−cut constraint, we need to use the linear constraints described in (2). So, we propose here an integer programming formulation for finding $E^{\mathfrak{R}}_{m.n.c}$:

$$\max \sum_{i \in V} x_i + \sum_{i \in H} y_i$$

$$\sum_{ij \in E} y_j + x_i = 1, \qquad \forall i \in H$$

$$\sum_{ij \in E} x_j + y_i = 1, \qquad \forall i \in V \qquad (5)$$

$$\sum_{i \in Cut_{(s,t)_j}} x_i + l(s,t)_j \le k+1, \qquad \forall (s,t)_j \in F_{s,t}$$

$$x_i \ge 0, y_j \ge 0 \qquad \forall i \in V, j \in H$$

$$x_i, y_j \ \ integer$$

To apply a solver like Cplex [6] or Glpk [12], program (5) must be posed as an LP. By dropping the integrality constraint and therefore using a LP solver, we obtain a solution for finding $E^{\mathfrak{R}}_{m.n.c}$. Let b be an integer equal to $|E^{\mathfrak{R}}_{m.n.c}|$. We then draw these b chords to divide $\mathfrak{R}$ into $b+1$ subpolygons in which no co-grid concave vertices are existed. In the next section, we describe an integer program for drawing horizontal cuts in these subpolygons.

4.2   Integer Linear Program for Drawing Maximum Horizontal Cuts

Suppose that the rectangle $\mathfrak{R}$ does not contains chords. Let $CV_{\mathfrak{R}}$ be the set of concave vertices in $\mathfrak{R}$. To each $i \in CV_{\mathfrak{R}}$, we associate a binary variable: $x_i = 1$ if a horizontal cut drawn from i is permissible, $x_i = 0$ otherwise, i.e a vertical cut drawn from i is permissible. Let l_i be the length of its horizontal cut. The objective function we use here is to obtain the maximal length of the horizontal cuts. Therefore, we have: $\max \sum_{i \in CV_{\mathfrak{R}}} l_i x_i$. Moreover, to satisfy the vertical k−cut constraint, we reuse the constraints described in program (2). So, an integer program for drawing maximum horizontal cuts is proposed as follows:

$$\max \sum_{i \in CV_{\mathfrak{R}}} l_i x_i$$

$$\sum_{i \in Cut_{(s,t)_j}} x_i + l(s,t)_j \le k+1, \qquad \forall (s,t)_j \in F_{s,t} \qquad (6)$$

$$x_i \in \{0,1\}$$

Before using a solver for LP, the integrality constraints $x_i \in \{0,1\}$ should be dropped (i.e. replacing it with $x_i \in [0,1]$). When a fractional solution is obtained, then we can use branch and bound algorithm to fixe this variable. For more details of this algorithm, the readers can refer to [17].

5   Application on Memory Protection Mechanism for Embedded Systems

When performances of embedded applications are strongly demanded because of low memory footprint or highly constrained for utilization of embedded memory protection devices, then a method for memory protection mechanism that can be applied here is to take advantage of physical descriptors of MMU or MPU [1]. Given a task t, a physical descriptor used by t is defined from a range of contiguous memory addresses on which identical access rights for t are associated. We call such a range of contiguous memory addresses a *memory section*. Nonetheless, setting up those protection devices is costly. Moreover, to ensure a minimal overhead of the memory protection update, an optimal design of both the memory mapping sections and the descriptors is required. The optimality what we discuss here comes from the limitations of memory protection resources, i.e. total number of available physical descriptors and time required to load necessarily physical descriptors when a task is activated. The first limitation enforces to reorganise descriptors within a given execution context in order to use as few as possible of them at a given time. So, two neighbouring memory sections on which a given context requires identical memory rights can be covered by the same descriptor. The second limitation involves to reorganize descriptors that can be used by several execution contexts on a given memory section, so that less descriptors have to be loaded when switching contexts. Therefore, optimizing the number of descriptors allows for a quicker task communication and better performances of the application. For that, a memory section on which two execution contexts require identical memory rights can be covered by the same descriptor. Our motivation described in this section presents a generic method for finding an optimal mapping of these physical descriptors. This mapping is obtained from an optimal configuration of logical descriptors in simulation machine with an appropriate format, this configuration is obtained through an offline procedure. We described here this procedure through three principal steps:

Step 1– The first one consists of spatially separating the execution contexts (application tasks, user's or supervisor's execution mode, . . .) by means of compiling the source files into object files (task X's stack, task Y's code, . . .). The advantage of this step allows to define sufficiently the memory access rights for each execution context on the resources that are required for its proper functioning. This preparation phase gives us an access control matrix A with n columns and m lines where each matrix entry a_{ij} presents the access rights for the task or execution context t_j on a logical memory section i. This matrix is called Lampson matrix [20]. An example of Lampson matrix A is illustrated in Fig. 9(a).

Step 2– With an access control matrix input A, the second step proceeds to reorganize lines and columns of this matrix such that the sum of absolute differences (Manhattan distances) between neighboring rows and between neighboring columns is minimized. This steps can be performed by means of clustering algorithms. For more details, the readers can refer to [14]. Description and application of these algorithms are not the main goal of this paper. The output of this step

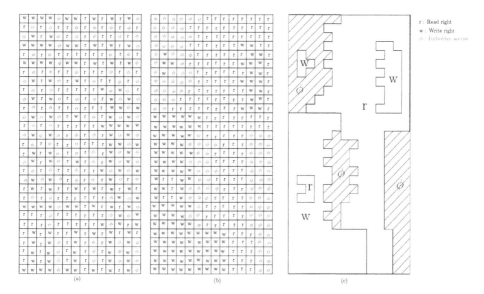

Fig. 9. (a) a Lampson matrix input with 14 columns (14 execution contexts) and 29 lines (29 memory sections), (b) an optimal permutation matrix obtained from the 2th step, (c) a corresponding rectangle with orthogonal polygons defined from r read right areas and w write right areas, the holds defined from ϕ forbidden access areas.

offers several optimal matrices which are the input data for the next step. An example for an optimal matrix of A is presented in Fig. 9(b).

Step 3– With several optimal matrices obtained from the $2nd$ step whose the number can be bounded by an positive integer set by the user. The last step is an application of our algorithms described in Section 4 to find an optimal mapping for logical descriptors. Each optimal matrix A^1 can be considered as a rectangle $\mathfrak{R}$ in which each orthogonal polygon in $\mathfrak{R}$ defines an access right area in A^1. For instance, in Fig. 9(c), we have 5 access right areas offering 5 orthogonal subpolygons. The constraint on the total number of available physical descriptors can be seen as the vertical $k-$cut constraint in our mathematical modeling. The goal of minimizing the number of physical descriptors loading when switching contexts can explain the choice of our objective function: finding a partition for $\mathfrak{R}$ whose total length of the partitioning horizontal line segments is maximum.

6   Conclusion

In this paper, we have proposed two integer programs to solve the problem of partitioning the rectangle $\mathfrak{R}$ containing a finite number of polygons with nondegenerate holes into a minimum number of rectangular partitions with respects to vertical cut constraints. We found such a partition by reducing the geometry problem to an $st-$graph problem for expressing mathematically vertical cut constraints. After, we reused the basic ideas for solving the matching problem of bipartite graph to find a maximum number of nonintersecting chords, we reformulated it by means of IP formulation. Finally, the $k - AROP$ problem can be solved with $O(n)$ linear constraints. In future works, we want to develop algorithms described in [14,3] to find good permutations for access control matrix on which our algorithm will be applied.

References

1. ARM: Technical reference manual - arm1176jz-s (2009)
2. Cormen, T., Leiserson, C.E., Rivest, R.L., Stein, C.: Introduction to Algorithms, 2nd edn. McGraw-Hill Book Company (2001)
3. DiMaggio, P.A., McAllister, S.R., Floudas, C.A., Feng, X., Rabinowitz, J., Rabitz, H.: Biclustering via optimal re-ordering of data matrices in systems biology: rigorous methods and comparative studies. BMC Bioinformatics (2008)
4. Ferrari, L., Sankar, P.V., Sklansky, J.: Minimal rectangular partition of digitized blobs. Comput. Vision, Graph. Image Proc. 28, 58–71 (1984)
5. Goemans, M.X.: Lecture notes on bipartite matching (February 2009)
6. IBM: Cplex optimizer (2012)
7. Imai, H., Asano, T.: Efficient algorithms for geometric graph search problems. SIAM J. Computer 15, 478–494 (1986)
8. Keil, J.M.: Polygon decomposition. In: Sack, J.R., Urrutia, J. (eds.) Handbook of Computational Geometry, pp. 491–518. Elsevier, North Holland, Amsterdam (2000)
9. Lipski, I.l.l.W., Lodi, J.E., Luccio, F., Muganai, C., Pagli, L.: On two dimensional data organization 11: Fundamenta informaticae. SIAM J. Computer 2, 245–260 (1979)
10. Lien, J., Amato, N.M.: Approximate convex decomposition of polygons. Comput. Geom. Theory Appl. 35(1), 100–123 (2006)
11. Lipski, W.: Finding a manhattan path and related problems. Networks 13, 399–409 (1983)
12. Makhorin, A.: Glpk: GNU GNU Linear Programming Kit (2000–2012)
13. Mead, C., Conway, L.: Introduction to vlsi systems. Addison-Wesley Pub. Co. (1979)
14. Miklos, I., Somodi, I., Podani, J.: Rearrangement of ecological data matrices via markov chain monte carlo simulation. Ecology 86(12), 530–536 (2005)
15. Muller-Hannemann, M., Weihe, K.: Quadrangular refinements of convex polygons with an application to finite-element meshes. International Journal of Computational Geometry & Applications 10(01), 1–40 (2000)
16. Nahar, S., Sahni, S.: Fast algorithm for polygon decomposition. IEEE Transactions on Computer-Aided Design 7, 473–483 (1988)
17. Nemhauser, G.L., Wolsey, L.A.: Integer and combinatorial optimization, vol. 18. Wiley, New York (1988)
18. Ohtsuki, T.: Minimum dissection of rectilinear regions. In: Proc. IEEE Symp. on Circuits and Systems, pp. 1210–1213 (1982)
19. Papadimitriou, C.H., Steiglitz, K.: Combinatorial Optimization: Algorithms and Complexity. Prentice Hall Inc., New Jersey (1982)
20. Lampson, B.W.: Protection. SIGOPS Oper. Syst. Rev. 8(1), 18–24 (1974)

The Confrontation of Two Clustering Methods in Portfolio Management: Ward's Method Versus DCA Method

Hoai An Le Thi[1], Pascal Damel[1], Nadège Peltre[2], and Nguyen Trong Phuc[3]

[1] L.I.T.A. Université de Lorraine, France
{hoai-an.lethi,pascal.damel}@univ-lorraine.fr
[2] C.E.R.E.F.I.G.E. Université de Lorraine, France
nadege.peltre@univ-lorraine.fr
[3] Ecole supérieur de transport et communication,
Hanoi Vietnam
Phuc.NguyenTrong@nttdata.com

Abstract. This paper presents a new methodology to cluster asset in the portfolio theory. This new methodology is compare with the classical ward cluster in SAS software. The method is based on DCA (Difference of Convex functions), an innovative approach in nonconvex optimization framework which has been successfully used on various industrial complex systems. The cluster can be used in an empirical example in the context of multi-managers portfolio management, and to identify the one that seems to best fit the objectives of portfolio management of a fund of funds or funds. The cluster is useful to reduce the choice of asset class and to facilitate the optimization of Markowitz frontier.

Keywords: clustering methods, portfolio management, DCA, Ward's method.

1 Introduction

In this paper, we will examine the portfolio optimization problem that occurs when a style constraint is introduced in a multi-manager portfolio management ; the funds included in the managed portfolio or fund of funds should not only achieve an optimal portfolio return, but they also need to implement the target benchmark. In that perspective, we propose to use a classification methodology which enables to make up fund packages that the manager can use to build up a portfolio corresponding to his investment goals. This article tends to confront two clustering methods: the very known and standard hierarchical ascendant clustering method, with Ward's measure, and the DCA (Difference of Convex functions Algorithm), an innovative approach in nonconvex programming framework.

In the second party, we will describe these two kinds of clustering methods, in the third party we will focus on the methodologies used to determine an optimal number

T.V. Do et al. (eds), *Advanced Computational Methods for Knowledge Engineering*,
Advances in Intelligent Systems and Computing 282,
DOI: 10.1007/ 978-3-319-06569-4_6, © Springer International Publishing Switzerland 2014

of clusters, in the fourth party, the database and the methodology of our study will be exposed, and in the last party, we will present the results and draw a conclusion about the use of these two clustering methods for portfolio management.

2   Types of Cluster Analyses

2.1   The Standard Clustering Methods

Automatic clustering problems have been addressed through several authors ([2, 5, 7] and so on). Those books show that there are two main kinds of standard clustering techniques: the non hierarchical methods (the method of leader, the K-means method and the dynamic clouds method) and the hierarchical methods.

We can distinguish two kinds of hierarchical clustering: the ascendant hierarchical clustering methods and the descendant hierarchical clustering methods. The ascendant hierarchical clustering methods are generally more used. The ascendant hierarchical clustering [7] consists in performing a series of regroupings to form increasingly fine classes, in aggregating at every stage objects or groups of objects closest. Among the ascendant hierarchical classification methods, The most known and used and even for some authors, the more efficient is certainly the Ward's method. We tested the nine ascendant hierarchical classification methods proposed by the statistical software SAS in a precedent conference [2] and the more efficient was incontestably the Ward's   method [12][1] .

At each step, the two clusters that merge are those whose "distance" is the lowest. The question is to find a good definition of what is meant by "distance" between two groups of points. There are many definitions of this distance, but the one used in the Ward's method is the squared Euclidean distance.

The principle of Ward's method is to obtain, at each iteration, a local minimum of intra-class inertia or a maximum of inter-class inertia.

The index of dissimilarity between two classes (or level of aggregation of these two classes) is equal to the loss of inter-class inertia resulting from their grouping.

2.2.   An Alternative Approach at the Standard Methods: The DCA

DC programming and DCA were introduced by Pham Dinh Tao in their preliminary form in 1985. They have been extensively developed since 1994 by Le Thi Hoai An and Pham Dinh Tao and become now classic and more and more popular (see, e.g. [8,9,10,13,14], and references therein). DCA has been successfully applied to many large-scale (smooth or non-smooth) non-convex programs in various domains of applied sciences, in particular in data analysis and data mining and machine learning for which it provided quite often a global solution and proved to be more robust and efficient than standard methods (see [8,9,10,13,14],   and references therein).

---

[1] Examining the dendogrammes elaborated according to the various methods, the one of the Ward's method alone was showing a classification including relatively distinct classes.

General DC Programs

Consider the general DC program

$$\alpha = inf\{f(x) := g(x) - h(x) : x \in R^n\}(P_{dc})$$

with g, h being proper lower semi-continuous convex functions on R^n. Such a function f is called DC function, and $g - h$, DC decomposition of f while the convex functions g and h are DC components of f.

Difference of Convex functions Algorithms (DCA)

The main idea behind DCA is to replace, at the current point x^k of iteration k, the concave part $-h(x)$ with its affine majorization defined by

$$h_k(x) := -h(x^k) - \langle x - x^k, \gamma^k \rangle, \gamma^k \in \partial h(x^k)$$

to obtain the convex program of the form

$$inf\{g(x) + h_k(x) : x \in R^n\} \Leftrightarrow inf\{g(x) - \langle x, \gamma^k \rangle : x \in R^n\}. (P_k)$$

Algorithm
Let $x^0 \in R^n$ be an initial guess. Set $k := 0$
REPEAT
    $\gamma^k \in \partial h(x^k)$.
    $x^{k+1} \in argmin\{g(x) - \langle x, \gamma^k \rangle : x \in R^n\}$.
    $k = k + 1$
UNTIL convergence.

DCA schemes have the following properties ([22], [34]):
  i)   the sequence $\{g(x^k) - h(x^k)\}$ is decreasing,
  ii) if the optimal value α of problem(P_{dc}) is finite and the infinite sequences $\{x^k\}$ is bounded, then every limit point $\tilde{x}$ of the sequence $\{x^k\}$ is a critical point of $g - h$.

Observe that a DC function has infinitely many DC decompositions and there are as many DCA as there are DC decompositions which have crucial impacts on the qualities (speed of convergence, robustness, efficiency, and globality of computed solutions) of DCA. Hence, the solution of a nonconvex program by DCA must be composed of two stages: the search of an appropriate DC decomposition and that of a good initial point.

DCA for solving the clustering problem via MSSC (Minimum Sum of Squares Clustering) formulation

An instance of the partitional clustering problem consists of a data set $A := \{a^1, ..., a^m\}$ of m points in R^n, a measured distance, and an integer k; we are to choose k members x^l $(l = 1, ..., k)$ in R^n as "centroid" and assign each member of A to its closest centroid. The assignment distance of a point $a \in A$ is the distance from a to the centroid to which it is assigned, and the objective function, which is to be minimized, is the sum of assignment distances. If the squared Euclidean distance is used, then the corresponding optimization formulation is expressed as ($\|.\|$ denotes the Euclidean norm) a so called MSSC problem

$$min \left\{ \sum_{i=1}^{m} \min_{l=1,...,k} \|x^l - a^i\|^2 : x^l \in R^n, l = 1, ..., k \right\}. (MSSC)$$

The DCA applied to (MSSC) has been developed in [10]. For the reader's convenience we will give below a brief description of this method.

To simplify related computations in DCA for solving problem (MSSC) we will work on the vector space $R^{k \times n}$ of $(k \times n)$ real matrices. The variables are then $X \in R^{k \times n}$ whose i^{th} row X_i is equal to x^i for $i = 1, ..., k$. The Euclidean structure of $R^{k \times n}$ is defined with the help of the usual scalar product

$$R^{k \times n} \ni X \leftrightarrow (X_1, ..., X_k) \in (R^n)^k, X_i \in R^n, (i = 1, ..., k),$$

$$\langle X, Y \rangle := Tr(X^T Y) = \sum_{i=1}^{k} \langle X_i, Y_i \rangle$$

and its Euclidean norm $\|X\|^2 := \sum_{i=1}^{k}\langle X_i, X_i \rangle = \sum_{i=1}^{k} \|X_i\|^2$ (Tr denotes the trace of a square matrix). We will reformulate the MSSC problem as a DC program in the matrix space $R^{k \times n}$ and then describe DCA for solving it.

Formulation of (MSSC) as a DC Program

According to the property

$$\min_{l=1,...,k} \|x^l - a^i\|^2 = \sum_{l=1}^{k} \|x^l - a^i\|^2 - \max_{r=1,...,k} \sum_{l=1,l\neq r}^{k} \|x^l - a^i\|^2$$

and the convexity of the functions

$$\sum_{l=1}^{k} \|x^l - a^i\|^2, \max_{r=1,...,k} \sum_{l=1,l\neq r}^{k} \|x^l - a^i\|^2,$$

we can say that (MSSC) is a DC program with the following DC formulation:

$$MSSC \Leftrightarrow min\{F(X) := G(X) - H(X) : X \in R^{k \times n}\}, \tag{1}$$

where the DC components G and H are given by

$$G(X) := \sum_{i=1}^{m} \sum_{l=1}^{k} G_{il}(X), G_{il}(X) = \frac{1}{2} \|X_l - a^i\|^2 \, for \, i = 1, \dots, m, l = 1, \dots, k$$

and

$$H(X) := \sum_{i=1}^{m} H_i(X), H_i(X) = \max_{j=1,\dots,k} H_{ij}(X); H_{ij}(X) = \sum_{l=1,l \neq j}^{k} \frac{1}{2} \|X_l - a^i\|^2 \, for \, i$$
$$= 1, \dots, m \tag{2}$$

It is interesting to note that the function G is a strictly convex quadratic form. More precisely we have, after simple calculations:

$$G(X) = \frac{m}{2} \|X\|^2 - \langle B, X \rangle + \frac{k}{2} \|A\|^2 \tag{3}$$

where $A \in R^{m \times n}, B \in R^{k \times n}$ are given by

$$A_i := a_i \, for \, i = 1, \dots, m$$

$$B_l := a = \sum_{i=1}^{m} a^i \, for \, l = 1, \dots, k.$$

In the matrix space $R^{k \times n}$, the DC program (1) then is minimizing the difference of the simplest convex quadratic function (3) and the nonsmooth convex one (2). This nice feature is very convenient for applying DCA, which consists in solving a sequence of approximate convex quadratic programs whose solutions are explicit.

DCA for Solving (1)

According to the description of DCA, determining the DCA scheme applied to (1) amounts to computing the two sequences $\{X^{(p)}\}$ and $\{Y^{(p)}\}$ in $R^{k \times n}$ such that

$$Y^{(p)} \in \partial H(X^{(p)}), X^{(p+1)} \in \partial G^*(Y^{(p)}).$$

The computation of $\partial H(X)$ and $\partial G^*(Y)$ leads to the following algorithm:

Description of DCA to solve the MSSC problem (1)

Initialization: Let $\varepsilon > 0$ be given, $X^{(0)}$ be an initial point in $R^{k \times n}$, set $p: = 0$;

Repeat

1. Calculate $Y^{(p)} \in \partial H(X^{(p)})$ by using (4)

$$Y^{(p)} = mX^{(p)} - B - \sum_{i=1}^{m} e_{j(i)}^{[k]} \left(X_{j(i)}^{(p)} - a^i \right). \tag{4}$$

2. Calculate $X^{(p+1)}$ according to (5)

$$X^{(p+1)} := \frac{1}{m} \left(B + Y^{(p)} \right). \tag{5}$$

3. Set $p \leftarrow p + 1$

Until
$$\|X^{(p+1)} - X^{(p)}\| \leq \epsilon(\|X^{(p)}\| + 1) \vee |F(X^{(p+1)}) - F(X^{(p)})| \leq \epsilon(|F(X^{(p)})| + 1).$$

Remark 2. *The DC decomposition (1) gives birth to a very simple DCA. It requires only elementary operations on matrices (the sum and the scalar multiplication of matrices) and can so handle large-scale clustering problems.*

3   Methodologies Determining an Optimal Number of Clusters

The ascendant hierarchical classification led to the construction of a hierarchical tree, called dendrogram, showing the partition of the population into subgroups according to a series of combinations. This dendrogram helps to visualize the hierarchy of partitions obtained by successive truncations. The more the truncation is at the top of the dendrogram, the less the partition includes classes. Consequently, a truncation made below the first node of the tree leads to what each class does not contain any individual. In contrast, a truncation made beyond the level of the root of the dendrogram results in a single class containing all individuals.

The search for the appropriate number of classes is then always a critical element in building a classification of data, but it is long and often ambiguous.

For the clustering hierarchical ascendant standard methods, the search for the appropriate number of classes is then always a critical element in building a classification of data, but it is long and often ambiguous. There is indeed no formula to calculate this number from the data but there are numerous articles outlined in the statistics literature detailing potential criteria for detecting the appropriate number of clusters [11,12].

The three criteria that performed best in these simulation studies with a high degree of error in the data were a pseudo-F statistics developed by Calinski and Harabasz [1], the pseudo-t2 statistics developed by Duda and Hart and the cubic clustering criterion developped for the statistical software SAS [16].

The CCC

The Cubic Clustering Criterion (CCC) was developed for the statistical software SAS [4] as a comparative measure of the deviation of the clusters from the distribution expected if data points were obtained from a uniform distribution.

The Pseudo F

Another very popular criteria to determine the appropriate number of classes is the pseudo-F.

The Pseudo t^2

The statistics pseudo-t^2 [2] is known as the ratio Je(2) /Je(1), where Je(2) is the within cluster sum of squares error when the data are divided into two clusters, Je(1) is the sum of square error before division.

It may be advisable to look for consensus among the three statistics, that is, local peaks of the CCC and pseudo-F statistics combined with a small value of the pseudo-t2 statistics and a larger pseudo-t2 for the next cluster fusion. It must be emphasized that these criteria are appropriate only for compact or slightly elongated clusters, preferably clusters that are roughly multivariate normal.

We can finally notice that, if some authors think that the CCC may be incorrect if clustering variables are highly correlated, others assess the correlation structure of the variables does not affect much the performance of the stopping rules ([12]).

For the DCA method, as the use of this method requires the number of clusters, we first apply DCA for some given value of k, the number of clusters. Then basing on the Calinski Harabasz (CH) Index, which is also called the pseudo-F criterium, we chose the DCA's result corresponding to this index.

4   Data and Methodology

Our study was realized on the database constituted of the monthly returns of 551 funds made of European stocks values, on the period from October 2002 to December 2007. For each fund, we calculated, in a first time, 17 performance, risk, and risk-adjusted performance indicators, on a rolling basis of 3 months, 1 year, and 3 years. Our intermediate database was then constituted of 51 variables.

The Computation of the Database's Indicators (Lipper's Database)
Some Indicators:

Suppose that we know N+1 indices of the fund $P_0, P_1, ..., P_N$ and N+1 indices of the benchmark fund $P_0^m, P_1^m, ..., P_N^m$. We also know the free rates $R_1^{rf}, R_2^{rf}, ..., R_N^{rf}$.

Table 1. The 17 financial indicators of the database

Return measurement	• Absolute performance • Relative performance (to benchmark)
Risk measurement	• Absolute risk (standard deviation) • Tracking Error (relative risk) • Max drawdown • Negative periods (%) • Positive periods (%) • VaR Cornish Fisher
Performance measurement adjusted to the risk	• Alphas of Jensen • Betas against the benchmark • Beta bull • Beta bear • Information ratio • Sharpe ratio • Sortino ratio • Treynor ratio • Black Treynor ratio

The returns (daily) of the fund and of the benchmark are calculated as follows:

$$R_i = \frac{P_i - P_{i-1}}{P_{i-1}}.100\%$$

$$R_i^m = \frac{P_i^m - P_{i-1}^m}{P_{i-1}^m}.100\%.$$

The Semi Parametric C.F. Value at Risk

The market Value-at-Risk (VaR) was popularized by J.P.Morgan's RiskMetrics group with the tool Risk Metrics. VaR is defined as the minimum level of loss at a given, sufficiently high, confidence level for a predefined time horizon. The recommended confidence levels are 95% and 99%. The parametric VaR needs a normal distribution of the return of asset. However, the returns on many financial assets are heavy tailed and skewed. We can measure better the market risk estimated VaR by accounting for the deviations from normality. Zangari [45] provides a Modifed VaR MVaR ou Cornish Fisher calculation that takes the higher moments of non-normal distributions (skewness, kurtosis). The value at risk corresponding the Cornish fisher law is calculated by

$$VaR = \overline{R} - \sigma.z_{cf}$$

where σ is the standard deviation and

$$z_{cf} = q_p + \frac{q_p^2 - 1}{6} S + \frac{q_p^3 - 3q_p}{24} K - \frac{2q_p^3 - 5q_p}{36} S^2.$$

Here, S is the skewness

$$S = \frac{\sum (R_i - \overline{R})^3}{(N-1)\sigma^3}$$

K is the kurtosis

$$K = \frac{\sum (R_i - \overline{R})^4}{(N-1)\sigma^4}$$

with $q_p = 1.65$.

We computed in a second time the four moments (mean, standard deviation, skewness and kurtosis) of each indicator, which witness the 204 variables of our final database. We have thus applied the principles mentioned in the second party to determine a satisfactory number of classes in our study. We have therefore retained 7 classes.

5   Results and Conclusion

If all the 204 variables take part to the building of clusters, their contribution to the characteristic propriety of each cluster is different. We actually calculated the coefficients of variation related to each variable in order to identify those who seem to better contribute to the profile definition of the funds making up the classes, that is to say, those with the lowest variation coefficients (variation coefficients lesser than 0.2).The main characteristics of the clusters are synthesized in the table 2.

We can observe that the clusters constituted with the Ward's method are a little more homogenous than the clusters constituted by the DCA. With the DCA, there are one big-size cluster (class 4) which regroups 77% of the population, one middle-size cluster (cluster 5) and 5 little-size clusters. With Ward's method there are 3 big-size clusters (clusters 1, 2 and 3), one middle-size cluster (cluster 4) and 3 little-size clusters (clusters 5, 6 and 7).

The aim of a clustering method is to get as much separation in the data as possible, thus we do not want the largest cluster to be too large.   Yet it is the case with the DCA in which the class 4 contains 77 % of the population.

But the Ward's method, however, has here a major disadvantage: no variable significantly contributes to the building of the cluster 2, that is to say, we can't describe this cluster with any variable. in the frame of portfolio management, this is a great disadvantage because this means that all funds of this class can't be described. This excludes a set of funds of the portfolio building, these set representing here almost one third (exactly 31%) of the population.

Table 2. The Results of the DCA and Ward's methods

DCA Method				Ward'sMethod			
Cluster	Number of contributive variables in the cluster	Number of funds in the cluster	% of funds in the cluster	Cluster	Number of contributive variables in the cluster	Number of funds in the cluster	% of funds in the cluster
DCA1	28	6	1	W1	8	188	34
DCA2	9	19	3	W2	0	172	31
DCA3	57	4	1	W3	67	131	24
DCA4	18	422	77	W4	79	53	10
DCA5	23	71	13	W5	94	4	1
DCA6	11	28	5	W6	44	2	0
DCA7	16	1	0	W7	31	1	0
Total		551	100	Total		551	100

We also studied the nature of the contributing variables to classes, for both methods, on the one hand versus time (indicators calculated on three months, one year and three years basis), on the other hand on the time used (mean, variance, skewness and kurtosis).

It shows that the period on which the indicator is calculated does not seem to be discriminating in choosing a method. The contributive variables are distributed quite evenly between the indicators at 3 months, 1 year, and three years, although the method DCA seems slightly to bring out the short-term indicators (at three months) and the Ward's method the long term ones (at three years).

Nevertheless, the two methods differ greatly in the timing indicators: indeed, the significant variables in the Ward's method are mainly (55%) the moments of order 3 and 4 (skewness and kurtosis), which are difficult to interpret, while the significant variables in the   DCA are overwhelmingly (74%) the moments of order 1 and 2 (the means and variances), much more easily interpretable.

We can conclude here that if Ward's method is the more known and used clustering method, it may be not well appropriate for portfolio management because it can build some clusters and thus a more or less big set of funds, that cannot be described, and consequently not used in the building of portfolios.

But the other drawback is to construct classes mainly on indicators of order 3 (skewness) and order 4 (kurtosis), which are difficult to interpret; this can conduct to a lack of interpretation of some classes.

We then choose the DCA, for which the table below details the interpretation of the seven classes constructed with this clustering method:

Table 3. The description of the classes constructed by DCA

Classes	Interpretation of the classes
DCA1	The performance of this class is lower than the sample average yield. All funds of this class have fewer negative periods than the average of the sample, but the absolute risk is much higher than the average. With market betas and bull betas above the sample mean, those funds do not behave well on the downside.
DCA 2	The average performance of funds of this class is much higher than that of the sample. This class include funds with positive Jensen alphas, and so an outperformance relative to the market. There were no significant risk indicators for the construction of this class. These funds have a larger variance of the absolute performance and a lower variance for the market beta.
DCA3	The fund's performance of this class is equivalent to that of the sample. This class mainly includes defensive funds, with weaker relative risk and VaR. Nevertheless, we can note that these funds have more negative periods than the sample.
DCA4	With this class, we are in the benchmark sample with a yield which is very close to the average yield of the sample. The risk of all of these funds is slightly lower than that of the sample (Absolute Risk VaR and somewhat lower). We can also note that we did not find any beta or indicators of relative risk among the significant variables.
DCA5	This class includes funds with a lower yield than the average of the sample. But indicators of absolute and relative risks are also lower. As against the VaR is much stronger, which explains the negative return. The risk is poorly estimated by indicators based on the average and standard deviation because returns are clearly nonnormal.
DCA6	The funds in this class have a better than average performance of the sample. This class is just the opposite of the class DCA5, in which the funds have higher aboslute and relative risks, but a lower VaR. Here the non-normality plays for the fund's performance.
DCA7	This class is composed of a single fund. We note that the performance is generally very negative and risk indicators very high

References

1. Calinski, T., Harabasz, J.: A dendrite method for cluster analysis. Communications in Statistics. Theory and Methods 3, 1–27 (1974)
2. Damel, P., Peltre, N., Razafitombo, H.: Classification, Performance and fund selection. In: Applied Econometrics Association 93th International Conference Exchange Rate and Risk Econometrics, Athènes Pireaus University (2006)
3. De Leeuw, J.: Applications of convex analysis to multidimensional scaling, Recent developments. In: Barra, J.R., et al. (eds.) Statistics, pp. 133–145. North-Holland Publishing Company, Amsterdam (1997)
4. Gordon, A.D.: How many clusters? An investigation of five procedures for detecting nested cluster structure. In: Hayashi, C., Ohsumi, N., Yajima, K., Tanaka, Y., Bock, H., Baba, Y. (eds.) Data Science, Classification, and Related Methods. Springer, Tokyo (1996)
5. Griffiths, A., Robinson, L.A., Willett, P.: Hierarchic Agglomerative Clustering Methods for Automatic Document Classification. Journal of Documentation 4(3), 175–205 (1984)
6. Kaufman, L., Rousseeuw, P.J.: Finding Groups in Data. Wiley & Sons, New York (1990)

7. Lance, G.N., Williams, W.T.: A general theory of classificatory sorting strategies 1. Hierarchical systems. The Computer Journal 9(4), 373–380 (1967)
8. Le Thi, H.A.: DC programming and DCA, `http://lita.sciences.univ-metz.fr/~lethi/index.php/dca.html`
9. Le Thi, H.A., Pham Dinh, T.: The DC (difference of convex functions) programming and DCA revisited with DC models of real world nonconvex optimization problems. Ann. Oper. Res. 133, 23–46 (2005)
10. Le Thi, H.A., Belghiti, T., Pham Dinh, T.: A new efficient algorithm based on DC programming and DCA for Clustering. Journal of Global Optimization 37, 593–608 (2007)
11. Lozano, J.A., Larranaga, P., Grana, M.: Partitional cluster analysis with genetic algorithms: Searching for the number of clusters. In: Hayashi, C., Ohsumi, N., Yajima, K., Tanaka, Y., Bock, H., Baba, Y. (eds.) Data Science, Classification, and Related Methods. Springer, Tokyo (1996)
12. Mingoti, S.A., Felix, F.N.: Implementing Bootstrap in Ward's Algorithm to estimate the number of clusters. Revista Eletrônica Sistemas & Gestão 4(2), 89–107 (2009)
13. Pham Dinh, T., Le Thi, H.A.: DC optimization algorithms for solving the trust region subproblem. SIAM J. Optim. 8, 476–505 (1998)
14. Pham Dinh, T., Le Thi, H.A.: Convex analysis approach to d.c. programming: Theory, Algorithms and Applications. Acta Mathematica Vietnamica (dedicated to Professor Hoang Tuy on the occasion of his 70th birthday) 2(1), 289–355 (1997)
15. Podani, J.: Explanatory variables in classifications and the detection of the optimum number of clusters. In: Hayashi, C., Ohsumi, N., Yajima, K., Tanaka, Y., Bock, H., Baba, Y. (eds.) Data Science, Classification, and Related Methods. Springer, Tokyo (1996)
16. Sarle, W.S.: Cubic Clustering Criterion, SAS Technical Report A-108. SAS Institute Inc., Cary (1983)

Approximating the Minimum Tour Cover with a Compact Linear Program

Viet Hung Nguyen

Sorbonne Universités, UPMC Univ Paris 06, UMR 7606, LIP6
4 place Jussieu, Paris, France

Abstract. A tour cover of an edge-weighted graph is a set of edges which forms a closed walk and covers every other edge in the graph. The minimum tour cover problem is to find a minimum weight tour cover. This problem is introduced by Arkin, Halldórsson and Hassin (Information Processing Letters 47:275-282, 1993) where the author prove the NP-hardness of the problem and give a combinatorial 5.5-approximation algorithm. Later Könemann, Konjevod, Parekh, and Sinha [7] improve the approximation factor to 3 by using a linear program of exponential size. The solution of this program involves the ellipsoid method with a separation oracle. In this paper, we present a new approximation algorithm achieving a slightly weaker approximation factor of 3.5 but only dealing with a compact linear program.

1 Introduction

Let $G = (V, E)$ be an undirected graph with a (nonnegative) weight function $c : E \Rightarrow \mathbb{Q}_+$ defined on the edges. A *tour cover* of G is a subgraph $T = (U, F)$ such that

1. for every $e \in E$, either $e \in F$ or F contains an edge f adjacent to e, i.e. $F \cap N(e) \neq \emptyset$ where $N(e)$ is the set of the edges adjacent to e.
2. T is a closed walk.

A tour cover is hence actually a tour over a vertex cover of G. The *minimum tour cover problem* consists in finding a tour cover of minimum total weight :

$$\min \sum_{e \in F} c_e,$$

over subgraphs $H = (U, F)$ which form a tour cover of G.

The minimum tour cover problem were introduced by Arkin, Halldórsson and Hassin [1]. The motivation for their study comes from the close relation of the tour cover problem to vertex cover, watchman route and traveling purchaser problems. They prove that the problem is NP-hard and provide a fast combinatorial algorithm achieving an approximation factor of 5.5.

Improved approximations came from Könemann, Konjevod, Parekh, and Sinha [7] where they use an integer formulation and its linear programming relaxation

T.V. Do et al. (eds.), *Advanced Computational Methods for Knowledge Engineering*,   99
Advances in Intelligent Systems and Computing 282,
DOI: 10.1007/978-3-319-06569-4_7, © Springer International Publishing Switzerland 2014

to design a 3-approximation algorithm. However, as the linear programming relaxation is of exponential size, their algorithm needs an ellipsoid method and a separation oracle to solve it.

Several problems are closely related to the minimum tour cover problem. First, if instead of a tour, we need a tree then this is the *minimum tree cover problem*. Second, if we need just a edge subset over a vertex cover then this is the *minimum edge dominating set problem*. These two problems are all NP-hard. Approximation algorithms [1],[7],[5] has been designed for the minimum tree cover problem and the current best approximation factor is 2 [5]. Similarly, approximation algorithms for the minimum edge dominating set problem have been discussed in [2], [6] and the current best approximation factor is also 2 [6]. Another related problem is the well known *minimum edge cover problem* which consists in finding a minimum weight edge subset which covers every vertex of G. This problem can be solved in polynomial time and we know a complete linear programming for it [4].

In this paper, we present a new approximation algorithm achieving a factor 3.5 which is slightly weaker than the factor 3 obtained by Könemann et al. but our algorithm needs only to solve a compact linear program. Precisely, we use a compact linear relaxation of the formulation in [7]. From an optimal solution of this relaxation, we determine a vertex cover subset and use a reduction to the edge cover problem to find a forest F spanning it. Finally, to obtain a tour cover, we apply the Christofides heuristic [3] to find a tour connecting the connected components of F and eventually duplicate edges in each connected component of F. We prove that the weight of such a tour cover is at most 3.5 times the weight of the minimum tour cover.

The idea of reduction to the edge cover problem is first given by Carr et al. [2] in the context of the minimum edge dominating set problem. We borrow their idea here to apply to the minimum tour cover problem.

Let us introduce the notations which will be used in the paper. For a subset of vertices $S \subseteq V$, we write $\delta(S)$ for the set of edges with exactly one endpoint inside S et $E(S)$ for the set of edges with both endpoints inside S. If $x \in \mathbb{R}^{|E|}$ is a vector indexed by the edges of a graph $G = (V, E)$ and $F \subseteq E$ is a subset of edges, we use $x(F)$ to denote the sum of values of x on the edges in the set F, $x(F) = \sum_{e \in F} x_e$.

The paper is organized as follows. First, we present the first three steps of the algorithm and we explain the idea of the reduction to the edge cover problem. Second and lastly, we describe the last step and we give a proof for the approximation factor of 3.5.

2   A 3.5-Approximation Algorithm

2.1   First Three Steps of the Algorithm

Let x be a vector in $\mathbb{R}^E$ which is an incidence vector of a tour cover $\mathcal{C}$ of G. Let $e = uv$ be an edge of G. We can see that e can belong or not to $\mathcal{C}$, but in the two cases, there must be at least two edges in $\delta(\{u, v\})$ belonging to $\mathcal{C}$, i.e. $|\delta(\{u, v\}) \cap \mathcal{C}| \geq 2$.

Hence, the vector x satisfies the following constraint: $x(\delta(\{u,v\})) \geq 2$. The following linear program which consists of all these constraints applying for every edge in G :

$$\min \sum_{e \in E} c_e x_e$$

s.t.
$$x(\delta(\{u,v\})) \geq 2 \qquad \text{for all edge } uv \in G, \qquad (1)$$
$$0 \leq x_e \leq 2 \text{ for all } e \in E.$$

This linear program (1) is thus a linear programming relaxation of the tour cover problem. We can see that this is a compact linear program, its size is even linear since the number of constraints is in $O(|E|)$. Note that the linear programming relaxation used in [7] has, in addition of the box constraints $0 \leq x_e \leq 2$ for all $e \in E$,

$$x(\delta(S)) \geq 2 \qquad \text{for all } S \subset V \text{ s.t. } E(S) \neq \emptyset, \qquad (2)$$

as constraints and then is clearly of exponential size and the set of the inequalities(1) is the subset of the inequalities in (2) having $|S| = 2$.
Now let us consider the following

$$\min \sum_{e \in E} c_e x_e$$

s.t.
$$x(\delta(\{u,v\})) \geq 1 \qquad \text{for all edge } uv \in G, \qquad (3)$$
$$0 \leq x_e \leq 1 \text{ for all } e \in E.$$

It is clear that an optimal solution of (1) is two times a optimal solution of (3) and a solution of (3) is a half of a optimal solution of (1).
Let x^* be an optimal solution of (3) found by usual linear programming techniques. Consequently $2x^*$ is an optimal solution of (1). Let $V_+ \subseteq V = \{u \in V \mid x^*(\delta(u)) \geq 1/2\}$ and let $V_- = V \setminus V_+$. It is not difficult to prove the following lemma.

Lemma 1. *The set V_+ is a vertex cover of G.*

Hence, a tour in G containing all vertices of V_+ is thus a tour cover. Our algorithm will build such a tour by building first an edge subset $D_+ \subseteq E$ covering V_+ (i.e. V_+ is a subset of the set of the end vertices of the edges in D_+) with weight not greater than $2 \sum_{e \in E} c_e x_e^*$ and after finding a tour connecting the connected components of D_+ by Christofides algorithm. The idea of build the edge set D_+ is borrowed from [2] where the authors apply it for designing an $2\frac{1}{10}$-approximation algorithm for the minimum edge dominating problem. Let us examine it in details.

Let V'_- be a copy of V_- where $v \in V_-$ corresponds to $v' \in V'_-$ and E' be the set of zero-weight edges, one between each $v \in V_-$ and its copy $v' \in V'_-$. We construct then the graph $\bar{G} = (\bar{V} = V \cup V', \bar{E} = E \cup E')$. We have the following lemma.

Lemma 2. *There is a one-to-one weight preserving correspondence between the edge subsets which cover V_+ in G and the edge cover subsets (that cover $\bar{V}$) of $\bar{G}$.*

Proof. If D_- is an edge cover of $\bar{G}$, then $D_+ = D_- \cap E$ must be an edge set of equal weight covering all the vertices in V_+. Conversely, if D_+ is an edge set covering all the vertices in V_+, then $D_+ \cup E'$ is an edge cover of $\bar{G}$ of equal weight, since the edges in E' cost nothing.

We can now describe the first three steps of the algorithm which can be stated as follows:

Step 1. Compute an optimal solution x^* of (3).
Step 2. Compute V_+.
Step 3. Build the graph $\bar{G}$. As the minimum edge cover problem can be solved in polynomial time [4], compute a minimum-weight edge cover D_- in $\bar{G}$ and set $D_+ = D_- \cap E$.

2.2   Analysis on the Quality of D_+

It is known that the minimum edge cover problem in $\bar{G}$ can be formulated by the following linear program [4]:

$$\min \sum_{e \in E} c_e x_e,$$

$EC(\bar{G})$ s.t.

$$x(\delta(u)) \geq 1 \ u \in \bar{V}, \tag{4}$$

$$x(E(S)) + x(\delta(S)) \geq \frac{|S|+1}{2} \ \ S \subseteq \bar{V}, |S| \geq 3 \text{ odd}, \tag{5}$$

$$0 \leq x_e \leq 1 \ e \in \bar{E}. \tag{6}$$

Theorem 1. *Le point $2x^*$ which is an optimal solution of (1) is feasible pour $EC(\bar{G})$.*

Proof. Let $y^* = 2x^*$. Suppose u is a vertex in $\bar{V}$. If $u \in V_+$, we have $x^*(\delta(u)) \geq \frac{1}{2}$, otherwise $u \in V_- \cup V'_-$, and we have $x^*_e = 1$ for all $e \in E'$, so in either case

$$y^*(\delta(u)) \geq 1, \tag{6}$$

So y satisfies the constraints (4). As x^* is a solution of (3), hence y^* satisfies

$$y^*(\delta(u)) + y^*(\delta(v)) \geq 2 + 2y^*_{uv}. \tag{7}$$

Suppose that S is a subset of $\bar{V}$ of odd cardinality; let $s = |S|$. When $s = 1$, the constraints (4) are trivially satisfied by y^*, so suppose that $s \geq 3$. By combining (6) and (7) we see

$$y^*(\delta(u)) + y^*(\delta(v)) \geq \begin{cases} 2 + 2y_{uv}^* & \text{if } uv \in \bar{E}, \\ 2 & \text{otherwise.} \end{cases}$$

Summing the approriate inequality above for each pair $\{u, v\}$ in $S \times S$, where $u \neq v$, we get

$$(s - 1)y^*(\delta(S)) + 2(s - 1)y^*(\bar{E}(S)) = (s - 1) \sum_{u \in S} y^*(\delta(u))$$

$$= \sum_{\{uv \in S \times S | u \neq v\}} y^*(\delta(u)) + y^*(\delta(v))$$

$$\geq s(s - 1) + 2y^*(\bar{E}(S)).$$

Isolating the desired left hand side yields

$$y^*(\delta(S)) + y^*(\bar{E}(S)) \geq \frac{s(s-1)+(s-3)y^*(\delta(S))}{2s-4} \geq \frac{s(s-1)}{2s-4}, \text{ for } s \geq 3.$$

We can see that for $s \geq 3$, $s(s - 1) > (s + 1)(s - 2)$, thus

$$\frac{s(s-1)}{2(s-2)} > \frac{(s+1)(s-2)}{2(s-2)} > \frac{s+1}{2}.$$

Hence y^* satisfies (5).

Corollary 1. *The weight of D_+ is a lower bound for the weight of an optimal solution of (1).*

2.3   Last Step

We can see that D_+ forms a forest.

Step 4. The last step of the algorithm consists of shrinking the connected compponents of D_+ along its edges into vertices to obtain a contracted graph G'. The contraction of a graph G along a set of edge D_+ produces a graph $G' = (V', E')$ and a set of vertices $S \subseteq V'$ defined as follows: For each edge in D_+, merge its two endpoints into a single vertex, whose adjacency list is the union of the two adjacency lists. If parallel edges occur, retain the edge with smaller weight. The resulting graph is G'. The set S consists of the vertices formed by edge contraction (i.e., the nodes in $V' \setminus V$). We now proceed to find an approximation of the optimal tour T' going through all the vertices of S in G' with distances modified (if necessary) to the shortest paths distances. This can be done by using Christofides heuristic [3]. Map this solution back to a partial tour T of G. For each component formed by the edges in D_+, form an Eulerian walk from the entry point to the exit point of Q in the component by duplicate some of its edges. Output the tour Q formed by T and the Eulerian walk.

Let APX be the weight of Q and OPT be the weight of an optimal tour cover in G.

Theorem 2. $APX \leq 3.5 \times OPT$.

Proof. We can see in the worst case, Q contains the tour T and 2 times the edge set D_+. Any tour cover in G should take visit to all the connected component of D_+, hence as the weight of T can not be worse than $1.5 \times OPT$ [3]. By Corollary 1, the weight of D_+ is less than OPT, then 2 times the edge set D_+ is of weight at most $2 \times OPT$. Thus, overall $APX \leq 1.5 \times OPT + 2 \times OPT = 3.5 \times OPT$.

References

1. Arkin, E.M., Halldórsson, M.M., Hassin, R.: Approximating the tree and tour covers of a graph. Information Processing Letters 47, 275–282 (1993)
2. Carr, R., Fujito, T., Konjevod, G., Parekh, O.: A 2 1/10-approximation algorithm for a generalization of the weighted edge-dominating set problem. Journal of Combinatorial Optimization 5(3), 317–326 (2001)
3. Christofides, N.: Worst-case analysis of a new heuristic for the travelling salesman problem. Technical Report 388, Graduate School of Industrial Administration, Carnegie Mellon University (1976)
4. Edmonds, J., Johnson, E.L.: Matching: A Well-solved Class of Integer Linear Programs. In: Combinatorial Optimization - Eureka, You Shrink!, pp. 27–30. Springer-Verlag New York, Inc., New York (2003)
5. Fujito, T.: How to trim a mst: A 2-approximation algorithm for minimum cost-tree cover. ACM Trans. Algorithms 8(2), 16:1–16:11 (2012)
6. Fujito, T., Nagamochi, H.: A 2-approximation algorithm for the minimum weight edge dominating set problem. Discrete Applied Mathematics 118(3), 199–207 (2002)
7. Konemann, J., Konjevod, G., Parekh, O., Sinha, A.: Improved approximations for tour and tree covers. Algorithmica 38(3), 441–449 (2003)

Part II

Queueing Models and Performance Evaluation

A New Queueing Model for a Physician Office Accepting Scheduled Patients and Patients without Appointments

Ram Chakka[1], Dénes Papp[2], and Thang Le-Nhat[3]

[1] RGMCET, RGM Group of Institutions, Nandyal, India
[2] Department of Networked Systems and Services
Budapest University of Technology and Economics
Budapest, Hungary
[3] Posts and Telecommunications Institute of Technology,
Hanoi, Vietnam

Abstract. Family physicians play a significant role in the public health care systems. Assigning appointments to patients in an educated and systematic way is necessary to minimize their waiting times and thus ensure certain convenience to them. Furthermore, careful scheduling is very important for effective utilization of physicians and high system efficiency.

This paper presents a new queueing model that incorporates important aspects related to scheduled patients (patients and clients are synonymous throughout this paper), patients without appointments and the no-show phenomenon of clients.

Keywords: health care efficiency, queueing model, appointments, patient waiting times.

1 Introduction

Queueing models provide fast and versatile tools to evaluate the performance of real systems such as telecommunication networks, cellular mobile networks and computer systems [1–15]. The performance evaluation studies help service providers to dimension and operate their systems in an efficient way. Therefore, intensive research using queueing theory to model health care systems has already been initiated [16, 17] and progressing well.

Family medical care plays a significant role in the public health system of the World. Family medical care is the first layer of the medical system to provide examination and treatment or redirection to other health services depending on the condition of patients. Usually a physician is assigned to a number of individuals and families from a particular region in order to provide disease prevention and to improve the health of families. In recent years patient satisfaction came into focus as one of the major indicators related to the quality of medical services. Excessive waiting times for examination can lead to low patient satisfaction decreases the overall quality. A measure to reduce waiting times of patients can

T.V. Do et al. (eds.), *Advanced Computational Methods for Knowledge Engineering*, 107
Advances in Intelligent Systems and Computing 282,
DOI: 10.1007/978-3-319-06569-4_8, © Springer International Publishing Switzerland 2014

be applied as follows. An arrangement is beneficial for both patients and the medical office since the patient can anticipate a lower waiting time upon her/his arrival and the physician can plan the daily activities in a more efficient way. However, this kind of schemes may fail if patients do not show up at the prearranged time. The no-show phenomenon is not rare [18] in practice. The larger the distance between the reservation instant and the prearranged time is, the more likely a patient will not show up at the appointment time. Of course, a pre-appointment is encouraged, but not mandatory in the family health care in some countries. As a consequence, a physician has to treat a significant proportion of the patients arrive at the reception without any prior appointments or notifications.

In order to ensure convenience to patients through minimizing the waiting time, patients are assigned appointments in a systematic and educated way. Scheduling of patients appointments at appropriate times after receiving requests, is a very important aspect in health care systems because it ensures the convenience of patients through minimizing their waiting time in hospitals. Furthermore, careful scheduling is crucial for the efficient utilization of physicians and other important resources in public health care systems. Some works [17,19,20] dealt with scheduled customers (patients).

Recently, the phenomenon of cancellations [18] and no-shows in health care systems have raised attention in the researcher community because this leads to inefficient utilization of public health service [16,17]. Hassin and Mendel [17] investigated a single-server queueing system with scheduled arrivals and Markovian service times. In their work customers [17] with a fixed probability of showing up were considered. Green and Savi [16] proposed the application of the M/M/1/K and the M/D/1/K queues to model the phenomenon of no-shows. However, these works [16,17] do not take into consideration the arrival process of patients without appointments which is rather important. The arrival of these patients (without appointments) is quite typical in case of the family medical care because there are frequently unexpected health problems (e.g., sudden sickness, accute illnesses) of people assigned to any specific family medical doctor.

In this paper, we propose a novel queueing model that incorporates both the phenomenon of no-shows and the impact of unscheduled arrivals, along with considering the scheduled patients.

The rest of this paper is organized as follows. In Section 2, our new queueing model is proposed. In Section 3 presents numerical results of this new analytical model. Finally, Section 4 concludes the paper.

2   The Proposed Queueing Model

2.1   Generalized Queueing Model

A generalized model for the queueing problem pertaining to the physician's office is illustrated in Figure 1. As observed, the proposed queueing model consists of

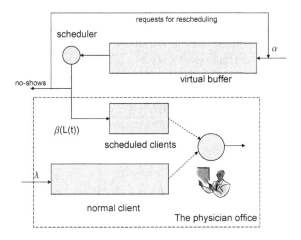

Fig. 1. A generalized queueing model

two "service stations". The first service station stands for the scheduling process, where patients get appointments (e.g., can be done through a telephone call and a scheduler can be a nurse or an assistant in the physician office).

The second station is the representation of the physician, which is modeled as the single server queue. Two categories of patients are distinguished in the physician office - normal and scheduled. The first category includes patients who arrive without appointment, while patients with an appointment belong to the second category. The normal category contains persons who need to see the physician on the day of arrivals and are willing to wait for more time than patients with appointments. The second category includes people who arrive for a regular check or who receive vaccinations.

It is worth emphasizing that scheduled patients have a priority over normal patients. However, the service of normal patients is not interrupted by the arrival of scheduled patients.

A Virtual Queue of Patients Requesting an Appointment. Although patients may not be physically present in the first station, a "virtual" queue for scheduling appointments is formulated as follows.

- The time spent by the ith patient in the virtual queue is $t_i - ta_i$, where t_i is the appointment time assigned to patient i with the arrival time ta_i (for the convenience $t_0 = 0$ by definition).
- The "service time" of patient i is defined as $t_{i+1} - t_i$. Let $F(x)$ and $f(x)$ denote the CDF and the probability density function of the service times of the virtual queues.

Therefore, the "virtual" queue for scheduling appointments is of type $M/G/1$ for an exponentially distributed inter-arrival times of requests for appointments with rate α. The number of patients in the virtual queue $L(t)$ is an embedded Markov process at the scheduled time epochs $(t_1, t_2, \ldots)$. For the stationary probability $p_l^{(1)} = lim_{t\to\infty} \Pr(L(t) = l)$, the Pollaczek-Khinchin formula [1] is valid

$$\sum_{l=0}^{\infty} z^l p_l^{(1)} = \frac{(1-\rho)F^*(\alpha - \alpha z)(1 - z))}{F^*(\alpha - \alpha z) - z}, \tag{1}$$

where

$$\rho = \alpha \int_0^{\infty} x f(x) dx,$$

$$F^*(s) = \int_0^{\infty} e^{-sx} f(x) dx.$$

Due to various reasons, only a portion of patients with appointments arrives at the physician office and quite a significant percentage of patients do not show up (or cancel the appointment, due to various reasons as explained before). The no-show probability depends on the appointment backlog too. Green and Savi [16] proposed the no-show function $\gamma(k) = \gamma_{max} - (\gamma_{max} - \gamma_0)e^{-k/C}$, where k is the appointment backlog in the first queue, γ_{max}, γ_0 and C are coefficients.

A Queue in the Physician Office. The server representing the physician accepts normal as well as scheduled patients. Normal patients arrive according to a Poisson process with the arrival rate λ. Scheduled patients arrives upon the completion of service from the virtual queue. We assume that the service times of normal and scheduled patients are exponentially distributed with parameter μ.

Let $J(t)$ be the number of normal patients in the physician office at time t (this also includes the normal client under service).

2.2   An Analytical Solution

To obtain an analytically tractable model we make some additional assumptions as follows.

- Let N denote the maximum number of scheduled patients in the physician office at any time, which is explained by the fact that the number of scheduled patients in the physician office should be limited to ensure a low value for waiting time in the day.
- Similarly, the maximum value of $J(t)$ ($J(t)$ is the number of normal patients in the office at any time) and $L(t)$ is denoted by M and K, respectively.
- The "service times" of the virtual queue are exponentially distributed. The show-up rate is expressed as $\beta(k) = (1 - \gamma(k))/ \int_0^{\infty} x f(x) dx.$

Table 1. Nomenclature

α	the arrival rate of patients requesting appointments
t_i	assigned time for patient i
$L(t)$	the number of patients in the virtual buffer
λ	the arrival rate of normal patients
$J(t)$	the number of patients without an appointment (normal patients)
μ	the service rate of patients
$\gamma(k)$	the no-show probability
$\beta(L(t))$	the show-up rate of scheduled patients
$I(t)$	description of the state of the physician and the number of scheduled patients
$f(x)$	the probability density function of the service times of the virtual queues
$F(x)$	the CDF of the service times of the virtual queues

With the assumption that the physician does not rest when there is a patient in the office, the states of the server (the physician) are classified as follows:

- free with no patient is in the office,
- busy servicing a normal patient with one patient with an appointment waiting for service,
- busy with the service of a normal patient and there is no a patient with an appointment,
- serving a patient with an appointment.

To describe the interaction of the physician and the patients in the physician office, the random variable $I(t)$ is introduced with the following values:

- 0 if the server is free or if the server is busy in serving a normal client and no scheduled client in the system.
- $2n - 1$, $n = 1, \ldots, N$ if the physician serves a normal client and there are n scheduled clients waiting in the office.
- $2n$, $n = 1, \ldots, N$ if the server is busy in serving a scheduled patient and the total number of scheduled clients in the system is n.

The system is modeled by a three-dimensional Continuous Time Markov Chain (CTMC) $X = (I(t), J(t), L(t))$. The following types of transitions are possible among the states of the CTMC X:

- $(0, 0, l) \Rightarrow (2, 0, l - 1)$: this transition is due to the arrival of a scheduled patient when no client is in the physician office.
- $(2i, 0, l) \Rightarrow (2i + 2, 0, l - 1)$ for $1 < i < N$: these transitions take place due to the show-up of a scheduled patient when only scheduled clients are in the physician office.

- $(0, j, l) \Rightarrow (1, j, l-1)$ for $1 \leq j \leq M$: these transitions are due to the show-up of a scheduled patient while the physician examines a normal client.
- $(2i - 1, j, l) \Rightarrow (2i + 1, j, l - 1)$ for $1 \leq j \leq M$ and $i = 1, \ldots, N - 1$: these transitions happen due to the show-up of a scheduled patient when the physician examines a normal client.
- $(2i, j, l) \Rightarrow (2i + 2, j, l - 1)$ for $0 \leq j \leq M$ and $i = 1, \ldots, N - 1$: these transitions happen due to the show-up of a scheduled patient when the physician examines a scheduled client.
- $(0, j, l) \Rightarrow (0, j - 1, l)$ for $1 \leq j \leq M$: these transitions occur due to the completion of the service of a normal patient.
- $(2i - 1, j, l) \Rightarrow (2i, j - 1, l)$ for $1 \leq j \leq M$ and $1 \leq i \leq N$: these transitions take place due to the completion of the service of a normal patient.
- $(2i, j, l) \Rightarrow (2i - 2, j,)$ for $1 \leq i \leq N$ and $0 \leq j$: these transitions happen due to the completion of the service of a scheduled client.
- $(i, j, l) \Rightarrow (i, j, l+1)$: these transitions take place due to the arrival of patients to the virtual queue.

The transitions between the states are organized as follows:

- The matrices A_l, $0 \leq l \leq K$, contain the transition rates from state (i_1, j_1, l) to state (i_2, j_2, l).
- The matrices B_l ($l = 0, \ldots, K - 1$) include the ransition rates from state (i_1, j_2, l) to state state $(i_2, j_2, l + 1)$.
- The elements of the matrices C_l ($l = 1, \ldots, K$) are the transition rates from state (i_1, j_1, l) to state $(i_1, j_1, l - 1)$.

As a consequence, the generator matrix of the CTMC is written as

$$
Q_X = \begin{bmatrix}
A_0^{(1)} & B_0 & 0 & 0 & \cdots & \cdots \\
C_1 & A_1^{(1)} & B_1 & 0 & \cdots & \cdots \\
0 & C_2 & A_2^{(1)} & B_2 & \cdots & \cdots \\
\vdots & \vdots & \vdots & \ddots & \cdots & \cdots \\
0 & 0 & \cdots & C_{K-1} & A_{K-1}^{(1)} & B_{K-1} \\
0 & 0 & \cdots & & C_K & A_K^{(1)}
\end{bmatrix}
$$

where

$$
A_j^{(1)} = \begin{cases}
A_0 - D^{A_0} - D^{B_0} & \text{if } j = 0, \\
A_j - D^{A_j} - D^{B_j} - D^{C_j} & \text{if } 0 < j < K, \\
A_K - D^{A_K} - D^{C_K} & \text{if } j = K.
\end{cases}
$$

Note that D^Z, $Z = A_l, B_l, C_l$ is a diagonal matrix whose diagonal element is the sum of all elements in the corresponding row of Z.

Let $p_{i,j,l}(t) = \Pr(I(t) = i, J(t) = i, L(t) = l)$ denote the probability that the CTMC X is in state (i, j, l) at time $t \geq 0$. The steady state probabilities are defined by $p_{i,j,l} = \lim_{t \to \infty} \Pr(I(t) = i, J(t) = i, L(t) = l)$. Correspondingly, we define vectors $\pi_l(t) = [p_{0,0,l}(t), \ldots, p_{2N,M,l}(t)]$ and $\pi_l = [p_{0,0,l}, \ldots, p_{2N,M,l}]$.

The balance equations can be expressed as follows:

$$\pi_0 A_0^{(1)} + \pi_1 C_1 = 0, \tag{2}$$

$$\pi_{k-1} B_{k-1} + \pi_k A_1^{(k)} + \pi_{k+1} C_{k+1} = 0, \quad 1 \leq k < K, \tag{3}$$

$$\pi_{K-1} B_{K-1} + \pi_K A_K^{(1)} = 0. \tag{4}$$

To compute the steady state probabilities, we shall proceed as follows. Let us define matrices R_k's such that $\pi_k = \pi_{k-1} R_k$ holds for $k = 1, 2, \ldots, M$. Then, based on equations (3) and (4), the matrices R_k's can be recursively computed using

$$R_K = -B_{K-1}(A_K^{(1)})^{-1}, \tag{5}$$

$$R_k = -B_{k-1}[A_1^{(k)} + R_{k+1} C_{k+1}]^{-1}, \quad k = K - 1, \ldots, 1. \tag{6}$$

This means, that π_k, $k = 1, 2, \ldots, K$, can be expressed in terms of π_0 and the already computed R_k's by

$$\pi_k = \pi_0 \prod_{i=0}^{k} R_k, \quad k = 0, \ldots, K, \tag{7}$$

where R_0 denotes by definition the identity matrix. Equation (2) can be rewritten as

$$\pi_0 (A_0^{(1)} + R_1 C_1) = 0. \tag{8}$$

Finally, we can utilize the normalization equation $\sum_{k=0}^{K} \pi_k \mathbf{e}_k = 1$ and equation (8) to compute π_0 and then the stationary probabilities.

3   Numerical Results

In Figures 2, 3 and 4, we plot the utilization of the physician and the average number of patients waiting in the physician office versus λ for $1/\mu = 15$ minutes, $1/\alpha = 77$ minutes. We compare the impact of the no-show and always-show on the performance of the physician office. Note that the no-show function is applied with $\gamma_{max} = 0.51$, $\gamma_0 = 0.15$ and $C = 9$ (see [16]).

From the curves, 18% relative decrease on the utilization of the physician is observed for low traffic ($\lambda = 0.01$ and appointment interval equal to 60 minutes –i.e., patients are scheduled every 60 minutes). However, for the physician with a high utilization (> 0.7), the no-show phenomenon only reduces the utilization less than 5%. It is worth emphasizing that the reduction minimally depends on the appointment interval for high loads.

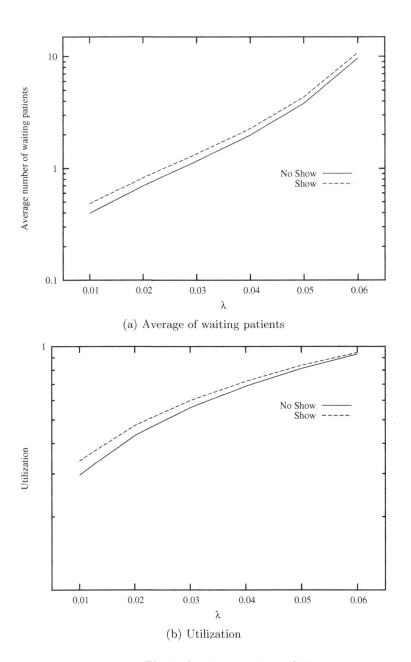

(a) Average of waiting patients

(b) Utilization

Fig. 2. Appointment interval 15 m

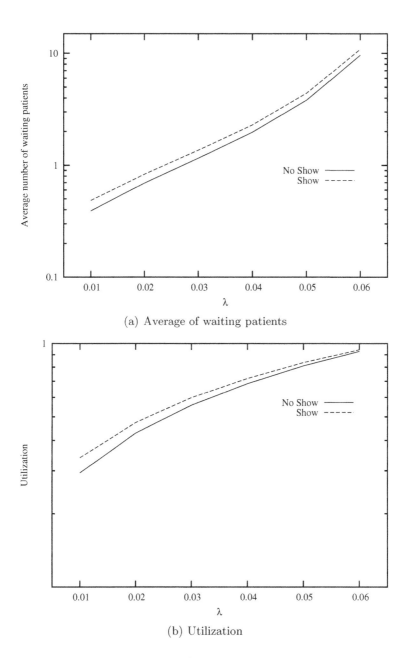

(a) Average of waiting patients

(b) Utilization

Fig. 3. Appointment interval 30 m

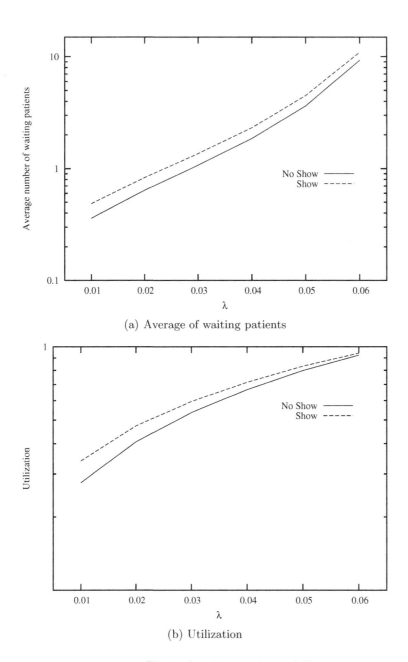

(a) Average of waiting patients

(b) Utilization

Fig. 4. Appointment interval 60 m

4   Conclusions

We have proposed a novel queueing model that incorporates the phenomenon of no-shows of patients with appointments and the impact of no scheduled arrivals, along with the scheduled patient arrivals. The queueing model can be used to obtain quantitative performance measures related to the operation of physician offices. Thus, this analytical tool can be of substantial help to determine the parameter for the scheduling policy applied in the physician's office.

References

1. Kleinrock, L.: Queueing Systems. Vol I: Theory. John Wiley & Sons, Inc. (1975)
2. Gelenbe, E., Mitrani, I.: Analysis and synthesis of computer systems. Academic Press (1980); Japanese translation: Ohm-Sha Publishing Co., Tokyo (1988); Expanded 2nd edn. World Scientific, Singapore (2009)
3. Latouche, G., Ramaswami, V.: Introduction to Matrix Analytic Methods in Stochastic Modeling. ASA-SIAM Series on Statistics and Applied Probability (1999)
4. Neuts, M.F.: Matrix Geometric Soluctions in Stochastic Model. Johns Hopkins University Press, Baltimore (1981)
5. Van Do, T., Krieger, U.R.: A performance model for maintenance tasks in an environment of virtualized servers. In: Fratta, L., Schulzrinne, H., Takahashi, Y., Spaniol, O. (eds.) NETWORKING 2009. LNCS, vol. 5550, pp. 931–942. Springer, Heidelberg (2009)
6. Do, T.V.: An Efficient Solution to a Retrial Queue for the Performability Evaluation of DHCP. Computers & OR 37(7), 1191–1198 (2010)
7. Do, T.V.: A new computational algorithm for retrial queues to cellular mobile systems with guard channels. Computers & Industrial Engineering 59(4), 865–872 (2010)
8. Do, T.V.: An efficient computation algorithm for a multiserver feedback retrial queue with a large queueing capacity. Applied Mathematical Modelling 34(8), 2272–2278 (2010)
9. Do, T.V.: Solution for a retrial queueing problem in cellular networks with the fractional guard channel policy. Mathematical and Computer Modelling 53(11-12), 2058–2065 (2011)
10. Do, T.V.: A new solution for a queueing model of a manufacturing cell with negative customers under a rotation rule. Performance Evaluation 68(4), 330–337 (2011)
11. Do, T.V., Rotter, C.: Comparison of scheduling schemes for on-demand IaaS requests. Journal of Systems and Software 85(6), 1400–1408 (2012)
12. Do, T.V.: M/M/1 retrial queue with working vacations. Acta Informatica 47(1), 67–75 (2010)
13. Do, T.V.: An initiative for a classified bibliography on G-networks. Performance Evaluation 68(4), 385–394 (2011)
14. Do, T.V.: Bibliography on G-networks, negative customers and applications. Mathematical and Computer Modelling 53(1-2), 205–212 (2011)
15. Do, T.V., Do, N.H., Zhang, J.: An enhanced algorithm to solve multiserver retrial queueing systems with impatient customers. Computers & Industrial Engineering 65(4), 719–728 (2013)

16. Green, L.V., Savin, S.: Reducing Delays for Medical Appointments: A Queueing Approach. Operations Research 56(6), 1526–1538 (2008)
17. Hassin, R., Mendel, S.: Scheduling Arrivals to Queues: A Single-Server Model with No-Shows. Management Science 54(3), 565–572 (2008)
18. Galucci, G., Swartz, W., Hackerman, F.: Impact of the wait for an initial appointment on the rate of kept appointments at a mental health center. Psychiatric Services 56(3), 344–346 (2005)
19. Mercer, A.: A queuing problem in which arrival times of the customers are scheduled. Journal of the Royal Statistical Society, Series B 22(1), 108–113 (1960)
20. Mercer, A.: A queuing problem in which arrival times of the customers are scheduled. Journal of the Royal Statistical Society, Series B 35(1), 104–116 (1960)

Usability of Deterministic and Stochastic Petri Nets in the Wood Industry: A Case Study

Ádám Horváth

Institute of Informatics and Economics, University of West Hungary
9400 Sopron, Bajcsy-Zs. u. 9., Hungary

Abstract. Deterministic and stochastic Petri nets (DSPNs) are commonly used for modeling processes having either deterministically, or exponentially distributed delays. However, DSPNs are not widespread in the wood industry, where the leaders of a company try to make decisions based on common sense instead of using high-level performance evaluation methods.

In this paper, we present a case study, in which we demonstrate the usability of DSPN models in the wood industry. In the case study, we model the production of wooden windows of a Hungarian company [2]. Using the model, we can simply determine the bottleneck of the manufacturing process and show how to eliminate it.

Keywords: wooden window manufacturing, DSPN, modeling, simulation.

1 Introduction

In many manufacturing processes, the work can be divided into different disjoint phases having deterministic holding times. Besides, there are some exponentially distributed variables even in these cases, like the inter-arrival times of the orders and the revealing of the material defects.

A process, in which the delays of the transitions are either exponentially, or deterministically distributed, can be appropriately described by a DSPN [4]. DSPNs are similar to stochastic Petri nets (SPNs [5]), except that deterministically delayed transitions are also allowed in the Petri Net model.

Although DSPNs have greater modeling strength than SPNs, there are some restrictions in the model evaluation phase. Namely, analytical solution using the underlying stochastic behavior can be obtained only if there is only *one* enabled deterministic transition in each marking. Otherwise, simulation must be used to get the steady state or the transient solution of the DSPN. In this paper, we model a window manufacturing process, where the production phases are deterministic, and work in parallel. Therefore, the traditional analytical approach cannot be used, so we will present some simulation results.

The rest of the paper is organized as follows. In Section 2, the formalism of DSPNs is described. In Section 3, we provide a short summary of the manufacturing process of wooden windows to help the understanding of the latter

T.V. Do et al. (eds.), *Advanced Computational Methods for Knowledge Engineering,*   119
Advances in Intelligent Systems and Computing 282,
DOI: 10.1007/978-3-319-06569-4_9, © Springer International Publishing Switzerland 2014

sections. Section 1 describes our model for the wooden window manufacturing process. In Section 5, we present the evaluation of the model. Finally, Section 6 concludes the paper.

2   DSPN Formalism

In this section, we give a short introduction to DSPNs, while a detailed description can be found, e.g. in [1] or [4].

DSPNs are bipartite directed graphs consisting two types of nodes: places and transitions. The places correspond to the state variables of the system, while the transitions correspond to the events that can induce a state change. The arcs are connected to places and vice versa expressing the relation between states and event occurrence. Places can contain tokens, and the state of a DSPN, called marking, is defined by the number of tokens in each place. We use the notation M to indicate a marking in general, and we denote by $M(p)$ the number of tokens in place p in marking M. Now we recall the basic definitions related to DSPNs.

Definition 1. *A DSPN system is a tuple*

$$(P, T, I, O, H, M_0, \tau, w),$$

where:

- *P is the finite set of* places. *A marking $M \in \mathbb{N}^{|P|}$ defines the number of tokens in each place $p \in P$.*
- *T is the set of* transitions. *T can be partitioned into the following disjoint sets: T^Z is for the set of immediate transitions, T^E is for the set of exponential transitions, and T^D is for the set of deterministic transitions. Note that $P \cap T = \emptyset$.*
- *$I, O, H : \mathbb{N}^{|P|} \to \mathbb{N}$ are the multiplicities of the* input arc *from p to t, the* output arc *from t to p, and the* inhibitor arc *from p to t, respectively.*
- *$M_0 \in \mathbb{N}^{|P|}$ is the* initial marking *of the net.*
- *$\tau : \mathbb{N}^{|P|} \to \mathbb{R}^+$ is the mean delay for $\forall t \in T^E \cup T^D$ (note that τ may be marking-dependent).*
- *$w : \mathbb{N}^{|P|} \to \mathbb{R}^+$ is the firing weight for $\forall t \in T^Z$ (note that w may be marking-dependent).*

The graphical representation of a place p is a circle, with the number of tokens in it is written inside (or illustrated with the corresponding number of black dots). A transition t is drawn as a box, a thin bar for the transitions in T^Z, an empty box for transitions in T^E, and a filled box for transitions in T^D. Input and output arcs end with an arrowhead on their destination, while inhibitor arcs have a small circle on the transition end. The multiplicity of an arc is written on the arc (the default value of multiplicity is one).

A transition is "enabled" in a marking if each of its input places contains the "necessary" amount of tokens where "necessary" is defined by the input functions

I and H. Formally, transition t is enabled in marking M if for all places p of the net we have $M(p) \geq I(t, p)$ and $M(p) < H(t, p)$. An enabled transition can fire and the firing removes tokens from the input places of the transition and puts tokens into the output places of the transition. The new marking M' after the firing of transition t is formally given $M'(p) = M(p) + O(t, p) - I(t, p), \forall p \in P$. Note that function H does not influence $M'(p)$.

The firing of a transition in T^E occurs after a random delay, while the transitions in T^D after a deterministic delay. The random delay associated with a transition in T^E has exponential distribution whose parameter depends on the firing intensity of the transition and on the current marking. In this paper, we assume that the number of tokens in a place does not influence the firing delay. This concept is called single server model, which practically corresponds to the real operation of the manufacturing process (other server policies also exist, for details see [5]).

When a marking is entered, a random delay is chosen for all enabled transitions by sampling the associated delay distribution. The transition with the lowest delay fires and the system changes the marking.

3   The Manufacturing Process of Wooden Windows

In this section, we describe the manufacturing process of wooden windows in detail.

From the company's point of view, the wooden window manufacturing process lasts from the arriving of an order to the delivery of the accomplished windows. After an order is accepted, the company starts manufacturing the sufficient number of windows. Each single window can be considered unique, since a building typically contains windows in several various sizes and types, so the quantity production cannot be carried out in this area.

In the first part of the production of a single window, the frame members are produced separately. Typically, a wooden window consists of 8 frame members: 4-4 frame members compose the frame and the casement frame, respectively. Since the manufacturing process cannot be started in the absence of raw material, the companies try to appropriately manage the quantity of the raw material in their store. The raw material typically means 6 meter long lumber.

The first work phase is cutting the raw material to size using circular saw. After cutting to size, the frame members must be planed with a planing machine. The third work phase is molding. In this work phase, a computer numerical control (CNC) machine is applied to bring the frame members into their final form. CNC machines are fully automated and programmable, and their use for molding ensures high level of precision and speed.

After molding, the frame members are glued into a frame. From this point, the basic unit of the manufacturing is the frame. The next work phase is dipping, which is a surface treatment for protecting the wooden material from damages caused by insects and fungi. Then, the frames are sanded with a sanding machine to obtain a smooth surface. After sanding, priming and final coating give the final color of the frame and ensure further protection for the wooden material. The last work phase is fixing the hardware parts. In this phase, the frame and the casement frame is supplied with the required metal components and the glass. Finally, the frame and the casement frame is joined, and the window is ready for transportation.

The peculiarity of the manufacturing process is that material defects can be revealed in any work phases before the final coating. If a defect is revealed before gluing, the single frame member can easily be re-produced, while after gluing, the glued frame must be decomposed, and all work phases must be repeated with the frame, when the re-produced single frame member is available. In both cases, the re-production has priority over the other works.

4   An Operation Model for the Manufacturing Process of Wooden Windows

In this section, we describe the operation model of a Hungarian wooden window manufacturing company.

From modeling point of view, the events of the production can be divided into two categories, based on the distribution of the transitions modeling the given event. Accordingly, we model the orders and the occurrence of material defects with exponentially delayed transitions, while we describe the different production phases (cutting, planing, molding, gluing, dipping, sanding, priming, final coating and fixing the hardware parts) with deterministic transitions (Fig. 1). Since the number of waiting window frames and frame members do not influence the intensity of a work phase in the model, we apply the exclusive server model [5] in this work.

In this example, there are three typical types of orders:

– for specific single window,
– for all windows of a family house, a windows on the average,
– for windows of a big building, b windows on the average.

According to the types of orders, the arrival process consists of transitions *arriveSingle*, *arriveAverage* and *arriveBig* in Fig. 1. In this case study, a and b are considered 18 and 65, respectively.

After the order was accepted[1], the 8 frame members of each window must be cut to size (each frame has 4 members and each casement frame has also 4 members in most cases[2]). The cut frame members must be planed, then molded (see transitions *cut*, *plane* and *mold* in Fig. 1). Until this point is reached, the frame members are individual entities each belonging to a specific window. Therefore, if a material defect is discovered, the given single member can be re-cut, re-planed and re-molded without causing bigger delay in the process (see transitions *md1*, *md2* and *md3* in Fig. 1).

---

[1] In Fig. 1, we took only the accepted orders into account
[2] Having the typical case is a modeling simplification.

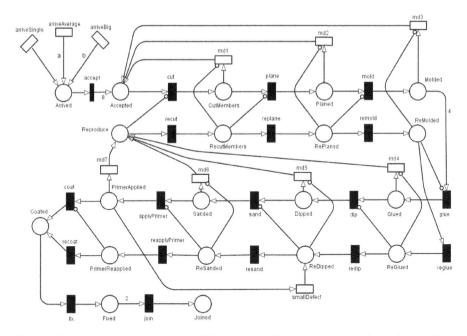

Fig. 1. The proposed operation model for the manufacturing process of wooden windows

After gluing the 4 frame members, one token denotes a frame in the model. The next operations after gluing are the dipping of the frame, then the sanding, the priming and the final coating (see transitions *glue, dip, sand, applyPrimer* and *coating* in Fig. 1). Note that a material defect can be discovered in *any* production phase before the final coating. However, when the members compose a frame, the delay is much bigger, since the frame must be recomposed, and the single member must be re-produced (see transitions *md4, md5, md6* and *md7* in Fig. 1). Therefore, we assume that if a material defect is discovered after the members compose a frame, further defects will not be discovered. This assumption is based on the thorough investigation, which follows when a frame must be decomposed to prevent further delay.

In our model, there is a dedicated path for this incident: since 3 frame members wait for the re-production of the fourth one, the transitions on the dedicated path have higher priority than the concurrent ones (see transitions *recut, replane, remold, reglue, redip, resand, reapplyPrimer* and *recoat*). In the DSPN model, inhibitor arcs ensure the priority. After priming, the smaller defects can be corrected by re-sanding, what is denoted by transition *smallDefect* in Fig. 1.

The last two steps of the production are fixing the hardware parts and join the casement frame and the frame (transitions *fix* and *join* in 1). The transitions with the corresponding delays are collected in Table 1[3] (in case of the exponentially delayed transitions, the expected delay is shown).

---

[3] The values of Table 1 were set based on a personal discussion with the representative of the company.

Table 1. The transitions of the DSPN model

Transitions	Description	Delay
arriveSingle	A single wooden window is ordered.	16.64
arriveAverage	18 wooden windows are ordered.	12.07
arriveBig	65 wooden windows are ordered.	1392
md1	A flaw is discovered after cutting.	3.125
md2	A flaw is discovered after planing.	15
md3	A flaw is discovered after molding.	50
md4	A flaw is discovered after gluing.	10
md5	A flaw is discovered after dipping.	60
md6	A flaw is discovered after sanding.	33.33
md7	A flaw is discovered after priming.	22.22
smallDefect	A frame is sent back to sand.	33.33
cut/recut	Cutting/re-cutting of a frame member.	0.0556
plane/replane	Planing/re-planing of a frame member.	0.0104
mold/remold	Molding/re-molding of a frame member.	0.05
glue/reglue	Gluing/re-gluing of the frame members.	0.1667
dip/redip	Dipping/re-dipping of a frame.	0.0333
sand/resand	Sanding/re-sanding of a frame.	0.2
applyPrimer/reapplyPrimer	Priming/re-priming of a frame.	0.1111
coat/recoat	Coating/re-coating of a frame.	0.1111
fix	Fixing hardware parts.	0.3333
join	Join the casement frame and the frame.	0.01

5   Evaluation of the Production Process

In this section, we investigate the production process using transient simulation. So, we can determine the throughput of the system for one year, and the bottlenecks of the production process. Moreover, we can investigate the effects of the material defects. However, several other aspects could be investigated, too.

For the performance evaluation of DSPNs, several tools exist, such as the ones presented in [3,6]. Since the analytical approach cannot be applied in our case, we used TimeNet [6] to obtain simulation results, which reflect 99% confidence level with 1% maximal error rating.

Table 2 shows the expected token distribution after one year. We can observe that the company can produce more than 3000 windows a year, while the bottleneck of the system is the fixing of the hardware parts, since there are about 585 frames on the *Coated* place (as a matter of fact, the employees in this session must work overtime to accomplish the orders until their deadline).

Table 2. The token distribution of the DSPN model after one year

Places	Expected number of tokens
Accepted	189.81933
CutMembers + RecutMembers	0.13467
Planed + RePlaned	0.63745
Molded + ReMolded	3.25834
Glued + ReGlued	0.10666
Dipped + ReDipped	0.70611
Sanded + ReSanded	0.35967
PrimerApplied + PrimerReapplied	0.35056
Coated	584.55711
Fixed	0.52422
Joined	3021.378

The bottleneck (fixing the hardware parts) is a limiting factor in the production if the company wants to increase the number of accepted orders. Fig. 2 shows that as we increase the intensity of the *arriveAverage* transition, the production is throttled by the fixing of the hardware parts work phase. In Fig. 2, we also show the case when the company detects the material defects always *before* gluing (e.g. by applying quality control) to prevent the major overhead caused by these defects. Technically, we deleted the transitions *md4*, *md5*, *md6* and *md7* from the DSPN model. We can observe that eliminating the late detection of the material defects does not affect the total throughput. However, it can obviously cause a delay of certain jobs. On the other hand, the elimination of the mentioned transitions increased the number of tokens in the bottleneck place when the system was heavily loaded.

The obvious solution for the bottleneck problem is to expand the critical production contingent (i.e. by employing more workers). Therefore, we investigated the throughput of the system as a function of the delay of transition *fix*, while the delay of transition *arriveAverage* remains on the lowest value of the previous simulation (the exact value is 7.606709638). Fig. 3 shows that decreasing the delay of transition *fix* from 0.33 hours to 0.22 hours is enough to move the

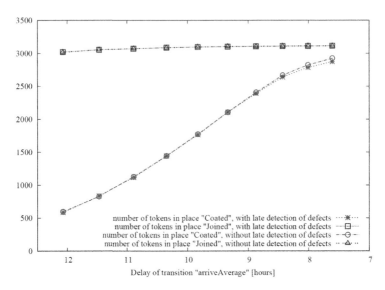

Fig. 2. The number of tokens in places "Coated" and "Joined"

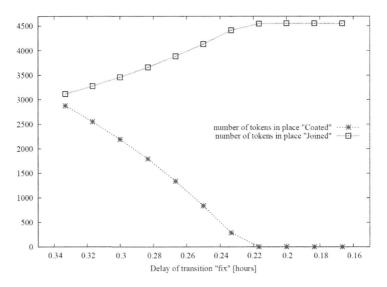

Fig. 3. The number of tokens in places "Coated" and "Joined"

bottleneck in the system, and gives the chance to increase the throughput by about 50%. In other words, the maximal throughput can be increased by 50%, if the delay of fixing the hardware parts onto one frame could be reduced from 20 minutes to 13.2 minutes.

6 Summary

In a production process, where the different work phases are deterministic processes, a DSPN is an appropriate modeling tool. In this paper, we presented the operation model of a wooden window production company. Obtaining the solution of the model, we identified the main bottleneck of the production process. Moreover, we determined the measure of improving the given work phase in order to eliminate the main bottleneck from the process. With this example, we demonstrated that the stochastic models can be applied in the wood industry.

Since our work is a pioneer one in the wood industry, several other investigations can be made. As a result, the whole manufacturing process can be optimized, what offers us a good opportunity in this area in the future.

Acknowledgments. The author thanks Roland Matuszka, the representative of Holz-Team Ltd., for his support during the preparation of this work.

This research was supported by the European Union and the State of Hungary, co-financed by the European Social Fund in the framework of TÁMOP 4.2.4. A/2-11-1-2012-0001 "National Excellence Program".

References

1. Ciardo, G., Lindemann, C.: Analysis of deterministic and stochastic Petri nets. In: Proc. 5th International Workshop on Petri Nets and Performance Models, pp. 160–169 (1993)
2. Holz-Team Ltd.: The web page of Holz-Team Ltd., `http://www.holzteam.hu/en` (last visited on November 28, 2013)
3. Lindemann, C.: DSPNexpress: A software package for the efficient solution of Deterministic and Stochastic Petri Nets. In: Proc. 6th Int. Conf. on Modelling Techniques and Tools for Computer Performance Evaluation, Edinburgh, Great Britain, pp. 15–29 (1992)
4. Marsan, M.A., Chiola, G.: On Petri nets with deterministic and exponentially distributed firing times. In: Rozenberg, G. (ed.) APN 1987. LNCS, vol. 266, pp. 132–145. Springer, Heidelberg (1987)
5. Marsan, M.A., Balbo, G., Conte, G., Donatelli, S., Franceschinis, G.: Modelling with Generalized Stochastic Petri Nets. John Wiley and Sons (1995)
6. Zimmermann, A., Knoke, M.: TimeNET 4.0: A software tool for the performability evaluation with stochastic and colored Petri nets; user manual. TU, Professoren der Fak. IV (2007)

A New Approach for Buffering Space in Scheduling Unknown Service Time Jobs in a Computational Cluster with Awareness of Performance and Energy Consumption

Xuan T. Tran[1,2] and Binh T. Vu[1]

[1] Department of Networked Systems and Services,
Budapest University of Technology and Economics, Hungary
[2] Faculty of Electronics and Communication technology,
Thai Nguyen University of Information and Communication technology, Vietnam

Abstract. In this paper, we present a new approach concentrating on buffering schemes along with scheduling policies for distribution of compute – intensive jobs with unknown service times in a cluster of heterogeneous servers. We utilize two types of ADM Operton processors of which parameters are measured according to SPEC's Benchmark and Green500 list. We investigate three cluster models according to buffering schemes (server-level queue, class-level queue, and cluster-level queue). The simulation results show that the buffering schemes significantly influence the performance capacity of clusters, regarding the waiting time and response time experienced by incoming jobs while they retain energy efficiency of system.

Keywords: heterogeneous resource, computational grid, scheduling policy, buffering scheme, server –level, class –level, cluster –level.

1 Introduction

In contemporary Grid computing systems, the resource heterogeneity, along with trade-off between energy efficiency and performance, makes job scheduling more challenged. If it is said that general job scheduling is NP-complete [1], job scheduling in a grid system is more challenging and multi-criteria in nature, presented in [2].

In order to advancing the efficient exploitation of scheduling algorithms in this such complex system, the computational grid can be divided into multiple levels as a tree-based model: Grid, Cluster, Site, and computing elements [3]. The scheduling algorithms can be also classified and applied to these four levels, but dominantly applied at two levels: grid and cluster, presented in [4, 5]. They can be also classified according to several different requirements of grid systems such as: job characteristics [6-8], QoS [9], energy consumption [10, 15], load balancing [3,11], e.t.c.

Beside the effective scheduling, saving energy in computing grid has become crucial due to the rapid increase of grid size and the goal of green network in the recent years. The most exploitable techniques to reduce power consumption in computing system are Dynamic Power Management (DPM) that can set the operating component to on/off state with different power levels or lower the operating voltage/frequency to make trade off performance for reducing power consumption [12-14].

T.V. Do et al. (eds.), *Advanced Computational Methods for Knowledge Engineering*,
Advances in Intelligent Systems and Computing 282,
DOI: 10.1007/978-3-319-06569-4_10, © Springer International Publishing Switzerland 2014

The heterogeneity and large-scale sharing of resources could be used as a new strategy dealing with energy saving problem, even different resources could be available in a single administrative domain owned by a specific service provider, such as a computational cluster. It is reliable to utilize a bunch of new resources instead of the old for purpose of enhancing performance or energy efficiency. This approach was presented by Zikos and Karatza in [15] where the authors proposed three policies applied for scheduling of compute-intensive jobs in heterogeneous cluster with awareness of performance and energy.

For the best of our knowledge, however, in traditional computational system considered in the literature studies, each computing component is responsible for serving jobs in one queue, without considering the organization of waiting space for jobs that are not immediately served upon their arrival instant. In our study, we propose a new approach to investigate new buffering schemes, co-operating with existing job scheduling policies. Motivated from heterogeneous cluster model and policies from [15], we examine buffering space at three levels, so-called Server-, Class-, and Cluster-level. Note that the Server-level queue based model plays a role of traditional configuration as studied in [15], Class-level and Cluster-level queue based models are our proposals.

The rest of paper is organized as follows. In Section 2, the architecture of buffering schemes along with the algorithms of distributing jobs are described. Section 3 describes simulation parameters and examined energy and performance metrics. The numerical results are presented in Section 4. Lastly, Section 5 concludes this study.

2   System Modelling

2.1   Related Work

We studied the system model along with three scheduling policies proposed in Zikos and Karatza research [15], therein two AMD Opteron processor [16, 17] types are utilized, so-called EE and HP types, in which EE type deals with energy efficiency under lower performance speed while HP type performs better but consumes more energy. The Opteron 2386 SE was chosen as HP processor, and Opteron 2376 HE was presented as EE processor. The main parameters used and computed in [15] of these processors are given in table 1.

Table 1. Main computed parameters of two processor types

Processor type	Frequency clock (GHz)	Voltage (V)	Idle power (W)	Active power (W)	Efficiency Ratio (FLOPS/Watt)
Opteron2376 HE (EE type)	2.3	1.2	105	160	$40.6*10^6$
Opteron 2386 SE (HP type)	2.8	1.35	125	240	$21.85*10^6$

According to evaluation performance capacities and energy efficiency of these processors, two ratios between two types of processor were taken into account:

+ Let PR denote performance ratio between EE and HP processor,

$$PR = \frac{EE\ speed}{HP\ speed} = \frac{2.3}{2.8} = 0.8$$

(1)

+ Let ER be energy efficiency ratio using metric FLOPS/Watt [18]:

$$ER = \frac{EE\_ER}{HP\_ER} = \frac{40.6 \times 10^{6}}{26.6 \times 10^{6}} = 1.53 \qquad (2)$$

From equation (1) and (2), we observe that EE processor shows a lower performance of about 20%, but greater about 1.5 times of energy efficiency than HP type. The model considered is a cluster of 16 servers equipped with processors of two described types and grouped into two classes, each of which consists of 8 processors with the same energy and performance characteristics.

In order to allocate jobs to servers, the authors presented three scheduling policies applied by the local scheduler of cluster (LS) which is aware of processors' characteristics as follows:

- *SQEE (Shortest Queue with Energy Efficiency priority):* LS monitors the queue lengths of all servers. When a job arrives, it will be routed to the server with shortest queue. In case of several shortest queue servers, an EE server is preferred to HP one.
- *SQHP (Shortest Queue with High Performance priority):* SQHP policy is very similar to SQEE, expect if there are several shortest queue servers, a HP server is preferred.
- *PBP - SQ (Performance-based probabilistic–shortest queue):* This policy is also performance oriented, but it is based on selection probabilities. LS computes the total computational capacity to create the selection probability for each group. When a job arrives, a group of servers is selected probabilistically and then shortest queue policy will be applied to allocate job to the best chosen server.

As concluded by the authors, SQEE yields the most energy efficiency, SQHP is preferred when system allows to trade more energy consumption for higher performance capacity. PBP-SQ was evaluated as the worst performance policy with medium energy consumption. Hence, in our study, we only apply SQEE and SQHP in our models to evaluate both energy and performance capacity.

2.2   System Models and Buffering Algorithms

In order to make these models obvious and easy for comparison and evaluation, we utilize the same types of processors to the related work [15], and also construct model as a cluster of 16 servers that are classified into two 8-server classes. The local scheduler (LS) is responsible for allocating jobs based on buffering schemes and applied policy (either SQEE or SQHP). Note that LS is aware of energy and performance capacity characteristics of all processors. Following [15], we also assume that jobs

-     - can be served by any server,
-     - are attended to by First Come First Served (FCFS) service policy,
-     - are non-preemptible, which means jobs cannot be suspended until completion,
-     - and have service times unknown by local scheduler.

Based on organization of buffering spaces, the corresponding models are described as follows:

- *Server-level queue based model*: each server has responsibility for serving jobs in its local queue, illustrated in figure 1. When an incoming job arrives LS, LS routes it to server which is chosen by applied policy. Job can be served immediately if server is idle or stored in queue in case server is busy. Job will be assigned to server as soon as server becomes idle.
- *Class-level queue based model*: Each class of servers has a local queue. When a job arrives LS, LS routes it to idle server if found according to which policy is applied. If all servers in cluster are busy, job can be stored in shortest class queue and taken into serving as soon as any server of this class become idle. Figure 2 shows this assumed model.
- *Cluster-level queue based cluster*: There is only one common queue to store jobs as presented in figure 3. When job arrives, if LS find that all servers in cluster are busy, it stores job in the queue until there is server ready to serve. We should note that in this case of model, shortest queue policy plays no role because there is only one queue. Routing job is based on either EE or HP priority that is applied by LS.

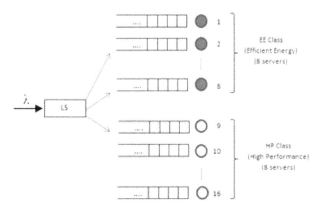

Fig. 1. Server-level queue based model

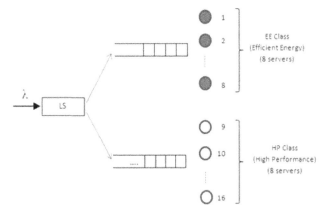

Fig. 2. Class-level queue based model

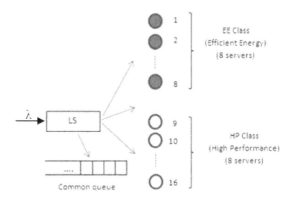

Fig. 3. Cluster-level queue based model

3   Simulation Parameters

3.1   System Load

We assume that arrival of jobs follows Poisson distribution with rate λ, and service rate of system denoted by μ is exponentially distributed with mean $1/\mu$.

Let denote μ_1 and μ_2 to be mean service rate of HP and EE processor. We assume mean service rate of HP processor to be equal to 1 ($\mu_1 = 1$) as a reference, according to the performance ratio (eq. 1), μ_2 will be equal to 0.8.

Denote N_{HP} and N_{EE} to be the number of HP and EE processors. The number of jobs able to be served in one time unit at full load is given by:

$$\mu_{sys} = N_{HP} * \mu_1 + N_{EE} * \mu_2 = 8*1 + 8*0.8 = 14.4 \qquad (3)$$

We estimate the system with utilization given by λ/μ_{sys} from medium to high load U = 50%, 60%, 70%, 80%, 90%. Therefore, the respectively corresponding values of arrival rate are derived as the following: $\lambda = 7.2, 8.64, 10.08, 11.52, 12.96$.

3.2   Performance Measures and Energy Metrics

We take into account energy consumption and performance, presented by system response time, as the evaluating parameters.

Response time r_j of job j is the time period from its arrival to the LS up to instant it leaves the system after served. Waiting time w_j of job j is the time in queue of job before service, and service time s_j of job j is time period taking server to execute job. In definition, $r_j = w_j + s_j$.

Let n denote the number of completed jobs. Let WT, ST, and RT consecutively be mean waiting time, mean service time and mean response time up to the time of n competed jobs. They can be defined as:

$$WT = \frac{1}{n} \sum_{i=1}^{n} w_i \qquad (4)$$

$$ST = \frac{1}{n} \sum_{i=1}^{n} s_i \tag{5}$$

$$RT = \frac{1}{n} \sum_{i=1}^{n} r_i \tag{6}$$

In order to evaluate energy conservation, we calculate energy consumed by server during the whole simulation time, following the definition:

$$E = P*t \tag{7}$$

where E, P, and t denote energy, power and time, respectively.

Note that depending on the state (idle or active), power consumption of each server are different, presented in table 1. Thus we define $E_{idle,k}$ as energy consumed when server k is in idle state with idle period time t_{idle}, and $E_{processing,k}$ as energy consumed when server k is processing jobs with busy period time $t_{processing}$, calculated by equation (7). The operating energy consumption ($E_{operating,k}$) at the server k is defined as the energy consumed during whole simulation time; therefore, it is sum of energy consumed in idle state and processing energy consumption of that server.

Let TO_E, TP_E and TI_E consecutively be total operating energy, total processing energy and total idle energy of whole 16-server cluster, hence be sum of $E_{operating}$, $E_{processing}$ and E_{idle} given by each server.

The performance measures and energy metrics are summarized in table 2:

Table 2. Notations of the parameters

w_j	Waiting time of job j in queue
WT	Mean waiting time
s_j	Service time of job j in serving
ST	Mean service time
r_j	Response time of job j in system
RT	Mean response time
$E_{idle,k}$	Energy consumed in idle state of server k
TI_E	Total idle energy
$E_{processing,k}$	Energy consumed in busy state of server
TP_E	Total processing energy
$E_{operating,k}$	Energy consumed by server k during simulation
TO_E	Total operating energy

4   Numerical Results

We build simulation software developed in C programming. In the related work [15], the authors implemented simulation 100 times and showed the results as the average derived from all simulation times. Each simulation ended when 16000 jobs' executions were completed. Herein, we implement one long – term simulation which ends after completion of 50 million jobs' executions. Using the statistical module developed by Politecnico di Torino for statistical analysis of the simulation results, our simulation is ensured with confident level of 98% and the accuracies (i.e. the ratio of the half-width of the confident interval and the mean of collected observations) of examined variables are at most 0.02.

We should note here that job distribution performed by LS into each class of servers is based on which policy is applied. In our collected results compared to related work [15], because there are no differences in job distribution of three buffering scheme based systems, we don't consider that issue as a evaluating parameter in our study.

4.1   Performance Capacity

Mean service times STs of three models presented in figure 4 show that at the low system load (50%), where the arrival rate is significantly less than service rate of system, ST is highest regarding to SQEE policy, since the EE processors with low performance rate are preferred for serving jobs; on the contrary, regarding to SQHP policy, ST is lowest due to HP processors are preferred, and increases with increasing of system load. It is also important to point out that in comparison between buffering scheme based models, STs are mostly equal due to utilizing the same types and organization of processors in every model. The anticipation that service times are not impacted by queuing schemes is confirmed in figure 4.

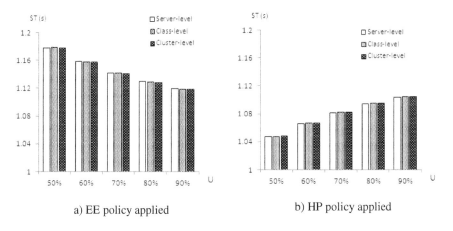

a) EE policy applied   b) HP policy applied

Fig. 4. Mean service time vs. system load

We observe the performance of buffering schemes in term of mean waiting time, WT. Figure 5 presents the mean waiting time of one job taking in queue. As the first point of view, WTs are insufficient at low load and increase with increasing system load regardless scheduling policies and buffering schemes. It is obviously seen that Class-level and Cluster-level queue schemes sufficiently reduce waiting time of jobs in queue, compared to the traditional one. This indicates that the more servers serve for a queue simultaneously, the sooner jobs can leave the queue.

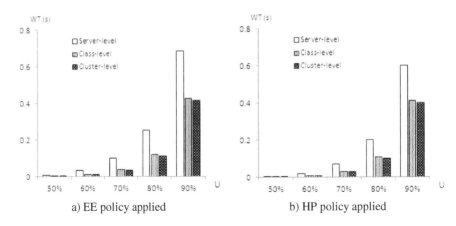

a) EE policy applied   b) HP policy applied

Fig. 5. Mean waiting time vs. system load

As the results obtained from above parameters, mean response times RTs are shown in figure 6. The observation presents that our proposed models with applying buffering schemes outperform traditional model, under both policies and in every system load.

Specifically, the Cluster –level model performs best, and Class-level model does slightly worse, especially in high system load. According to increasing system load, no significant differences among three system performances can be discerned at 50% when the delay in queue is ignored, but there are noticeable gaps at higher loads.

Based on terms of ST, WT, and RT, we can conclude that the efficient queuing schemes advance the system performance behaviour respectably under both cases of policies.

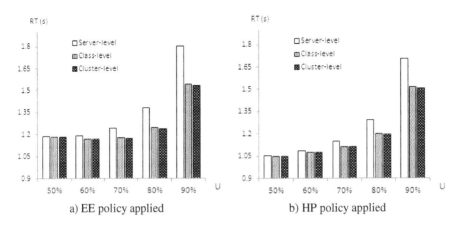

a) EE policy applied   b) HP policy applied

Fig. 6. Mean response time

4.2   Energy Conservation

In order to evaluate the energy efficiency, we take into account three parameters in terms of total idle energy consumption (TI_E), total processing energy consumption (TP_E), and total operation energy consumption (TO_E).

The total idle energy consumption (TI_E) of system during simulation is illustrated in figure 7. It can be observed that the energy consumption in idle state always shows a decreasing tendency, since a higher system load brings a greater number of incoming jobs leading to servers are required more in active state and less idle time.

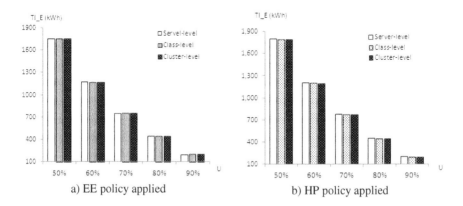

Fig. 7. TI_E vs. system load

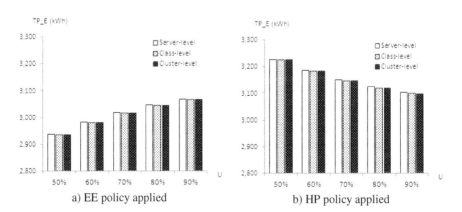

Fig. 8. TP_E vs. system load

Figure 8 and 9 present the total energy consumed in active state to process jobs (TP_E) and the total energy consumed from beginning up to the end of simulation (TO_E) under both policies, regarding to system load. It should be noted that, the energy conservation behaviours Zikos and Karatza showed in the previous study are confirmed in our current study. More specifically, in term of TP_E, there are sufficient differences between two policies' influence at low load (50%), but the differences decrease with increasing of system load and converge to value of 3100 (kWh) at 90% system load. That is because with higher system load, both types of servers are taken into active state more equally.

In addition, TO_Es decrease when the system load increases, due to the reducing of energy consumed in idle state. We would like emphasize an important point also indicated in related work that SQEE yields significant energy efficiency at the low load, but at 90% of system utilization, the efficiency of SQEE is not sufficient in comparison with SQHP policy.

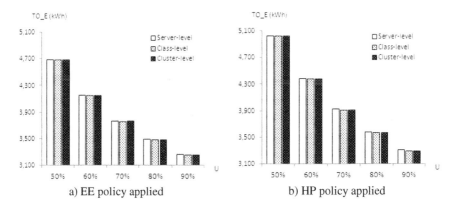

a) EE policy applied   b) HP policy applied

Fig. 9. TO_E vs. system load

Lastly, compared to traditional system, figures 7- 9 show that our proposals give the identical results of energy conservation compared to traditional model, independently of scheduling policies.

5   Conclusions

In this paper, we investigate different buffering schemes in a computational cluster, which consists of two eight-server classes corresponding two types of processors, along with applying existing scheduling policies (SQEE and SQHP). Breaking down model based on levels of buffering, we construct three models: Server-, Class-, and Cluster-level queue models.

The numerical results show that, our proposed models (Class-level and Cluster-level) sufficiently outperform the traditional system (Server-level queue model) due to reduction of local queue in use in new queuing schemes, while retaining the energy efficiency under both SQEE and SQHP policy.

References

1. El-Rewini, H., Lewis, T., Ali, H.: Task Scheduling in Parallel and distributed Systems. Prentice Hall, Englewood Cliffs (1994)
2. Gkoutioudi, K.Z., Karatza, H.D.: Multi-Criteria Job Scheduling in Grid Using an Accelerated Genetic Algorithm. Journal of Grid Computing, 311–323 (March 2012)

3. Yagoubi, B., Slimani, Y.: Dynamic load balancing strategy for grid computing. World Academy of Science, Engineering and Technology 19 (2006)
4. Terzopoulos, G., Karatza, H.D.: Performance evaluation of a real-time grid system using power-saving capable processors. The Journal of Supercomputing, 1135–1153 (2012)
5. Zikos, S., Karatza, H.D.: Communication cost effective scheduling policies of nonclairvoyant jobs with load balancing in a grid. Journal of Systems and Software, 2103–2116 (2009)
6. Zikos, S.: Helen D. Karatza: A clairvoyant site allocation policy based on service demands of jobs in a computational grid. Simulation Modelling Practice and Theory, 1465–1478 (2011)
7. Zikos, S., Karatza, H.D.: The impact of service demand variability on source allocation strategies in a grid system. ACM Trans. Model. Comput. Simul., 19:1–19:29 (November 2010)
8. He, Y., Hsu, W., Leiserson, C.: Provably efficient online non-clairvoyant adaptive scheduling, pp. 1–10 (March 2007)
9. Wang, T., Zhou, X.-S., Liu, Q.-R., Yang, Z.-Y., Wang, Y.-L.: An Adaptive Resource Scheduling Algorithm for Computational Grid, pp. 447–450 (December 2006)
10. Opitz, A., König, H., Szamlewska, S.: What does grid computing cost? Journal of Supercomputing, 385–397 (2008)
11. Kaur, P., Singh, H.: Adaptive dynamic load balancing in grid computing an approach. International Journal of Engineering Science and Advance Technology (IJESAT), 625–632 (June 2012)
12. Chedid, W., Yu, C., Lee, B.: Power analysis and optimization techniques for energy efficient computer systems
13. Zhuo, L., Liang, A., Xiao, L., Ruan, L.: Workload – aware Power Management of Cluster Systems, pp. 603–608 (August 2010)
14. Chedid, W., Yu, C.: Survey on Power Management Techniques for Energy Efficient Computer Systems, http://academic.csuohio.edu/yuc/mcrl/survey-power.pdf (retrieved)
15. Zikos, S., Karatza, H.D.: Performance and energy aware cluster-level scheduling of compute-intensive jobs with unknown service times. Simulation Modelling Practice and Theory, 239–250 (2011)
16. AMD Opteron Processor-Based Server Benchmarks (2010), http://www.amd.com/us/products/server/benchmarks/Pages/benchmarks-filter.aspx
17. SPEC's Benchmarks and Published Results (2010), http://www.spec.org/benchmarks.html
18. Feng, W., Cameron, K.: Power Measurement of High-End Clusters, The Green500 List, Version 0.1 (November 12, 2006)

Part III
Software Development and Testing

Supporting Energy-Efficient Mobile Application Development with Model-Driven Code Generation*

Imre Kelényi[1], Jukka K. Nurminen[2], Matti Siekkinen[2], and László Lengyel[1]

[1] Budapest University of Technology and Economics, Hungary
imre.kelenyi@aut.bme.hu
[2] Aalto University, Finland
jukka.k.nurminen@aalto.fi

Abstract. Energy-efficiency is a critical attribute of mobile applications, but it is often difficult for the developers to optimize the energy consumption on the code level. In this work we explore how we could use a model and code library based approach to assist the developer. Our vision is that developers can specify the operation on a high level and the system automatically converts the model to an appropriate software pattern. In this way, the developer can focus on the actual functionality of the app. We exemplify our approach with several energy-efficient software patterns, which focus on wireless data communication which is one of the biggest energy hogs with typical mobile applications. We discuss the pros and cons of different implementation alternatives and suggest open questions needing further exploration.

Keywords: Energy-efficiency, modeling, code generation.

1 Introduction

Mobile devices are becoming increasingly important as platforms for Internet applications. To deal with the constrains and more limited resources, application developers need new skills. One of the main issues is energy-efficiency of applications. High energy consumption may spoil an otherwise useful application. Increasing the burden of developers with yet another non-functional requirement can slow down the development process and increase the complexity of the resulting applications. Furthermore, it can effect the performance of the applications, i.e. while energy is important, it is usually hard to convince a developer to sacrifice performance for efficiency. When developers have to trade things off, they

* This work was partially supported by the European Union and the European Social Fund through project FuturICT.hu (grant no.: TAMOP-4.2.2.C-11/1/KONV-2012-0013) organized by VIKING Zrt. Balatonfred, and This work was partially supported by the Hungarian Government, managed by the National Development Agency, and financed by the Research and Technology Innovation Fund (grant no.: KMR_12-1-2012-0441).

do it to improve the performance of the application. Therefore, new ways are needed to facilitate the creation of quality energy-efficient mobile applications whit appropriate performance.

Research has shown that by developing applications in a smart way, major savings in energy consumption can be achieved. For instance, Hoque et al. [1] show that up to 65% of the energy consumed by wireless communication can be saved through traffic shaping when streaming video to a mobile device and Nurminen et al.[2] report 80% savings when data transfer happens during voice calls. These results show that when an application is properly constructed it is possible to fundamentally improve the energy-efficiency of at least some classes of mobile applications.

Unlike energy-efficiency improvements in hardware or low-level software, these application-level techniques require extra development work. Increasing the awareness of energy-efficient programming techniques could be one way but educating the vast mass of mobile developers to know the needed patterns and tricks is slower and tedious.

Increasing the abstraction level is a successful way to hide low-level details from the developers, which makes it possible to use complex data structures without worrying too much how they are actually implemented, or we can use a high-level communication protocol without worrying how the bits are actually moving the communication channel.

The key question we investigate in this paper is the following: How can we hide some of the complexity needed for energy-efficient operation behind a properly selected abstraction? Some related work exists that touches the interface between applications and power management of mobile devices: Anand et al. [3] propose specific kind of power management interfaces for I/O access that allow applications to query the power state and propose middleware for adaptive disk cache management. In [4], the authors propose to leverage application hints in power management. However, choosing the right level of abstraction in order to make the energy efficient programming techniques and patterns widely and easily available to the developers still seems to be an open problem and we feel that it requires further exploration. As far as we know, we are the first ones to investigate how the techniques and coding-patterns needed for energy-efficient software could be made widely and easily available to the developers.

An ideal solution would enable also distributing newly discovered patterns and techniques later on to the software developer community without requiring further efforts from the developers.

In summary, we propose a way to reduce the complexity of developing energy efficient mobile applications. The key contributions of this paper are:

- We identify energy-efficient software patterns that can reduce the cost of communication.
- We propose a way to hide most of the complexity of the patterns behind a high-level software abstractions and discuss what kind of issues arise with such an approach and what are the remaining open problems.

– We specifically advocate a model-driven approach that is one of the possibilities to address our goals, i.e. to effectively support the development of energy efficient mobile applications.
– In the discussion section, we collect the main questions and issues of the field and provide our thoughts about them.

The rest of this paper is organized as follows. Section 2 introduces the key concepts of model-driven approaches supporting mobile application development. We shortly discuss the key achievements of these type of solutions, compare them with our vision and highlight that currently there is no other solution supporting the modeling of energy efficient issues. Section 3 suggests the energy-efficient communication patterns. Section 4 provides a solution for modeling and generating energy efficient applications. We introduce the architecture of our Android implementation and the related library support, discuss the concept of model-based source code generation, and provide some examples. Section 5 contains our discussion about the topic. We collect the open issues and provide our thoughts about them.

2   Model-Driven Approaches in Mobile Application Development

Nowadays modeling is a key concept, which facilitates the system definition and supports the better communication on the appropriate abstraction level. Furthermore, system models are the first-class artifacts in model-based development. Domain-specific modeling facilitates that a system can be specified on a higher level of abstraction. Model processors automatically generate the lower level artifacts. Model-driven development approaches (e.g. Model-Integrated Computing (MIC) and OMG's Model-Driven Architecture (MDA) [5] emphasize the use of models at all stages of system development. They have placed model-based approaches to software development into focus.

In software development and domain engineering, a domain-specific language (DSL) [6] [7] is a programming language or specification language dedicated to a particular problem domain, a particular problem representation technique, and/or a particular solution technique. There are two aims of domain-specific modeling-based software development. First, raise the level of abstraction beyond programming by specifying the solution in a language that directly uses concepts and rules from a specific problem domain. Second, generate the final products, source code, configuration files, and other artifacts, from these high-level specifications. The automation of application development is possible because the modeling language, the code generator, and the supporting framework need to fit the requirements of a relatively narrow application domain. It is worth to apply the benefits of domain-specific modeling and model processing for the energy efficient aspects of mobile application development as well. In the current case the domain is the development of mobile applications focusing on the energy efficiency aspects of mobile apps.

Model-driven methods have a wide application field. Different solutions support one or more from the following scenarios: refining the design to implementation, transforming models into other domains, aspect weaving, analysis and verification, refactoring purposes, simulation and execution of a model, querying some information and providing a view for it, abstracting models, assigning concrete representation to model elements, migration, normalization, optimization, and synchronization [8]. There are several academic approaches supporting model-driven methods, some examples are: the Attributed Graph Grammar (AGG), AToM3, Fujaba, Generic Modeling Environment (GME), GReAT, VIATRA2, and Visual Modeling and Transformation System (VMTS). These solutions makes possible modeling and certain types of model processing. Fujaba supports editing UML diagrams and generating Java code from it. AToM3, GME and VMTS facilitate to define optional domain-specific languages and utilize them to define software artifacts. Model processors provide different ways to perform model transformations, for example some of them provide APIs to make the mobile editing easier (GME, VMTS), template-based processing (VMTS), and/or graph rewriting-based processing (AGG, GReAT, VIATRA2, and VMTS).

In the field of mobile application development, there are several cross-platform development frameworks and solutions that, within certain conditions, utilizes the domain-specific modeling concepts and support the rapid application development for different mobile platforms. We mention some of these approaches and compare their achievements with our goals. Also we highlight that currently there is no solution that provide support for modeling energy efficient issues of mobile applications.

PhoneGap is a mobile development framework enabling developers to build applications for mobile devices using standard web technologies, such as JavaScript, HTML5 and CSS3, instead of device-specific languages such as Java, Objective-C or C♯. The resulting applications are hybrid. Appcelerator Titanium is a platform that, similarly to PhoneGap, using web technologies supports developing mobile, tablet and desktop applications. Xamarin.Mobile is a library that exposes a single set of APIs for accessing common mobile device functionality across iOS, Android, and Windows platforms.

The above mentioned solutions produce final executable files, i.e. the applications that are ready to use which can only contain those functions that have the appropriate implementation or support in the mobile platform-specific libraries, namely in the supporting SDKs or APIs. In contrast, the goal of our approach is to speed up the development and not to eliminate the native programming. Furthermore, our goal is also to introduce the different aspects of energy efficiency on higher level, on the modeling level, which the existing solutions currently lack.

3   Energy-Efficient Communication Patterns

A mobile device has several hardware components that drains the battery. The three highest energy consumers are the wireless interfaces (3G/Wi-fi/LTE radio), the display and the application processor. These three are responsible for more than 70% of the devices energy consumption for most use cases. This research does not focus on user interface design or display technologies, thus the display factor can be neglected. The effect of the processor highly depends on the use case; however, considering most common applications, its share in the overall energy consumption is still much smaller than the radio's [9].

We next identify communication patterns that can achieve smaller energy consumption compared to the unregulated transfer of data. All of these methods assume that an IP-based communication channel is used over Wi-Fi or 3G cellular data. Table 1 summarizes the main characteristics of each discussed pattern. We note that other such patterns besides the ones we describe certainly exist and the purpose of the chosen ones is merely to exemplify our approach.

Table 1. Summary of the discussed energy-efficient communication patterns

Pattern	Requirements	Tradeoffs
Delayed	Smaller, impulse-like data transfers	Increased latency
Bursty	Larger transfer sessions and streamable data; regular data transfer must not utilize full network capacity to achieve energy saving	Increased latency
Compressed	Larger, compressible data	Energy and performance cost of compression/decompression

3.1 Delayed Transfer

Delayed transfer means that data transfer requests are queued and transmissions happen in a batch after a certain time has passed or the size of the data to be transferred is large enough. The energy saving in this case is the result of the fewer bursts of data which means less energy overhead from the switching of radio states. When using 3G access, the radio needs to enter the high energy transmission state only once a specific number of requests has been queued, which reduces the amount of so called tail energy[10]. With Wi-fi, although the energy saving can be less significant, performing several requests at a time still has a positive effect on the overall energy consumption [11].

This pattern can be used with smaller, impulse-like data transfers, such as generic HTTP request during the communication with a web application, excluding the larger media transfers. Nevertheless, a key characteristic of the method is the maximum delay allowed for certain requests. The queuing increases the latency which might not be tolerable for certain types of communication where real-time feedback is important, such as when sending and receiving non-critical notifications. A more concrete example could be a Twitter client queuing the requests to create some new tweets and update the user's news feed.

In this paper, we do not tackle the problem of how exactly to delay data transfers, which is a scheduling problem with potentially many jobs with differing deadlines. Existing scheduling algorithms and related work[12] can be reused here.

3.2   Bursty Transfer

Bursty transfer exploits the fact that faster communication incurs lower energy per bit cost with radio communication [13]. The main idea is that the two communicating partner reschedules the transfer session in a way that the data is transferred in high rate bursts instead of at a steady but slower rate. This differs from the delayed transfer pattern, where the complete data payload is always sent in one burst. Obviously, this only makes sense if one of the communication partners cannot receive data at maximum speed using normal communication. The lower the rate at which the devices communicate without such traffic shaping, the larger the achievable energy savings when using the bursty transfer.

This method can be used effectively if the amount of data is large enough to divide it into smaller bursts. Generally, software modification is required only on the uploading device to reschedule the transmission into bursts but it is also possible to generate bursty transfer at the receiver side by only periodically reading from the socket, which leads to a bursty transmission because of TCP flow control and limited TCP receive buffer space. Example applications include video streaming or transferring large files.

3.3   Compressed Transfer

Compressed transfer is the most generic solution to save energy[14]. It involves compressing the data to be transferred at the sending device and decompressing it at the recipient after the transfer session.

Compressed transfer is viable only if a high compression ration can be achieved with the data to be transferred. Different types of data might require different compression algorithms. Another issue is the computing and energy requirements of the algorithm, which should be taken into account, especially on mobile devices.

An example is transferring large amount of textual data. The compressed transfer can be combined with the bursty transfer if the requirements of both patterns are met.

4   Modeling and Generating Energy Efficient Applications

In order to support the aforementioned energy efficient patterns, we designed an architecture and implemented a prototype system for Android. Based on the model, which describes the energy aspects of the application, the generated code uses a library which interacts with our runtime system. The runtime system, also referred to as Energy-efficient Communication Service, implements the various energy efficient patterns and takes care of the related tasks. The following subsections describes the components of the approach supported by examples.

4.1 Mobile Runtime Supporting Energy Efficient Communication Pattern

The architecture of our reference Android implementation is illustrated in Figure 1. A key element of the system is the EnergyCommService service. This is a standard Android service component which is running in the background, waiting for requests from applications. The energy-efficient patterns are implemented in the form of manager classes hosted by the the service, each implementing the mechanics related to a certain pattern. For instance, the DelayedTransferManager class acts as a facade to the subsystem responsible for scheduling the delayed transfer requests and performing them at the right time.

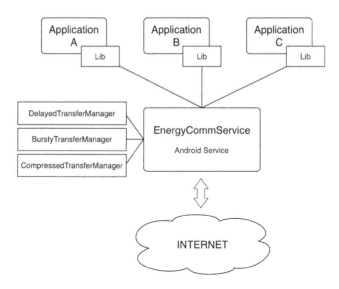

Fig. 1. Architecture for the Android implementation

A centralized service components was needed so that all communication related requests can be handled at a single place. This is important because the effectiveness of the patterns highly depend on how they are used concurrently. Any application that wants to use the service must implement a library, which contains the APIs that give access to the patterns.

The following API examples shows the pseudocode declarations of the methods responsible for initiating bursty and delayed transfer requests. In addition to the general parameters required for the transfer, such as the address and the data to be sent, the different energy-efficient patterns also require parameters.

```
/**
 Opens a TCP session to the given address and sends
 the data using the bursty communication pattern.
*/
void sendBursty(data, address, burstSize, burstDelay)

/**
 Opens a TCP session to the given address and sends
 the data using the delayed communication pattern.
 */
void sendDelayed(data, address, maximumDelay)
```

Choosing which pattern to apply depends on several variables, including the server-side support and the size and characteristics of the data to be sent. To support the programmer in this task, we introduce a model-based code generation solution.

4.2   Model-Based Code Generation

Our modeling and model transformation framework is the Visual Modeling and Transformation System (VMTS) [15] [16]. VMTS is a metamodeling environment which supports editing models according to their metamodels. Models are formalized as directed, labeled graphs. VMTS uses a simplified class diagram for its root metamodel ("visual vocabulary").

Also, VMTS is a model transformation system, which transforms models using both template-based and graph rewriting techniques. Moreover, the tool facilitates the verification of the constraints specified in the transformation step during the model transformation process. VMTS framework is one possible framework that can support the modeling of energy efficiency related aspect of mobile applications. Other frameworks can also adapt the suggested method.

The architecture of the mobile application modeling and generation process is depicted in Figure 2. Modeling and processing different aspects of mobile applications are performed in the VMTS framework. We use mobile-specific languages to define the required structure and application logic. Platform-specific model processors are applied to generate the executable artifacts for different target mobile platforms. The generated code is based upon the previously assembled mobile-specific libraries. These libraries provide energy efficient solutions for mobile applications through the runtime component introduced in Section 4.1.

Mobile-specific languages address the connection points and commonalities of the most popular mobile platforms. These commonalities are the basis of further modeling and code generation methods. The main areas, covered by these

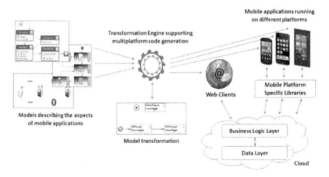

Fig. 2. Supporting multi-mobile platform development

domain-specific languages, are the static structure, business logic (dynamic behavior), database structure and communication protocol. Using these textual and visual languages, we are able to define several parts of the mobile applications.

The following example introduces a textual language-based sample for defining REST-based (Representational State Transfer) communication related interface. The solution can generate both the server and client-side proxy classes supporting the communication based on these type of interface definitions.

To indicate that this interface defines the API of a server application we denote it with the [RestApi] attribute above the interface. By finding this attribute the code generator will recognize that this interface should be treated as a REST API interface, and it should generate the client-side proxy class for that. This [RESTUrl] attribute specifies which Url to invoke. This way, the generated proxy will navigate to the *insertuser.php* page. The sample also covers the definition of REST method, command type, return format, REST parameter, and mapping type.

```
[RestApi]
public interface MyService
{
  [RestMethod(Url = "$client/insertuser.php", CommandType
    = CommandType.POST, ReturnFormat = FormatType.XML)]
  UserInfo InsertUser ([RestParam(Mapping =
    RestMethodMappingType.Custom)]string client,
    [RestParam(Name="usr", Mapping =
    RestMethodMappingType.Body)]string userName);
}
```

For each target platform, a separate transformation is realized since, at this step, we convert the platform-independent models into platform-specific executable code. The transformation expects the existence of the aforementioned frameworks and utilizes their methods. The generated source code is essentially a list of parameterized activities (commands) that certain functions of the mobile application should perform. This means that the core realization of the functions is not generated but utilized from mobile-platform specific frameworks, i.e., the generated code contains the correct function calls in an adequate order, and with appropriate parameters. The advantages of this solution are the followings:

- The software designer has easier task. The model processors are simpler. The model processing is quicker. The generated source code is shorter and easier to read and understand.
- We use prepared mobile-specific libraries. These libraries are developed by senior engineers of the actual mobile platform and subject. These solutions, applying the appropriate patterns, take into account the questions of energy efficiency as well.

In order to handle the energy related aspects of mobile applications also on the modeling level, we should introduce those attributes that enables to define the envisioned conditions and circumstances of the data transfer. This include the amount of the data, the frequency of data transfer, and the tolerated delay.

The above example is the textual view of the model, of course the modeling framework makes possible to provide a visual interface to edit these and other domain-specific models as well.

There are various solutions that can be utilized when it is about code generation. We have chosen the Microsoft T4 (Text Template Transformation Toolkit). T4 is a mixture of static texts and procedural code: the static text is simply printed into the output while the procedural code is executed and it may result in additional texts to be printed into the output.

The next section discusses some examples, where the method makes it possible to move the energy related aspects to higher abstraction level. Moving these configuration settings to the modeling level means that energy efficient settings are also available to generate the applications, analyze the models and verify the application.

4.3   Energy-Efficient Code Generation Examples

Figure 3 shows and example how different code is generated based on the model. The model in this case is code based, where the modeler defines the attributes of the data transfer, and the result is a generated method that can be used by the programmer to perform the actual transfer.

In the example on the left, the attributes define a transfer session with textual content, 5 KB default data size and 10 minutes maximum delay. The generated code performs a combined compressed and delay transfer since the model attribute values enable the utilization of both patterns. The example on right describes transferring a larger amount of data which makes the bursty pattern a suitable choice.

Figure 4 depicts another example in which some device parameters has also been taken into account in the modeling phase. Being aware that the device is going to use 3G to connect to the internet and operate on battery, the parameters

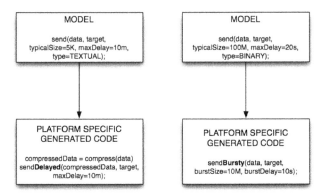

Fig. 3. Two examples of the model-based code generation approach

in the generated code can be refined: in this case, the high overhead (ramp and tail energy) of bursts with 3G makes using larger bursts more energy-efficient. Nevertheless, it is important to recognize that depending on such system parameters only make sense if the device is operated in a relatively static environment.

These are of course only some examples of the many possibilities such system could offer. Choosing which patterns to apply, how to combine the patterns and fine tuning the parameters of the various patterns are some of the tasks the modeling system needs to handle.

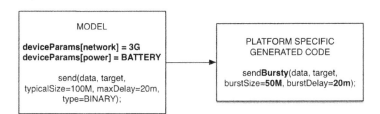

Fig. 4. Example of including system parameters in the model

5   Discussion

The first of the aspects to be considered is where to place the logic that chooses and configures the patterns to apply. One option is to integrate it into the client side library which, monitoring the current device environment, could then operate adaptively. The other option is that the pattern-related decisions are made during the code generation phase by the model processor. As usual, there are pros and cons for both approaches. Having the logic in the client side library is beneficial because it can be configured adaptively. On the other hand, the developer may provide a hint for the library and in this way the result can be a clearer and more understandable source code.

The next questions are the followings: Why is it important to increase the abstraction level in mobile application development? What are the benefits of modeling mobile applications? Why is it important to include the modeling of energy efficient issues of mobile applications as well? To raise the level of abstraction in model-driven development, both the modeling language and the model processor (generator) need to be domain-specific. The approach can improve both the performance of the development and the quality of the final products. The commonly identified benefits of the model-driven approach are improved productivity, better product quality, hidden complexity, and reuse of expertise.

Moving certain parts of mobile application definition to the modeling level, including the energy efficient settings, means that these aspects of the applications are also available on higher level. This makes it possible to gain an overview of the applications, to analyze them from different points of view, to verify or validate certain critical issues, and to generate the source code of the application from models.

We believe that it is necessary to involve the software developers in making apps more energy efficient at least to a certain extent. Solutions that delay traffic and are completely transparent to the developer, like the one described in [17] may not be optimal either because they might lead to unexpected behavior from the developer's viewpoint. From the perspective of a typical developer, the model-driven approach may sound unnecessarily cumbersome.

However, from the perspective of the product owner, the project manager, the software architect, and also the customer it is worthwhile to move all possible aspects of mobile applications to a higher level of abstraction because that is the level at which they typically operate. They cannot think on the level of the libraries (source code), because they are responsible for a whole system with different parts: server side, desktop client, web clients, mobile clients, cloud related things and so on. Still, they want to see the important aspects of each area. Modeling should concentrate on all key aspects of the mobile applications, including energy efficiency.

Energy efficiency is typically context dependent. This applies especially for data communication where the amount of Joules per bit spent depends on at least SNR (transmit power) and available bandwidth that vary with location and time. Therefore, regardless of whether model-driven or purely library-based approach is used, some logic should be executed at run time. For example, the energy utility of data transfer (J/bit) depends on e.g. SNR and available bandwidth that are location and time dependent which in turn impact the decision whether compressing specific type of content yields energy savings or not [18].

Our examples focus only on communication energy. Concerning other important sources of energy drain, such as CPU, display, and sensors, we can imagine ways to include those as well. For example, the developer could decide to execute a function that renders the display in the most energy efficient manner by adapting the colors [19]. As for sensors, the programmer could explicitly set the trade-off between accuracy and energy consumption.

In summary, we can say that applying model-driven solutions in mobile application development is the current trend, and in case of energy efficiency it is also advised to follow it. Obviously, there are scenarios where modeling (higher abstraction level) does not provide immediate advantages. We hope that this work could serve as a starting point of a discussion within the research community concerning how to effectively transfer the existing and novel energy efficiency techniques to the mobile application development at large.

References

1. Hoque, M., Siekkinen, M., Nurminen, J.K.: TCP receive buffer aware wireless multimedia streaming - an energy efficient approach. In: Proceedings of the 23rd ACM Workshop on Network and Operating Systems Support for Digital Audio and Video, NOSSDAV 2013, pp. 13–18. ACM (2013)
2. Nurminen, J.K., Nöyränen, J.: Parallel data transfer with voice calls for energy-efficient mobile services. In: Bonnin, J.-M., Giannelli, C., Magedanz, T. (eds.) Mobilware 2009. LNICST, vol. 7, pp. 87–100. Springer, Heidelberg (2009)
3. Anand, M., Nightingale, E.B., Flinn, J.: Ghosts in the machine: interfaces for better power management. In: Proceedings of the 2nd International Conference on Mobile Systems, Applications, and Services, MobiSys 2004, pp. 23–35. ACM (2004)
4. Anand, M., Nightingale, E.B., Flinn, J.: Self-tuning wireless network power management. Wirel. Netw. 11, 451–469 (2005)
5. OMG Model-Driven Architecture (MDA) Specification, OMG Document ormsc/01-07-01 (2001)
6. Fowler, M.: Domain-Specific Languages. Addison-Wesley Professional (2010)
7. Kelly, S., Tolvanen, J.-P.: Domain-Specific Modeling: Enabling Full Code Generation. Wiley-IEEE Computer Society Press (2008)
8. Amrani, M., Dingel, J., Lambers, L., Lucio, L., Salay, R., Selim, G., Syriani, E., Wimmer, M.: Towards a model transformation intent catalog. In: Proceedings of the First Workshop on the Analysis of Model Transformations (AMT 2012), New York, USA, pp. 3–8. ACM (2013)
9. Nurminen, J., Noyranen, J.: Energy-consumption in mobile peer-to-peer - quantitative results from file sharing. In: 5th IEEE Consumer Communications and Networking Conference, CCNC 2008, pp. 729–733 (2008)
10. Balasubramanian, N., Balasubramanian, A., Venkataramani, A.: Energy consumption in mobile phones: a measurement study and implications for network applications. In: Proceedings of the 9th ACM SIGCOMM Conference on Internet Measurement Conference, IMC 2009, pp. 280–293. ACM, New York (2009)
11. Xiao, Y., Cui, Y., Savolainen, P., Siekkinen, M., Wang, A., Yang, L., Yla-Jaaski, A., Tarkoma, S.: Modeling energy consumption of data transmission over wi-fi. IEEE Transactions on Mobile Computing 99, 1 (2013) (PrePrints)
12. Ra, M.-R., Paek, J., Sharma, A.B., Govindan, R., Krieger, M.H., Neely, M.J.: Energy-delay tradeoffs in smartphone applications. In: Proceedings of the 8th International Conference on Mobile Systems, Applications, and Services, MobiSys 2010, pp. 255–270. ACM, New York (2010)
13. Nurminen, J.: Parallel connections and their effect on the battery consumption of a mobile phone. In: 2010 7th IEEE Consumer Communications and Networking Conference (CCNC), pp. 1–5 (2010)

14. Barr, K.C., Asanovic, K.: Energy-aware lossless data compression. ACM Trans. Comput. Syst. 24, 250–291 (2006)
15. Angyal, L., Asztalos, M., Lengyel, L., Levendovszky, T., Madari, I., Mezei, G., Meszaros, T., Siroki, L., Vajk, T.: Towards a fast, efficient and customizable domain-specific modeling framework. In: Proceedings of the IASTED International Conference, Innsbruck, Austria, pp. 11–16 (2009)
16. VMTS: Visual modeling and transformation system, http://www.aut.bme.hu/vmts/
17. Qualcomm Inc., Managing background data traffic in mobile devices (2012), http://www.qualcomm.com/media/documents/ managing-background-data-traffic-mobile-devices
18. Xiao, Y., Siekkinen, M., Yla-Jaaski, A.: Framework for energy-aware lossless compression in mobile services: The case of e-mail. In: 2010 IEEE International Conference on Communications (ICC), pp. 1–6 (2010)
19. Dong, M., Zhong, L.: Chameleon: a color-adaptive web browser for mobile oled displays. In: Proceedings of the 9th International Conference on Mobile Systems, Applications, and Services, MobiSys 2011, pp. 85–98. ACM (2011)

Problems of Mutation Testing and Higher Order Mutation Testing

Quang Vu Nguyen and Lech Madeyski

Institute of Informatics, Wroclaw University of Technology, WybrzezeWyspianskiego 27,
50370 Wroclaw, Poland
{quang.vu.nguyen,Lech.Madeyski}@pwr.wroc.pl

Abstract. Since Mutation Testing was proposed in the 1970s, it has been considered as an effective technique of software testing process for evaluating the
quality of the test data. In other words, Mutation Testing is used to evaluate the
fault detection capability of the test data by inserting errors into the original
program to generate mutations, and after then check whether tests are good
enough to detect them. However, the problems of mutation testing such as a
large number of generated mutants or the existence of equivalent mutants, are
really big barriers for applying mutation testing. A lot of solutions have been
proposed to solve that problems. A new form of Mutation Testing, Higher Order Mutation Testing, was first proposed by Harman and Jia in 2009 and is one
of the most promising solutions. In this paper, we consider the main limitations
of Mutation Testing and previous proposed solutions to solve that problems.
This paper also refers to the development of Higher Order Mutation Testing and
reviews the methods for finding the good Higher Order Mutants.

Keywords: Mutation Testing, Higher Order Mutation, Higher Order Mutants.

1 Introduction

According to IEEE Std 829-1983 (IEEE Standard Glossary of Software Engineering
Terminology), software testing is the process of analyzing a software item to detect
the differences between existing and required conditions and to evaluate the features
of the software items. In other words, software testing is execution a program using
artificial data and evaluating software by observing its execution in order to find
faults or failures. It is worth mentioning that "testing can only show the presence of
errors, not their absence" which is often referred as Dijkstra's law. Where, according
to IEEE Std 829-1983, error is a human action that produces an incorrect result, fault
is an incorrect step, process, or data definition in a computer program and failure is
the inability of a system or component to perform its required functions within specified performance requirements.

Software testing is always one of the important activities in order to evaluate the
software quality. However, the quality of the set of testcases is a problem to be discussed. In addition, there are many cases that testers have not mentioned in the set of

T.V. Do et al. (eds.), *Advanced Computational Methods for Knowledge Engineering*,
Advances in Intelligent Systems and Computing 282,
DOI: 10.1007/978-3-319-06569-4_12, © Springer International Publishing Switzerland 2014

testcases. Mutation Testing has been introduced as a technique to assess the quality of the testcases.

Mutation Testing (MT), a technique that has been developed using two basic ideas: Competent Programmer Hypothesis ("programmers write programs that are reasonably close to the desired program") and Coupling Effect Hypothesis ("detecting simple faults will lead to the detection of more complex faults"), was originally proposed in 1970s by DeMillo et al.[1] and Hamlet[2]. While other software testing techniques focus on the correct functionality of the programs by finding error, MT focuses on test cases used to test the programs. In other words, the purpose of software testing is to find all the faults in a particular program whilst the purpose of MT is to create good sets of testcases. A good set of testcases is a set which is able to discover all the faults. With MT, mutants of a program are the different versions of the program. More specifically, each of which is generated by inserting only one semantic fault into original program (Table 1 gives an example to mutant). That generation is called mutation and that semantic fault is called mutation operator. It is a rule that is applied to a program to create mutants, for example modify expressions by replacing operators and inserting new operators. Mutation operators depend on programming languages, but there are traditional mutation operators: deletion of a statement; replacement of boolean expressions; replacement of arithmetic; replacement of a variable.

Table 1. An example of mutant (First Order Mutant)

Program P	Mutant P'
| ```
. . .
while (hi<50) {
system.out.print(hi);
        hi = lo +hi;
lo = hi -lo;
        }
. . .
``` | ```
. . .
while (hi>50) {
system.out.print(hi);
        hi = lo +hi;
lo = hi -lo;
        }
. . .
``` |

The process of MT can be explained simply in following steps:

1. Suppose we have a program P and a set of testcases T
2. Produce mutant P1 from P by inserting only one semantic fault into P
3. Execute T on P and P1 and save results as R and R1
4. Compare R1 with R:
   4.1 If R1 ≠ R: T can detect the fault inserted and has killed the mutant.
   4.2 If R1 =R: There could be 2 reasons:
   + T can't detect the fault, so have to improve T.
   + The mutant has the same semantic meaning as the original program. It's
equivalent mutant (an example of equivalent mutant is showed in Table 2).

MT evaluates a set of testcases T by Mutation Score (MS), will be between 0 and 1, which is calculated by the following formula:

$$MS = \frac{\text{Number of killed mutants}}{\text{Total mutants} - \text{Equivalent mutants}}$$

A low score means that the majority of faults cannot be detected accurately by the test set. A higher score indicates that most of the faults have been identified with this particular test set. A good test set will have a mutation score close to 100%. When MS = 0, have no any testcase that can kill the mutants and when MS=1, we say that mutants are very easy to kill.

Table 2. An example of equivalent mutant

Program P	Mutant P'
...	...
`int a =2;`	`int a =2;`
`if (b==2) {`	`if (b==2) {`
`System.out.print(b);`	`System.out.print(b);`
`  b = a + b;}`	**`  b = a * b;}`**
...	...

In the next section, we summarize main limitations of mutation testing. Section 3 shows the previous proposed solutions for solving the limitations of mutation testing. Section 4 presents Higher Order Mutation Testing and its effectiveness. Section5 presents techniques to find good Higher Order Mutants. Section 6 presents conclusions and future work.

2   Main Limitations of Mutation Testing

Although MT is a high automation and effective technique for evaluating the quality of the test data, Mutation Testing has three main problems in our view.

The first limitation of mutation testing is **a large number of mutants**, because program may have a fault in many possible places and with only one inserted semantic fault we will have one mutant. Thus, a large number of mutants will be generated in the mutant generation phase of mutation testing. Typically, this is a large number for even small program. For example, a simple program with just a sentence such as return a+b (where a, b are integers) may be mutated into many different ways: a−b, a*b, a/b, a+b++, −a+b, a+−b, 0+b, a+0, etc. This problem leads to a **very high execution cost** because the test cases are executed on not only original program but also each mutants. For example, assume that we have a program under test with 150 mutants and 200 testcases, it requires (1+150)*200 = 30200 executions with their corresponding results.

The second limitation of mutation testing is **realism**. Mutations are generated by single and simple syntactic changes, hence they do not denote realistic faults. While according to Langdon et al.[9], 90 percent of real faults are complex. In addition, it is not sure that we have found a large proportion of real faults present even if we have killed all the killable mutants[4].This is one of big limitations for applying mutation testing.

The third limitation of mutation testing is **equivalent mutant problem**[68]. In fact, many mutation operators can produce equivalent mutants which have the same behavior as original program. In this case, there is no testcase which could "kill" that mutants and the detection of equivalent mutants often involves additional human effort. Madeyski et al.[68] manually classified 1000 mutants as equivalent or non-equivalent and confirmed the finding by Schuller and Zeller[89] that it takes about12 minutes to assess one single mutation for equivalence. Therefore, there is often a need to ignore equivalent mutants, which would mean that we are ready to accept the lower bound on mutation score named mutation score indicator MSI [98][99][100][101].

3   Solutions Proposed to Solve Problems of Mutation Testing

In order to collect a complete set of all the techniques to solve the problems of MT since the 1970s (after MT was proposed by DeMillo et al. [1] and Hamlet [2]) and sort them in chronologic order, with search terms which have either "mutation testing", "mutation analysis", "mutants + testing", "mutants + methods", "mutants + techniques", "mutants + problems", "mutation testing + improve", "equivalent mutants", "mutants*", "higher order mutants", we searched papers that were published in IEEE Explore, ACM Portal, Springer Online Library, Wiley Online Library, Citeseerx and journals or conference proceedings as IST (Information and Software Technology), JSS (Journal of Systems and Software), TSE (IEEE Transactions on Software Engineering), IET Software. After then, we continued to look at the references of each paper to find other articles related to our purpose. In addition, we also searched for Master and PhD theses that have contents related to "mutation testing". Figure 1 shows number of publications proposing techniques to solve the problems of Mutation Testing.

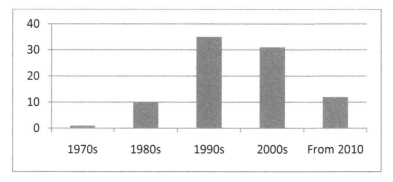

Fig. 1. Number of publications proposing techniques to solve the problems of MT

We searched and listed techniques to solve the problems of MT as reduce the number of mutants and execution cost, solve the realism problem and solve the equivalent mutant problem in chronologic order as follows:

3.1   Mutant and Execution Cost Reduction Techniques

Mutant Sampling approach[14-20] was first proposed by Acree[14] and Budd[15] in 1980. In this approach, a small subset of mutants are randomly selected from the entire set. Some studies implied that Mutant Sampling is valid with a x% value higher than 10% (x% is % of mutants were selected randomly).

Reduction number of mutation operators will leads reduction number of mutants. This is idea of Agrawal et al.[23] and Mathur[24] and it was called Selective Mutation by Offutt et al.[18][25]. This approach suggested finding a small operators set that generate a subset of all possible mutants without losing test effectiveness. There are many authors have been applied this approach to effectively reduce generated mutants[20][26-32].

Instead of selecting mutants randomly, Husain[21] in 2008 used clustering algorithms to choose a subset of mutants. There is a empirical study of this approach in the work of Li et al.[22].

Second Order Mutation Testing[68][76][83][84[97] in particular and Higher Order Mutation Testing[3][4][9][10][35] in general are the most promising solutions to reduce number of mutants. Basic idea of this approach is improvement of MT by inserting two or more faults into original program to generate its mutants. For example, by combining two First Order Mutants to generate Second Order Mutant[68][83][84[97] or by using Subsuming Higher Order Mutants algorithms[3][4], the number of generated mutants can be reduced to about 50%.

Weak Mutation[36-42][93][94] was first proposed by Howden[36] in 1982 proposed. This is the first technique to optimize the execution of Traditional Mutation Testing (called Strong Mutation). Suppose that a program P is constructed from a set of components C= $\{c_1,..,c_n\}$. And mutant m is made by changing c_m,. So, we need only compare immediately the result of mutant m with the result of c_m execution to say m be killed or not (immediately check after mutated component be executed).

In order to improve quality of Strong Mutation and Weak Mutation, Woodward et al. [43][44] suggested an approach named Firm Mutation in 1988. This approach provides a continuum of intermediate possibilities to overcome the disadvantages of weak and strong mutation. In firm mutation, components may be groups of statements and partial program executions may be considered rather than each separate component execution as in weak mutation or the complete program execution as in strong mutation. It combines the reduced execution cost of weak mutation with the greater transparency of strong mutation.

In group of Run-time Optimization techniques, there are 6 techniques have been proposed. Interpreter-Based technique was first proposed in 1987 by Offutt and King[45][18] to reduce execution cost with idea: the original program is translated into an intermediate form, and then mutants are generated from this intermediate code. With Compiler-Based technique[46-48], each mutant is first compiled into an executable program, and then execute testcases on the compiled mutant. It is faster because execution of compiled binary code takes less time than code interpretation. Whilst main idea of Compiler-Intergrated[49][50] is designing an instrumented compiler to generate and compile mutants instead of compiling each mutant   individually in the Compiler-Based technique. In order to improve the Interpreter-Based technique, Mutant Schema Generation[51-54][95] approach has been proposed. All mutants are encoded into a metaprogram, then the metaprogram is compiled and run with faster speed, basically, compared to compile and execute each individual mutation. Another technique is Bytecode Translation Technique[55-58] which was first proposed by Ma et al.[56] in 2005 to reduce execution cost of MT by using bytecode translation. Instead of from source code, mutants are generated from the compiled bytecode of original program and that generated mutants can be executed directly without compilation[55-58]. The last technique of this group is Aspect-Oriented Mutation which has been proposed by Bogacki and Walter[59][60]. In this approach, do not need to compile each mutant. An aspect patch is generated and each aspect will run programs twice: First for the original program and then for mutants.

There are also some techniques to improve computational cost base on computer architectures as SIMD (Single Instruction   Multiple Data)[61-63], MIMD (Multiple Instruction   Multiple Data)[64][65] and Parallel[66][67].

In 2013, Lisherness et al.[96] suggested a new approach by using coverage discounting for mutation analysis can reveal which functions are changed in the mutant, and in turn what is not being adequately tested if the mutation is undetected.

3.2   Techniques to Solve the Realism Problem

There are many works have been demonstrated that the majority of real faults are complex faults. E.g. in [34][35], it is known that 90%. Base on that suggestion, Second Order Mutation Testing[68][77][84][85[97] in particular and Higher Order Mutation Testing[3][4][9][10] in general have been considered as the most promising solutions to solve the realism problem. That approaches have been suggested using more complex faults to generate mutants by inserting two or more faults into original program. In addition, Langdon et al.[9][35] introduced a new form of mutation testing: Multi Objective Higher Order Mutation Testing (with Genetic Programming)[35] in order to find higher order mutants that   more realistic complex faults.

3.3   Techniques to Solve the Equivalent Mutant Problem

According to Madeyski et al.[68], the techniques to solve the equivalent mutant problem can be classified into 3 groups:

  - Applying techniques to detect equivalent mutants such as Compiler Optimization[69][70], Mathematical Constraints[71][90],Program Slicing[72], Semantic differences in terms of running profile[73],Margrave's change-impact analysis[74] and Lesar model-checker[75].
  - Applying techniques to avoid (or reduce) generating the equivalent mutants in the process of mutants generation such as Selective mutation[77], Program dependence

analysis[78], Co-evolutionary[79], Equivalency conditions[80], Fault hierarchy[81], Semantic exception hierarchy[82], Higher order mutation testing[3], [83], [84], [68].

- Applying techniques in order to suggest the equivalent mutants such as Using Bayesian-learning based guidelines[85], Examining the impact of equivalent mutants on coverage [86], Using the impact of dynamic invariants [87], Examining changes in coverage[88][89].

4   Higher Order Mutation Testing and Its Effectiveness

Second Order Mutation Testing in particular and Higher Order Mutation Testing in general is an approach for generating mutants by applying mutation operators more than once. The idea about Second Order Mutation Testing was first mentioned by Offut[76] in 1992. After then, Polo et al. [100] in 2008, Kintis et al. [83] and M. Papadakis and N. Mlevris [84] in 2010, and L. Madeyski et al.[68] in 2013 further studied to suggest their algorithms to combine First Order Mutants to generate Second Order Mutants (SOMs). With this approach, not only at least 50% mutants were reduced without loss of effectiveness of testing, but also the number of equivalent mutants can be reduced (the reduction in the mean percentage of equivalent mutants passes from about 18.66% to about 5%) and generated second order mutants can be harder to kill [97][84][85][68].

In 2009, Higher Order Mutation Testing (HOM Testing) was first proposed by Jia, Harman and Langdon[3][4]. According to them, mutants can be classified into two types: First Oder Mutants (FOMs – used in Traditional Mutation Testing) - are generated by applying mutation operators only once - and Higher Oder Mutants (HOMs- used in Higher Order Mutation Testing) – are constructed by inserting two or more faults. A simple example is presented as Table 3:

Table 3. An example of higher order mutant

Program P	FOM1
... ```while ((hi < 50) && (hi>lo)){` `System.out.print(hi);` ` hi = lo + hi;` ` lo = hi - lo; }``` ...	... ```while ((hi > 50) && (hi>lo)){` `System.out.print(hi);` ` hi = lo + hi;` ` lo = hi - lo; }``` ...
FOM2	**HOM (Createdfrom FOM1 and FOM2)**
... ```while ((hi < 50) && (hi<lo)){` `System.out.print(hi);` ` hi = lo + hi;` ` lo = hi - lo; }``` ...	... ```while ((hi > 50) && (hi<lo)){` `System.out.print(hi);` ` hi = lo + hi;` ` lo = hi - lo; }``` ...

The same as Second Order Mutants, the combination of faults of higher order mutation testing can reduces the number of generated mutants and limits the number of mutants and those generated mutants are harder to kill than any of the individual constituent faults. E.g. in case of subsuming Higher Order Mutants [3][4].A subsuming HOM was constructed from constituent FOMs and so that HOM is only killed by subset of the intersection of testcases which kill each constituent FOMs. Therefore, the quality of testcases will be better but the number of testcases will be reduced. Fewer number mutants and fewer number of test cases lead to a fewer execution cost. In addition, with the combination of faults, we also limit unrealistic and avoid generating equivalent mutants [3][68][76][83-84].

To be more clearly, let's consider the example given in Table 3. In this case, we have only one HOM created from FOM1 and FOM2 instead of FOMs. And there are TWO test cases which kill FOM1: {hi<50 && hi>lo} and {hi>50 && hi>lo}, TWO test cases which kill FOM2: {hi<50 && hi>lo} and {hi<50 && hi<lo}. But we need ONLY ONE test case which kills HOM: {hi<50 && hi>lo} and this test case also kills both FOM1 and FOM2. But the reverse is not true, the test case {hi>50 && hi>lo} does not kill HOM and the test case {hi<50 && hi<lo} also does not kill HOM. And so, with HOM, we can reduce number mutants generated and number of test cases without leading to loss of effectiveness during testing. Similarly, the combination of two or more errors to generate mutants will cause limited number of generated equivalent mutants and number of "easy to kill" mutants.

Unlike some other techniques presented in part 3 that only solve individual problems of traditional mutation testing, higher order mutation testing could help us to deal with three main problems of the traditional mutation testing at the same time. That is why we choose higher order mutation testing to study. In the next part, we will present more detail on this issue.

5   Finding the Good HOMs

5.1   Second Order Mutants (SOMs)

In 2008, with the results of their study, Polo et al.[97] believed that "mutant combination does not decrease the quality of the test suite". They suggested[97] 3 algorithms to generate second order mutants (mutants containing two simple faults are called second order mutants). With the LastToFirst algorithm, assume that we have the list of n mutants of First Order Mutants (FOMs), SOMs were generated by combining the first FOM with the FOM number n, the second-FOM with the FOM number (n-1), and so on. In the second one, DifferentOperators algorithm, SOMs were generated by combining two FOMs generated by different mutation operators. And the last algorithm, RandomMix, combines any two FOMs to generate SOMs. In their study, after FOMs were generated from 6 programs under test (Bisect with 31 LOC-Line of Code, Bub-54, Find-79, Fourballs-47, Mid-59 and TriTyp-61), the three combination algorithms were used to generate SOMs. Number of test cases corresponding to each program are 25, 256, 135, 96, 125, 216 were passed to the FOMs and the SOMs. Results showed that the number of SOMs decreased about 50% compared with FOMs, and the mean percentage of equivalent SOMs is only about 5% compared with 18,66% of equivalent FOMs [97].

Based on the algorithms of the Polo et al.[97], M. Papadakis and N. Malevris studied and proposed five new strategies to combine FOMs: First2Last , SameNode, SameUnit , SU_F2Last and SU_DiffOp[84]. They executed both of First Order Mutation Testing strategies (Strong Mutation, Rand 10%, Rand 20%, Rand 30%, Rand 40%, Rand 50%, Rand 60%) and Second Order Mutation Testing strategies (RandomMix, DifferentOperators, First2Last, SameNode, SameUnit, SU_F2Last and SU_DiffOp) with the same test programs[84]. With their empirical study[84], the mean number of equivalent mutants with the First Order Mutation Testing strategies is about 6237, and with the Second Order Mutation Testing strategies is about 2727. Meanwhile, Second Order Mutation Testing strategies reduced the number of generated SOMs by about 50%, because of SOMs also were generated combining two FOMs, and of course will lead to reduced execution cost.

In 2010, Kintis et al.[83] suggested two categories of Second Order Mutation Testing strategies: The Second Order Strategies category includes RDomF and SDomF strategy; The Hybrid Strategies category includes HDom(20%) and HDom(50%) strategy. The results of their study indicates that the number of generated equivalent mutants of Weak Mutation, RDomF, SDomF, HDom(20%) and HDom(50%) strategies reduced about 73%, 85.4%, 86.8%,81.4% and 65.5% in turn compared with the number of generated equivalent mutants of Strong Mutation. Meanwhile the Mutation Scores are 99.94%, 99.99%, 99.91%, 99.91% for the RDomF, SDomF, HDom(20%) and HDom(50%) strategies respectively. That mutation socres are higher than Mutation Score of Weak Mutation strategy (96.90%).

Most recently, in 2013, Madeyski et al.[68] introduced the JudyDiffOp algorithm and NeighPair algorithm. JudyDiffOp algorithm is a modification of the DifferentOperatorsalgorithm[100] with the idea that "both constituent FOMs were used only once for producing a SOM". And NeighPair algorithm was introduced with idea "generate SOM by combining FOMs which are as close to each other as possible". They experimented[68] 4 algorithms RandomMix, Last2First[97] and JudyDiffOp, NeighPair[68] on programs under test Barbecue (7.413 LOC – Line of Code), Commons IO (16.283 LOC), Commons Lang (48.507 LOC) and Commons Math (80.023 LOC) with the number of test cases are 21,43,88,221 respectively. That algorithms reduce the number of generated SOMs at least by half compared with generated FOMs, especially JudyDiffOp algorithms over 60%. The mean reductions of equivalent mutants number are about 47%, 58.5%, 66% for the RandomMix, Last2First and JudyDiffOp in turn. And the NeighPair algorithm is not good in terms of equivalent mutant reduction. However, "In most of the cases mutation score estimations were higher for the SOM strategies than for FOM"[68].

5.2   Subsuming HOMs

In 2009, Jia and Harman[3] introduced the term "Subsuming HOMs" as follows: The HOM is named "Subsuming HOM" if it is harder to kill than their constituent FOMs. And then they suggested some approaches to find the Subsuming HOMs by using some meta-heuristic algorithms[3]: Greedy Algorithm, Genetic Algorithm and Hill-Climbing Algorithm. In order to find the Subsuming HOMs more effectively, they introduced two definitions[3]: Fragility value, a value for measuring the ability of a FOM or HOM can be killed and Fitness Function is the ratio of the fragility of a HOM to the fragility of the constituent FOMs. In their approach, they used this Fitness Function to evaluate the fitness of HOMs. They experimented with 10 benchmark C programs under test (14850 LoC and 35473 test cases in total) and the results indicate that genetic algorithm is the most efficient algorithm for finding those subsuming HOMs, while the greedy algorithm and hill climbing algorithm can also be used to improve the quality of the results[3].

5.3   Multi-object Genetic Programming

In order to find higher order mutants that are hard to kill and more realistic complex faults, Langdon et al.[9][35] introduced a new form of mutation testing: Multi Objective Higher Order Mutation Testing (with Genetic Programming)[35]. They believed that although there are a lot of number of FOMs was generated but most are simply and do not denote realistic faults. So they suggested inserting faults that are semantically close to the original program instead of inserting faults that are syntactically close to the original program in order to find higher order mutants that are syntactically similar to the original program under test. With this, multi objective Pareto optimal genetic programming approach, the number of mutants grows exponentially with order but the number of equivalent mutants fall rapidly with number of changes made [35]. For example, with the chosen set of the C comparison operations (<,<=, ==, !=, >=, >), and the program under test that contains 17 comparison operators (i.e. the Triangle benchmark Triangle.c), there are 85 1st order mutants (17×5 programs with one change), 3400 2nd order mutants ($17 \times 5 \times 16 \times 5/2$ with two changes), 85000 3rd order mutants($17 \times 5 \times 16 \times 5 \times 15 \times 5/6$ with three changes), and 1487500 4th order mutants ($17 \times 5 \times 16 \times 5 \times 15 \times 5 \times 14 \times 5/24$ with four changes),whilst respectively only 8, 28, 56, 70 equivalent mutants that pass all test cases and 7, 55, 189, 371 mutants that fail just one test[35].

6   Conclusion and Future Work

Since was proposed in 1970s by DeMillo et al.[1] and Hamlet[2], Mutation Testing has been considered as a powerful technique for evaluating the quality of the test cases. Basically, there is still work to be done to improve the quality of mutation testing. This paper reviewed a range of strategies that were proposed to solve three main problems of mutation testing: a vast number of mutants (and also high execution cost), realism of faults and equivalent mutant problem. Each technique has its own advantages and disadvantages and we focus on Higher Order Mutation testing because this is not only a newest method but also a promising solution of three main problems of the traditional mutation testing at the same time. However, the number of mutants grows exponentially with order. So, in the future, we will research to improve and solve that problem for finding good HOMs by applying Multi-Object optimization algorithm. Specifically we are going to:

- Use Java language programming and Judy mutation testing tool for Java (http://madeyski.e-informatyka.pl/tools/judy/)[68][98].
- Search for strongly subsuming HOMs applying multi-objective optimization.
- Assess results, according to the criteria of solving the problems of traditional mutation testing (reduce the number of mutants and execution cost, realism and the equivalent mutant problem), and compare that results with the results of algorithms that have been proposed previously.

References

1. DeMillo, R.A., Lipton, R.J., Sayward, F.G.: Hints on test data selection: help for the practicing programmer. IEEE Computer 11(4), 34–41 (1978)
2. Hamlet, R.G.: Testing programs with the aid of a compiler. IEEE Transactions on Software Engineering SE-3(4), 279–290 (1977)
3. Jia, Y., Harman, M.: Higher order mutation testing. Information and Software Technology 51, 1379–1393 (2009)
4. Harman, M., Jia, Y., Langdon, W.B.: A Manifesto for Higher Order Mutation Testing. In: Third International Conf. on Software Testing, Verification, and Validation Workshops (2010)
5. Afzal, W., Torkar, R., Feldt, R.: A systematic review of search-based testing for nonfunctional system properties. Information and Software Technology 51(6), 957–976 (2009)
6. Harman, M.: The current state and future of search based software engineering. In: Briand, L., Wolf, A. (eds.) Future of Software Engineering 2007, pp. 342–357. IEEE Computer Society Press, Los Alamitos (2007)
7. Harman, M., Mansouri, A., Zhang, Y.: Search based software engineering: A comprehensive analysis and review of trends techniques and applications. Technical Report TR-09-03, Department of Computer Science, King's College London (2009)
8. Raiha, O.: A survey on search based software design. Technical Report Technical Report D-2009-1, Department of Computer Sciences, University of Tamper (2009)
9. Langdon, W.B., Harman, M., Jia, Y.: Efficient multi-objective higher order mutation testing with genetic programming. The Journal of Systems and Software 83 (2010)
10. Jia, Y., Harman, M.: Constructing Subtle Faults Using Higher Order Mutation Testing. In: Proc. Eighth Int'l Working Conf. Source Code Analysis and Manipulation (2008)
11. Jia, Y., Harman, M.: MILU: A Customizable, Runtime-Optimized Higher Order Mutation Testing Tool for the Full C Language. In: Proceedings of the 3rd Testing: Academic and Industrial Conference Practice and Research Techniques (TAIC PART 2008), pp. 94–98. IEEE Computer Society, Windsor (2008)
12. Offutt, A.J.: Investigations of the software testing coupling effect. ACM Transactions on Software Engineering and Methodology 1(1), 5–20 (1992)
13. Adamopoulos, K., Harman, M., Hierons, R.M.: How to Overcome the Equivalent Mutant Problem and Achieve Tailored Selective Mutation Using Co-evolution. In: Deb, K., Tari, Z. (eds.) GECCO 2004. LNCS, vol. 3103, pp. 1338–1349. Springer, Heidelberg (2004)
14. Acree, A.T.: On Mutation. PhD thesis, Georgia Inst. of Technology (1980)

15. Budd, T.A.: Mutation Analysis of Program Test Data. PhD thesis, Yale Univ. (1980)
16. DeMillo, R.A., Guindi, D.S., King, K.N., McCracken, W.M., Offutt, A.J.: An Extended Overview of the Mothra Software Testing Environment. In: Proceedings of the Second Workshop on Software Testing, Verification, and Analysis, pp. 142–151 (1988)
17. Sahinoglu, M., Spafford, E.H.: A Bayes Sequential Statistical Procedure for Approving Software Products. In: Proc. IFIP Conf. Approving Software Products, pp. 43–56 (1990)
18. King, K.N., Offutt, A.J.: A Fortran Language System for Mutation-Based Software Testing. Software: Practice and Experience 21(7), 685–718 (1991)
19. Mathur, A.P., Wong, W.E.: An Empirical Comparison of Mutation and Data Flow Based Test Adequacy Criteria. Technical Report, Purdue Univ. (1993)
20. Wong, W.E.: On Mutation and Data Flow. PhD thesis, Purdue Univ. (1993)
21. Hussain, S.: Mutation Clustering. Master's thesis, King's College London (2008)
22. Ji, C., Chen, Z., Xu, B., Zhao, Z.: A Novel Method of Mutation Clustering Based on Domain Analysis. In: Proc. 21st Int'l Conf. Software Eng. and Knowledge Eng. (2009)
23. Agrawal, H., DeMillo, R.A., Hathaway, B., Hsu, W., Krauser, E.W., Martin, R.J., Mathur, A.P., Spafford, E.: Design of Mutant Operators for the C Programming Language. Technical Report SERC-TR-41-P, Purdue Univ. (1989)
24. Mathur, A.P.: Performance, Effectiveness, and Reliability Issues in Software Testing. In: Proc. Fifth Int'l Computer Software and Applications Conf., pp. 604–605 (1991)
25. Offutt, A.J., Rothermel, G., Zapf, C.: An Experimental Evaluation of Selective Mutation. In: Proc. 15th Int'l Conf. Software Eng., pp. 100–107 (1993)
26. Wong, W.E., Mathur, A.P.: Reducing the Cost of Mutation Testing: An Empirical Study. J. Systems and Software 31(3), 185–196 (1995)
27. Offutt, A.J., Lee, A., Rothermel, G., Untch, R.H., Zapf, C.: An Experimental Determination of Sufficient Mutant Operators. ACM Trans. Soft. Eng. and Methodology (1996)
28. Mresa, E.S., Bottaci, L.: Efficiency of Mutation Operators and Selective Mutation Strategies: An Empirical Study. Software Testing, Verification, and Reliability 9(4), 205–232 (1999)
29. Barbosa, E.F., Maldonado, J.C., Vincenzi, A.M.R.: Toward the Determination of Sufficient Mutant Operators for C. Software Testing, Verification, and Reliability 11(2), 113–136 (2001)
30. Namin, A.S., Andrews, J.H.: Finding Sufficient Mutation Operators via Variable Reduction. In: Proc. Second Workshop Mutation Analysis, p. 5 (2006)
31. Namin, A.S., Andrews, J.H.: On Sufficiency of Mutants. In: Proc. 29th Int'l Conf. Software Eng., pp. 73–74 (2007)
32. Namin, A.S., Andrews, J.H., Murdoch, D.J.: Sufficient Mutation Operators for Measuring Test Effectiveness. In: Proc. 30th Int'l Conf. Software Eng., pp. 351–360 (2008)
33. Polo, M., Piattini, M., Garcia-Rodriguez, I.: Decreasing the Cost of Mutation Testing with Second-Order Mutants. Software Testing, Verification, and Reliability 19(2), 111–131 (2008)
34. Purushothaman, R., Perry, D.E.: Toward Understanding the Rhetoric of small source code changes. IEEE Transactions on Software Engineering 31(6) (2005)
35. Langdon, W.B., Harman, M., Jia, Y.: Multi Objective Higher Order Mutation Testing with Genetic Programming. In: Proc. Fourth Testing: Academic and Industrial Conf. Practice and Research (2009)
36. Howden, W.E.: Weak Mutation Testing and Completeness of Test Sets. IEEE Trans. Soft. Eng. 8(4), 371–379 (1982)

37. Girgis, M.R., Woodward, M.R.: An Integrated System for Program Testing Using WeakMutation and Data Flow Analysis. In: Proc. Eighth Int'l Conf. Software Eng. (1985)
38. Horgan, J.R., Mathur, A.P.: Weak Mutation is Probably Strong Mutation. Technical Report SERC-TR-83-P, Purdue Univ. (1990)
39. Woodward, M.R.: Mutation Testing-An Evolving Technique. In: Proc. IEE Colloquium on Software Testing for Critical Systems, pp. 3/1–3/6 (1990)
40. Marick, B.: The Weak Mutation Hypothesis. In: Proc. Fourth Symp. Software Testing, Analysis and Verification, pp. 190–199 (1991)
41. Offutt, A.J., Lee, S.D.: How Strong is Weak Mutation. In: Proc. Fourth Symp. Software Testing, Analysis and Verification, pp. 200–213 (1991)
42. Offutt, A.J., Lee, S.D.: An Empirical Evaluation of Weak Mutation. IEEE Trans. Software Eng. 20(5), 337–344 (1994)
43. Woodward, M.R., Halewood, K.: From Weak to Strong Dead or Alive? An Analysis of Some Mutationtesting Issues. In: Proc. Second Workshop on Software Testing, Verification, and Analysis, pp. 152–158 (1988)
44. Jackson, D., Woodward, M.R.: Parallel Firm Mutation of Java Programs. In: Proc. First Workshop on Mutation Analysis, pp. 55–61 (2000)
45. Offutt, A.J., King, K.N.: A Fortran 77 Interpreter for Mutation Analysis. ACM SIGPLAN Notices 22(7), 177–188 (1987)
46. Choi, B., Mathur, A.P.: High-Performance Mutation Testing. J. Systems and Software 20(2), 135–152 (1993)
47. Delamaro, M.E.: Proteum-A Mutation Analysis Based Testing Environment. Master's thesis, Univ. of Sao Paulo (1993)
48. Delamaro, M.E., Maldonado, J.C.: Proteum: A Tool for the Assessment of Test Adequacy for C Programs. In: Proc. Conf. Performability in Computing Systems (1996)
49. DeMillo, R.A., Krauser, E.W., Mathur, A.P.: Compiler-Integrated Program Mutation. In: Proc. Fifth Ann. Computer Software and Applications Conf., pp. 351–356 (1991)
50. Krauser, E.W.: Compiler-Integrated Software Testing. PhD thesis, Purdue Univ. (1991)
51. Untch, R.H.: Mutation-Based Software Testing Using Program Schemata. In: Proc. 30th Ann. Southeast Regional Conf., pp. 285–291 (1992)
52. Mathur, A.P.: CS 406 Software Engineering I. Course Project Handout (1992)
53. Untch, R.H., Offutt, A.J., Harrold, M.J.: Mutation Analysis Using Mutant Schemata. In: Proc. Int'l Symp. Software Testing and Analysis, pp. 139–148 (1993)
54. Untch, R.H.: Schema-Based Mutation Analysis: A New Test Data Adequacy Assessment Method. PhD thesis, Clemson Univ. (1995)
55. Offutt, A.J., Ma, Y.S., Kwon, Y.R.: An Experimental Mutation System for Java. ACM SIGSOFT Software Eng. Notes 29(5), 1–4 (2004)
56. Ma, Y.S., Offutt, A.J., Kwon, Y.R.: MuJava: An Automated Class Mutation System. Software Testing, Verification, and Reliability 15(2), 97–133 (2005)
57. Ma, Y.S., Offutt, A.J., Kwon, Y.R.: MuJava: A Mutation System for Java. In: Proc. 28th Int'l Conf. Software Eng., pp. 827–830 (2006)
58. Schuler, D., Dallmeier, V., Zeller, A.: Efficient Mutation Testing by Checking Invariant Violations. In: Proc. Int'l Symp. Software Testing and Analysis (2009)
59. Bogacki, B., Walter, B.: Evaluation of Test Code Quality with Aspect-Oriented Mutations. In: Abrahamsson, P., Marchesi, M., Succi, G. (eds.) XP 2006. LNCS, vol. 4044, pp. 202–204. Springer, Heidelberg (2006)

60. Bogacki, B., Walter, B.: Aspect-Oriented Response Injection: An Alternative to Classical Mutation Testing. In: Sacha, K. (ed.) Soft. Eng. Techniques: Design for Quality. IFIP, pp. 273–282. Springer, Boston (2007)

61. Mathur, A.P., Krauser, E.W.: Mutant Unification for Improved Vectorization. Technical Report SERC-TR-14-P, Purdue Univ. (1988)

62. Krauser, E.W., Mathur, A.P., Rego, V.J.: High Performance Software Testing on SIMD Machines. In: Proc. Second Workshop on Software Testing, Verification and Analysis (1988)

63. Krauser, E.W., Mathur, A.P., Rego, V.J.: High Performance Software Testing on SIMD Machines. IEEE Trans. Software Eng. 17(5), 403–423 (1991)

64. Offutt, A.J., Pargas, R.P., Fichter, S.V., Khambekar, P.K.: Mutation Testing of Software Using a MIMD Computer. In: Proc. Int'l Conf. Parallel Processing, pp. 255–266 (1992)

65. Zapf, C.N.: A Distributed Interpreter for the Mothra Mutation Testing System. Master's thesis, Clemson Univ. (1993)

66. Weiss, S.N., Fleyshgakker, V.N.: Improved Serial Algorithms for Mutation Analysis. ACM SIGSOFT Software Eng. Notes 18(3), 149–158 (1993)

67. Fleyshgakker, V.N., Weiss, S.N.: Efficient Mutation Analysis: A New Approach. In: Proc. Int'l Symp. Software Testing and Analysis, pp. 185–195 (1994)

68. Madeyski, L., Orzeszyna, W., Torkar, R., Józala, M.: Overcoming the Equivalent Mutant Problem: A Systematic Literature Review and a Comparative Experiment of Second Order Mutation. IEEE Transactions on Software Engineering (2013), http://dx.doi.org/10.1109/TSE.2013.44 (accepted)

69. Baldwin, D., Sayward, F.G.: Heuristics for determin-ing equivalence of program mutations. Yale University, New Haven, Connecticut, Tech. Report 276 (1979)

70. Offutt, A.J., Craft, W.M.: Using compiler optimization techniques to detect equivalent mutants. Software Testing, Verification and Reliability 4(3), 131–154 (1994)

71. Offutt, A.J., Pan, J.: Detecting equivalent mutants and the feasible path problem. In: Proc. Eleventh Annual Conf. 'Systems Integrity Computer Assurance COMPASS 1996 Software Safety. Process Security', pp. 224–236 (1996)

72. Hierons, R., Harman, M., Danicic, S.: Using program slicing to assist in the detection of equivalent mutants. Software Testing, Verification and Reliability (1999)

73. Ellims, M., Ince, D., Petre, M.: The Csaw C mutation tool: Initial results. In: Proceedings of the Testing: Academic and Industrial Conference Practice and Research Techniques-MUTATION, pp. 185–192. IEEE Computer Society Press, Washington, DC (2007)

74. Martin, E., Xie, T.: A fault model and mutation testing of access control policies. In: Proceedings of the 16th International Conference on World Wide Web. WWW 2007, pp. 667–676. ACM Press, New York (2007)

75. DuBousquet, L., Delaunay, M.: Towards mutation analysis for Lustre programs. Electronic Notes in Theoretical Computer Science 203(4), 35–48 (2008)

76. Offutt, A.J.: Investigations of the software testing coupling effect. ACM Transactions on Software Engineering Methodology 1, 5–20 (1992)

77. Mresa, E.S., Bottaci, L.: Efficiency of mutation operators and selective mutation strategies: An empirical study. Software Testing, Verification and Reliability (1999)

78. Harman, M., Hierons, R., Danicic, S.: The relationship between program dependence and mutation analysis. In: Wong, W.E. (ed.) Mutation Testing for the New Century, pp. 5–13. Kluwer Academic Publishers, Norwell (2001)

79. Adamopoulos, K., Harman, M., Hierons, R.M.: How to overcome the equivalent mutant problem and achieve tailored selective mutation using co-evolution. In: Deb, K., Tari, Z. (eds.) GECCO 2004. LNCS, vol. 3103, pp. 1338–1349. Springer, Heidelberg (2004)

80. Offutt, A.J., Ma, Y.S., Kwon, Y.R.: The class-level mutants of MuJava. In: Proceedings of the 2006 International Workshop on Automation of Software Test - AST 2006. AST 2006, pp. 78–84. ACM Press, New York (2006)

81. Kaminski, G., Ammann, P.: Using a fault hierarchy to improve the efficiency of DNF logic mutation testing. In: Proc. Int. Conf. Software Testing Verification and Validation, ICST 2009, pp. 386–395 (2009)

82. Ji, C., Chen, Z., Xu, B., Wang, Z.: A new mutation analysis method for testing Java exception handling. In: Proc. 33rd Annual IEEE Int. Computer Software and Applications Conf., COMPSAC 2009, vol. 2, pp. 556–561 (2009)

83. Kintis, M., Papadakis, M., Malevris, N.: Evaluating mutation testing alternatives: A collateral experiment. In: Proc. 17th Asia Pacific Soft. Eng. Conf. (APSEC) (2010)

84. Papadakis, M., Malevris, N.: An empirical evaluation of the first and second order mutation testing strategies. In: Proceedings of the 2010 Third International Conference on Software Testing, Verification, and Validation Workshops. ICSTW 2010, pp. 90–99. IEEE Computer Society (2010)

85. Vincenzi, A.M.R., Nakagawa, E.Y., Maldonado, J.C., Delamaro, M.E., Romero, R.A.F.: Bayesian-learning based guide-lines to determine equivalent mutants. International Journal of Soft. Eng. and Knowledge Engineering 12(6), 675–690 (2002)

86. Grün, B.J.M., Schuler, D., Zeller, A.: The impact of equivalent mutants. In: Proceedings of the IEEE International Conference on Software Testing, Verification, and Validation Workshops, pp. 192–199. IEEE Computer Society, Denver (2009)

87. Schuler, D., Dallmeier, V., Zeller, A.: Efficient mutation testing by checking invariant violations. In: Proceedings of the Eighteenth International Symposium on Software Testing and Analysis. ISSTA 2009. ACM Press, New York (2009)

88. Schuler, D., Zeller, A.: (Un-)covering equivalent mutants. In: Proceedings of the 3rd International Conference on Software Testing Verification and Validation (ICST 2010), Paris, France, pp. 45–54 (2010)

89. Schuler, D., Zeller, A.: Covering and uncovering equivalent mutants. Software Testing, Verification and Reliability 23(5), 353–374 (2012)

90. Offutt, A.J., Pan, J.: Automatically Detecting Equivalent Mutants and Infeasible Paths. Software Testing, Verification and Reliability 7(3), 165–192 (1997)

91. Clark, J.A., Dan, H., Hierons, R.M.: Semantic Mutation Testing. In: Third International Conf. on Software Testing, Verification and Validation Workshops (2010)

92. Dan, H., Hierons, R.M.: Semantic Mutation Analysis of Floating-point Comparison. In: IEEE Fifth International Conference on Software Testing, Verification and Validation (2012)

93. Boonyakulsrirung, P., Suwannasare, T.: A Weak Mutation Testing Framework for WS-BPEL. In: Eighth International Joint Conf. on Computer Science and Soft. Engineering (2011)

94. Durelli, V.H.S., Offutt, A.J., Delamaro, M.E.: Toward Harnessing High-level Language Virtual Machines for Further Speeding up Weak Mutation Testing. In: IEEE Fifth International Conference on Software Testing, Verification and Validation (2012)

95. Mateo, P.R., Usaola, M.P.: Mutant Execution Cost Reduction Through MUSIC (MUtant Schema Improved with extra Code). In: IEEE Fifth International Conference on Software Testing, Verification and Validation (2012)

96. Lisherness, P., Lesperance, N., Cheng, K.T.: Mutation Analysis with Coverage Discounting. In: Design, Automation and Test in Europe Conference and Exhibition (2013)

97. Polo, M., Piattini, M., Garcia-Rodriguez, I.: Decreasing the Cost of Mutation Testing with Second-Order Mutants. Software Testing, Verification, and Reliability 19(2), 111–131 (2008)
98. Madeyski, L., Radyk, N.: Judy - a mutation testing tool for Java. IET Software 4(1), 32–42 (2010),
    http://madeyski.e-informatyka.pl/download/Madeyski10b.pdf
99. Madeyski, L.: On the effects of pair programming on thoroughness and fault-finding effectiveness of unit tests. In: Münch, J., Abrahamsson, P. (eds.) PROFES 2007. LNCS, vol. 4589, pp. 207–221. Springer, Heidelberg (2007),
    http://madeyski.e-informatyka.pl/download/Madeyski07.pdf
100. Madeyski, L.: Impact of pair programming on thoroughness and fault detection effectiveness of unit test suites. Software Process: Improvement and Practice 13(3), 281–295 (2008),
    http://madeyski.e-informatyka.pl/download/Madeyski08.pdf
101. Madeyski, L.: The impact of test-first programming on branch coverage and mutation score indicator of unit tests: An experiment. Information and Software Technology 52, 169–184 (2010),
    http://madeyski.e-informatyka.pl/download/Madeyski10c.pdf

Realization of a Test System Framework

Tamás Krejczinger, Binh Thai Vu, and Tien Van Do

Budapest University of Technology and Economics,
Budapest, Hungary

Abstract. This paper proposes an architecture for a test system framework where the execution of the logic of test cases is decoupled from the communications between a system under a test and the proposed test framework. A proof of concept is demonstrated through an example.

1 Introduction

Telecommunication equipments and software products often go through intensive conformance and performance testing processes to minimize a faulty operation and bugs. Test processes often include the following activities: requirement specification [3], preparation and planning of test cases, preparation of a test environment, carrying out test cases, inspection of captured data. These activities can be specified with various tools and languages [1] where an approach to automate these processes can play a significant role in a cost reduction for vendors and operators.

The most widely used technology in testing is TTCN-3 (Testing and Test Control Nonation), which was standardized by the European Telecommunications Standards Institute (ETSI) [2,7]. TTCN-3 specifies the language itself, and the interfaces of the test execution system. There are a number of open source tools and commercial tools to execute test cases written in TTCN-3 (see [6]).

This paper presents a new approach for realizing a test system framework using advanced software technologies. In our approach, abstract process graphs are used to define the logic of test cases and embedded code snippets are applied to express system specific behaviors.

In order to demonstrate the proof of concept, emerging software technologies are exploited. To represent test cases, a business process management system (jBPM) is used. We follow the principles of the system oriented architecture, and apply an enterprise service bus solution to handle interactions. A prototype of a lightweight testing framework is implemented with the use of several tools. NetPDL language is selected for a protocol description, while the NetBee library is selected to generate coding and decoding entities.

The rest of the paper is organized as follows. In Section 2 we describe the architecture of a proposed test framework. In Section 3 we present the implementation details and demonstrate the operation of the realized prototype test system. Finally, the paper is concluded in Section 4.

T.V. Do et al. (eds.), *Advanced Computational Methods for Knowledge Engineering*,   173
Advances in Intelligent Systems and Computing 282,
DOI: 10.1007/978-3-319-06569-4_13, © Springer International Publishing Switzerland 2014

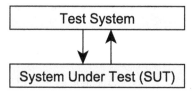

Fig. 1. General test configuration

2   A Proposed Test Framework

A general test configuration is depicted in Figure 1, where test cases can be planned and executed with the help of a test system. A test case describes the dynamic behavior of an intended test and the behavior of a System Under Test (SUT) when a test case is executed. Responses that are triggered by the execution of test cases from a System Under Test (SUT) are evaluated to to check whether the SUT satisfies specified requirements.

2.1   Functionality

We present required functionality in the planning and execution phases in Figure 2. In the planning phase the specification, the requirements (e.g., the conformance to specific standards and dynamic behaviors) and a high-level model of a system shall be considered to design abstract test cases. Normally, each ATS contains a number of test cases related to a particular function of the SUT. Each test case contains only the logic (the flow of tasks) in the abstract level. That is, each test case can be represented as a state machine workflow that contains the states and transitions to specify the logic of the state machine.

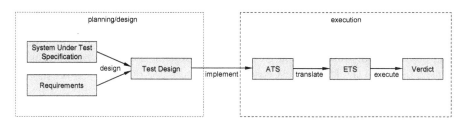

Fig. 2. Steps of the design and execution of tests

To execute the tests, the ATS is translated into an Executable Test Suite (ETS) by the test system with a system specific information that contains the details of interactions with the SUT (see Figure 3). The system specific information determines how information in the high level ATS should be transformed into an information that the SUT can interpret and includes a communication procedure between the ETS and the SUT. It is worth emphasizing that the decoupling of ATS and ETS enhances the reuse of test cases.

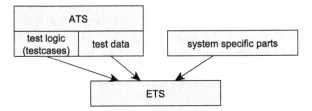

Fig. 3. ATS to ETS translation

2.2 Architecture

Based on the analysis presented in Section 2.1 and the aim of distributed testing capabilities, the following layers are proposed:

1. Test Scenario Logic Layer: This layer represents the modelling, creation, execution and management of test cases. To meet the requirements, stating that test logic should be decoupled from system specific implementations, and to represent the test cases as abstract as possible, we aim to handle the test cases as processes, described by graphs. The nodes of the process graphs should be configurable, which means that the concept of a keyword driven testing should be applied here: decouple the test logic from low level implementations at test case level, and use adaptor codes for specifying state machine nodes' operation.
2. Communication Layer: The main task of Communication Layer is to handle the messaging and interaction between system entities. This layer is also responsible for the test case based orchestration between the functionality provided by system entities: to deliver operations coming from the test cases to the proper underlying service or another test case.

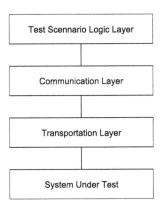

Fig. 4. Layers

3. Transportation Layer: The Transportation Layer is responsible for SUT specific communication and message transformation. This means the entities
composing this layer will be analogue to the TTCN-3 specific Codecs and
Adaptors. The Codecs handle the back and forth transformation of the test
data coming from test cases into the proper form, the SUT is capable handling, while the Adaptors are responsible for the realization of concrete communication with the SUT.

   The introduced layers are divided into several entities, where each entity has
its own role in the given layer (see Figure 5).

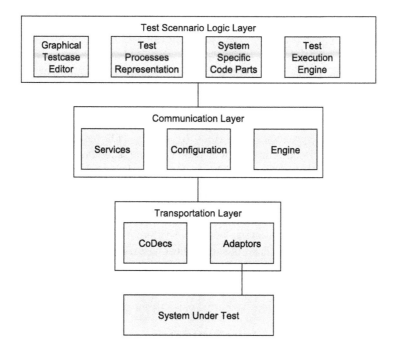

Fig. 5. Detailed layers of the proposed test framework

- Graphical Testcase Editor: Test process abstraction is achieved by using
  graphical testcase representation. This entity facilitates the modelling and
  creation of testcase graphs.
- Test Processes Representation: After graphical modelling, test processes are
  stored in a compact and representative format, which this entity represents.
- System Specific Code Parts: Test logic, and system specific parts are decoupled from each other. This entity represents the test specific actions.
- Test Execution Engine: Responsible for the execution of test processes, and
  controlling other entities of the underlyig layers.

The Communication Layer is responsible for messaging and interactions be-tween other entities, and the orchestration of test processes. The entities of the Communication Layer are:

- Services: The entity consists of the realized test system services. Services encapsulate the test system entities.
- Configuration: Configuration specifies the virtual topology of the services' connections.
- Engine: Represents the distributed execution engine.

Transportation Layer is responsible for the System Under Test (SUT) specific interactions. This layer consists of the language implementations of test system entities that are responsible for transportation tasks. There are two categories of SUT specific components: Codecs and Adapters.

- Codecs: The Codec entities are responsible for the translation between test system data and concrete SUT specific messages. The specific codecs are generated by separate codec generator tools. The tools usually create the source code of the codecs for a given protocol, then these codecs are compiled into executable codes.
- Adaptors: Adaptors handle the concrete communication between the test system and the SUT.

2.3   Choice of Technology

A number of decisions concerning the choice of software technology are taken as follows:

- The logic of test cases is represented using the Business Process Management (BPM) concept. The reason is the modeling and the execution phase of a business process is separated. The process model consists of a directed graph representing the different steps, completed using the system specific details of the operations performed by each step. An internal workflow engine is responsible for the execution of business processes, and the orchestration between underlying services based on the process logic itself.
  JBoss Business Process Management(jBPM) system, one of the most popular open source systems, is chosen. jBPM has detailed user guides for all versions, and a proper API documentation hosted on their web site [4]. The engine is created in Java as a library, and it supports several BPM languages: jPDL, BPML, Pageflow. It can be extended with further languages, but it is not an easy development task. jDPL language is highly customizable due to embeddable actions, events, and the possibility of adding custom nodes. Integration to ESB systems, such as Mule ESB, can be achieved by using Spring framework.
      Test Scenario Logic Layer (TSLL) is implemented by using a Business Process Management System, the jBPM. In jBPM, the modelling, jPDL, and execution of processes are distributed in separate steps.

– Bundle test system entities into services using the Service Oriented Architecture (SOA) concept. SOA is used to encapsulate the components of the underlying infrastructure, hiding their concrete implementation, permit the dynamic creation, and independent execution of their operations in the design of the test framework.
– Use an Enterprise Service Bus (ESB) framework to integrate the entities of the test system Communication and interactions between test logic and system components will be handled by an ESB subsystem, which will compose the Communication Layer.

The selection of the proper ESB solution was inspired by a comprehensive comparison of open source ESBs, presented in the book Open Source ESBs in Action [5]. Mule ESB is chosen because it supports the most complete ESB core functionalities and provides high level of flexibility through configuration capabilities. Integration capability comes from the great number of supported transport protocols, and the support for Spring framework enables integration with other solutions.

3   Illustration of Prototype Implementation

In this section the implementation details of the test system entities is presented through an example taken from [7].

The test case contains several steps to test a Domain Name System (DNS) in Internet:

1. first it sets up a DNS query, and sends it out to a DNS server (the SUT),
2. then waits for the response as a DNS answer,
3. it examines the content of the answer and checks if the queried domain name is resolved correctly.

3.1   Implementation

Figure 6 shows the process graph of the DNS test case. Proto_Start is the start node of the process, where the initialization of the DNS query is done. The task of Send_Message node is to send out the DNS query, invoking the necessary mechanisms in the underlying infrastructure of the test system. Receive_Message handles the decapsulation of the received DNS answer. The Expected? node is responsible for the decision, its task is to examine the received DNS answer, and check if the resolved IP address is correct. Depending on the result, it directs the execution of the test process into one of the Correct or Incorrect nodes. The Correct and Incorrect nodes only display a message for the user about the test result. In a more detailed test scenario, these states could be used to determine the verdict.

System Specific Code Parts. The execution of a node in jBPM consists of different stages: enter node, execute, and leave node. During the execution of a

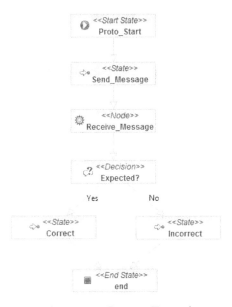

Fig. 6. Test Process Example

test process, the jBPM engine signals these stages using appropriate events. In the jPDL definition, the user-written actions can be assigned to these events.

To create any action, the `ActionHandler` interface of jBPM has to be implemented. The `execute()` method is associated to this interface. Developers can write the system specific parts of their system by implementing the `execute()` method of their actions. The action handler can take parameters from the jPDL process. The value of parameters are directly passed on to the member variables in the implementing class. In jPDL, parameters are specified as sub-elements of the action.

The evaluation can be implemented by the process developer as a decision handler. When the evaluation stage is reached, the `decide()` method of decision handler is called. This method returns the label of the transition.

To permit the proper execution of DNS Tester test case, the following action handlers are to be implemented:

- `MessageActionHandler`: Prints the received text to the standard output for debug purposes.
- `ProcessIncoming`: Processes the received parameter, and stores into a context variable of the test process. Used in the initialization stage of the process, which happens in Proto_Start.
- `org.mule.transport.bpm.jbpm.actions.SendMuleEvent`: This action handler is implemented by Mule ESB. The task of this handler is to send a message to the specified Mule endpoint.

– `org.mule.transport.bpm.jbpm.actions.StoreIncomingData`: It waits for a message from Mule ESB, and when received, stores the message into the specified process variable.

`Msg_Dec_Hndlr` is the decision handler implemented for the `"Expected?"` decision node. It decapsulates the answered IP address from the received DNS message, and checks against the decision parameter. Returns "Yes" for match, and "No" otherwise. The `"Expected?"` node handles the transition depending on which label is returned by the decision handler.

Communication Layer

Services. To support the execution of test processes, and to integrate system specific implementations to the test system, we created several services in Mule ESB. All of these services have their own task, and together they compose a communication system, which encapsulates system specific components (implemented in Transportation Layer), and transports the requests sent by the test logic to the appropriate system component.

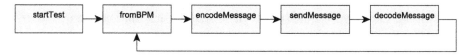

Fig. 7. Test system ESB services

Figure 7 shows the created services, and their relation to each other and to the jBPM process. The most important services are the following:

– `startTestcase`: This service is for initialization purposes.
– `fromBPM`: Handles the events sent to and from BPM processes.
– `encodeMessage`: It is responsible for encoding the messages sent by the test process, where the Encoder class of the TL layer executes the encoding operation. The service contains an instance of that class, as a Mule ESB component.
– `sendMessage`: This service is responsible for handling the communication with the SUT. It involves sending out the encoded message and waiting for the response, which is returned to Mule ESB. Also, encapsulates an instance of Communication_Adapter class.
– `decodeMessage`: Responsible for decoding the messages sent by the SUT. Similarly to the `encodeMessage` service, this service contains an instance of Decode class as a component too.

Configuration. Mule ESB configuration is set up in a separate XML file, which contains every detail, related to the communicating system.

Before defining services, and setting up their configuration, some global entities has to be set:

- Connectors: Connectors specify the mechanisms of connecting external systems to Mule ESB. Several connectors are provided by Mule, whose details can be configured in the configuration XML file. In the test system, a BPM connector is used for connecting to jBPM, and a File connector was applied to handle external file operations.
- Global endpoints: These elements are template endpoints, and their details are specified in concrete service declarations. In our system, a global endpoint was set up for endpoints that are connected to our jBPM test process. The name of the DNS tester process is a Prototype.
- Global transformers: Similar to global endpoints. These elements provide templates for transformers. The only transformer used in our test system is a file-to-string transformer.
- jBPM Spring Bean: Mule ESB and jBPM co-operation can be achieved using Spring framework, where jBPM environment is encapsulated as a spring bean.

In service definitions, the endpoints, transformers, inbound and outbound routers, components, and their configuration have to specified . Without configuring, Mule ESB will create a default service, which simply forwards all messages from its incoming endpoint to its outgoing endpoint.

Endpoints can be created either from global ones, or by full specification. Several properties can be set for an endpoint: its name, reference to the global endpoint from which it was created, URI address to identify it for other services, synchronous/asynchronous operation, connector reference, transformers to apply, and so on. In our approach, File, BPM and VM endpoints are used. BPM is a global endpoint for BPM connectivity. VM endpoint is a Mule ESB default using a virtual memory connector, and performs in-memory message communication. File endpoint uses a file connector, which handles file read and write operations.

Mule ESB provides a number of default transformers and routers. The only transformer used in our test system will be the globally defined FileToString transformer. As the configuration of our services is quite simple (see Figure 7), complex routing is not needed. This way, outbound pass through routers (passes messages to the outbound endpoint without any interaction) and an endpoint selector router (selects outgoing endpoint based on an expression) will be used.

Component is the part of a service, which processes and generates messages. It means, that these elements contain the logic required for service operations. The components of the test system will encapsulate Java classes.

Transportation Layer

Codecs. Figure 8 shows the realization of the Codec entity built on NetPDL tool.

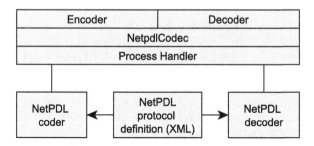

Fig. 8. Realization of Codec entity with NetPDL

NetPDL coder is a PDML-to-binary, NetPDL decoder is a binary-to-PDML message translator. These programs are implemented using the NetBee library, and compiled into binary executable codes. During their operation, they first parse the NetPDL protocol definition, then perform their task on the given binary or PDML input files according to the protocol rules.

`ProcessHandler` is a Java class, and its role is to invoke an executable, pass on command line parameters, handle its execution, and catch the output or error response of it.

`NetpdlCodec` class is responsible for the execution of coding and decoding. This class has 3 publicly accessible methods: `init()`, `encode()`, and `decode()`. The `init()` method is used to set up and configure the actually used encoder or decoder, `encode()` method executes the encoding, and `decode()` method handles the decoding based on the set up configuration.

Encoder and Decoder classes are the wrapper classes of the codec entity for Mule ESB services. Mule ESB invokes them via their implemented `onCall()` method, when they get a message from another service. This message contains the pdml file in case of the `Encoder`, or the binary PDU (Protocol Data Unit) file in case of the Decoder. The role of these classes is to properly configure the NetpdlCodec for the appropriate operation, pass the received files to process, and manage the execution of encoding/decoding respectively. After the execution, these classes return a calculated content for Mule ESB.

Adapters. A simple, straightforward adapter is created as part of our test suite: the `Communication_Adapter` class. This class realizes a simple UDP socket communication. The `onCall()` method is invoked by Mule ESB, it sends out the message which is passed on to the specific socket defined by parameters(ip address:port), and then it waits for the response, and returns the received message.

3.2   Execution

Initialization. Execution of the test scenario starts with the initialization of the test system framework. This process is controlled by the configuration XML files. The jBPM framework, ESB connectors, and ESB components are instantiated based on the configuration.

4   Conclusions

We have proposed the architecture of a new test system framework. We have discussed general principles (decoupling and distributed testing) to design and realize a test system framework. Three layers are proposed and advanced software technology are applied to achieve decoupling and distributed testing capabilities.

References

1. Adamis, G., Horváth, R., Pap, Z., Tarnay, K.: Standardized languages for telecommunication systems. Computer Standards & Interfaces 27(3), 191–205 (2005)
2. ETSI Standard: Methods for Testing and Specification (MTS); The Testing and Test Control Notation version 3; Part 1: TTCN-3 Core Language, ETSI ES 201 873-1 V3.1.1 (2005)
3. Jaskó, S., Dulai, T., Muhi, D., Tarnay, K.: Test aspect of requirement specification. Computer Standards & Interfaces 32(1-2), 1–9 (2010)
4. jBPM, jPDL official web page, http://docs.jboss.org/jbpm/
5. Rademakers, T., Dirksen, J.: Open Source ESBs in Action - Example Implementations in Mule and ServiceMix. Manning Publications Co. (2009)
6. TTCN-3 tools, http://www.ttcn-3.org
7. Willcock, C., Deiβ, T., Tobies, S., Keil, S., Engler, F., Schulz, S.: An Introduction to TTCN-3. Wiley (2005)

Part IV
Computational Methods for Knowledge Engineering

Processing Collective Knowledge from Autonomous Individuals: A Literature Review

Van Du Nguyen and Ngoc Thanh Nguyen

Institute of Informatics, Wroclaw University of Technology,
Wyb. St. Wyspianskiego 27, 50-370 Wroclaw, Poland
{van.du.nguyen,Ngoc-Thanh.Nguyen}@pwr.wroc.pl

Abstract. In recent years, Collective Intelligence is one of the major research subjects in Computer Science. Determining knowledge of a collective is one of the most important issues and has been considered as a subfield of Collective Intelligence. In this paper, we present a literature review on the research progress related to determining collective knowledge from autonomous indidviduals. For this aim, we will analyze some related works on determining representation of a collective, team or group knowledge, and collective intelligence. Finally, some conclusions and future directions are presented.

Keywords: collective knowledge, consensus theory, collective intelligence, knowledge integration.

1   Introduction

Nowadays processing knowledge originating from autonomous and distributed source is a very popular task not only for information retrieval systems like Internet search engines, but also for humans. For some specific subjects people tend to look for experts' opinions or information from the internet through forums. Thus he/she has to deal with some kind of collective knowledge, i.e. the knowledge determined on the basis of knowledge of members of a group (collective). This is useful in providing a proper opinion. However, it also causes a conflict because the opinions may be different from each other [41]. The tools for automatic knowledge integration are still poor and do not satisfy the requirements of their users. In addition, a very important aspect is related to the number of sources which should be enough for a subject and how the number of collective members can affect its knowledge. Owing to them one can know how many experts to ask for a given issue. The number of jury members for a competition or expert commissions would be set in a proper way, with clear rules.

As mentioned above, determining collective knowledge is considered as a subfield of Collective Intelligence. Firstly, we need to know what Collective Intelligence is. According to Newell [30], an intelligent collective is a social system, whether a small group or a large formal organization, that is capable to act, event approximately, as a single, rational agent. Collective Intelligence is considered as an intelligence that emerges from the collaboration and competition of many individuals – an intelligence

T.V. Do et al. (eds.), *Advanced Computational Methods for Knowledge Engineering*,
Advances in Intelligent Systems and Computing 282,
DOI: 10.1007/978-3-319-06569-4_14, © Springer International Publishing Switzerland 2014

that seemingly has a mind of its own [22,50]; it is the collective process by which a large number of people cooperate and act [56]; it is the general ability of a group to perform a wide variety of tasks [65]. In other work, Hand book of Collective Intelligence: Collective Intelligence is groups of individuals doing things collectively that seem intelligent[1]. According to that way, in [57], authors have stated that the term "*Collective Intelligence*" has appeared in recent years but maybe it has existed for a very long time. Families, companies, and countries are examples of groups of individual people doing things that at least sometimes intelligent. The other examples of collective intelligence on the social insects are: beehives, ant colonies, even birds, and shoal of fish.

Moreover, research works on collective intelligence are interdisciplinary. They are related to biology [3], sociological media [53], computer science [23], complexity sciences [52], psychological [66], cognitive studies [58], and etc. In addition, research progress related to collective intelligence have been investigated from purely theoretical [55], simulations [5], case studies [11], conceptual [25], and systems design [61]. Because of interdisciplinary and diverse characteristic of collective intelligence, in this work, we couldn't cover all of works related to collective intelligence. This aspect has been investigated by other authors: *Collective Intelligence on human* [51], *Swarm Intelligence in animals and humans* [19], and others. Instead, we present some basic concepts related to collective intelligence as well as its applications in business. Then, a general view about determining collective knowledge is considered as a subfield of collective intelligence which will be presented. We will give an overview about the research progress related to determining collective knowledge of autonomous individuals. That is the main problem what we investigate in this work and we hope that it is useful for future research on collective knowledge. For this aim, we analyze papers related to collective intelligence, team or group performance, collective knowledge and knowledge integration. These keywords could be considered as popular research directions in recent years.

The remaining part of this paper is organized as follows: section 2 presents an overview of related works to collective intelligence. In section 3, we present the research progress on collective knowledge. In this section, base on their approaches, we classify the related works to two categories as *non-structured* and *well-structured*. Finally, conclusions and future works are presented in section 4.

2   Collective Intelligence

As mentioned above, the needs for collective intelligence processing have existed for a long time for both human and animal environments. One of the previous attempts tended to predict group performance based on some statistic involving members' performances. According to Hackman, the interaction between a task and a team are important [12]. It is significant for improving team performance. Then, group

---

[1] Definition is from
  http://scripts.mit.edu/~cci/HCI/index.php?title=Main_Page

performance is an average of individual performances [1,60]. This statement was confirmed by Neuman and colleagues [29]. Another variable that is beneficial for team performance is collective efficacy [15]; this is the result of exploring the relationships among transactive memory system, collective mind, and collective efficacy and the impact of these variables on team performance by Hue-Chen. Similarly, through simulation Hutchins has confirmed that team knowledge can be greater than external or internal knowledge of team members, and it is the consequence of interactions of factors in the sociotechnical system [16]. In addition, team knowledge is greater than the collection of knowledge of individual team members and it emerges as a result of interactions among team members [18]. Woolley measured the collective intelligence in the performance of human groups [65]. They experimentally proved that collective intelligence is not strongly correlated with the average or maximum individual intelligence of group members but is correlated with the average social sensitivity of group members, the equality in distribution of conversational turn-taking, and the proportion of females in the group. In the work published in Nature [6], Conradt suggested that the accuracy of a decision often increases with the number of decision makers, a phenomenon exploited by betting agents, Internet search engines and stock markets.

Additionally, applications of collective intelligence have been implemented in the real world and brought significant benefits to business. One of the most popular features that applied in business is prediction markets that are markets mainly used for the purpose of making predictions. Some companies (such as: Hewlett Packard, Eli Lilly, ArcelorMittal, Microsoft, Best Buy and etc) have applied prediction markets to get more accurate predictions about future events such as: future sales, base-material supply, store openings and product success [57]. Agora Market is a platform that used prediction markets to improve the efficiency of internal information management and strategic decision-making. This platform has been developed by researchers at Experimental Economics Laboratory in the University Jaume I. Moreover, a nascent project is leading by Mark Klein in the Center for Collective Intelligence to help deal with the problems of global climate change. InnoCentive is the global leader in crowdsourcing innovation problems to the world's smartest people who compete to provide ideas and solutions to important business, social, policy, scientific, and technical challenges.

Furthermore, another intelligence form could be emerged from social insects, fish shoal, birds beehives, ant colonies and mammals [3]. This form has been named by "*Swarm Intelligence*" which is initially introduced by Beni and Wang [2]. Applications inspired from behavior of social insects have helped several companies, including Southwest Airline, Unilever, McGraw-Hill, and Capital One, to schedule factory equipment, divide tasks among workers, organize people, and even plot strategy in more efficient ways [4]. According to Conradt [6], fish also uses "*wisdom of the crowd*" effect. However, no mathematical model has been proposed to prove this hypothesis. In addition, both speed and accuracy of decision making increase with group size in fish shoal under predator threat [64]. In other work [14], authors have identified three key rules for the social interactions of mosquitofish: attraction forces are important in maintaining group cohesion; repulsion is mediated principally by changes in speed; although the positions and directions of all shoal members are highly correlated, individuals only respond to their single nearest neighbor. Through study on collective behavior in animal species, including humans through analyzing captured of 5 socials species – ants, fish, frogs, chickens, and humans [9]. ISOMAP (isometric mapping) – a machine learning method is used to analyze captured video alongside with human observation. The result showed that according to humans and ISOMAP: ants have ascribed the highest degree of interaction, while the least for frogs.

3   Collective Knowledge

Up to now, the idea for processing collective knowledge remains controversial [48]. There exist many conceptions for processing collective knowledge such as: collective knowledge is meant justified true belief or acceptance held or arrived at by groups as plural subjects [49]; the sum of shared contributions among community members [47]; the common state of knowledge of a collective as a whole [44], and etc. In the process of consumption and creation of collective knowledge, individual plays an important role in contributing the wealth of collective knowledge [24] while communities in the knowledge-building process aim at producing new collective knowledge by developing and improving ideas constantly [17]. In this section, we analyze some related works to collective knowledge and classify them in two categories by taking into account the structure of knowledge. They are *non-structured* and *well-structured*. We assume *non-structured* knowledge is knowledge that doesn't exist in any specified structure. In other words, this is knowledge that gains from forum, social network, Wikipedia, and etc. While *well-structured* knowledge aims to mention collective knowledge that could be described by a specific structure. Moreover, we can measure dissimilarity between components of knowledge based on distance functions. Each kind of collective knowledge has its own strengths and the combination of their strength could promise a novel result. Gruber [11] has proposed a class of applications named Collective Knowledge Systems by combining the best idea from the Social Web and Semantic Web. In what follows, we present two categories of collective knowledge which is classified based on knowledge structure.

3.1   Non-structured Knowledge

In this kind, first of all we consider conceptions of authors Gilbert [10], Wray [67], Rolin [49], and Ridder [48] about collective knowledge in scientific activities. Gilbert has argued that scientific communities have collective knowledge. While according to Wray, the necessary condition for groups to have collective knowledge is organic solidarity. In authors' view, research teams are capable of having collective knowledge but scientific communities are not because they are lack of organic solidarity. Then, in a similar research, Rolin has argued that collective knowledge is not limited in research teams. In addition, author also showed that scientific communities have an interest in collective knowledge by introducing a contextualist theory of epistemic justification. Collective knowledge is not merely research team because the theory enables scientific communities to establish a context of epistemic justification. In other work, through some characteristic features of contemporary science survey, Ridder has stated that scientific knowledge is collective knowledge, in a sense to be specified and defended.

In other area - organization theory [13], collective knowledge is considered as: shared knowledge, complementary knowledge, and knowledge embedded in collective artifacts. With the first conception, knowledge usually originates from common experiences and knowledge-sharing activities. It is considered as intersection of individuals' knowledge. While the second concept, collective knowledge is knowledge distributed among individuals interacting in a complementary way. Each individual holds specified knowledge and together collective knowledge is not only sum individuals knowledge and a "*surplus*" – this is knowledge inferred from the sum of individual knowledge. The last one is a strange assumption. Author has assumed collective knowledge which is existed in a tangible form such as documents and databases. This kind of collective knowledge is implicitly embedded in manufacturing technology, prototypes, products, formalized operating rules and organizing principles – instances of artifacts.

Mentioning to collective knowledge, the important issue we couldn't ignore is the role of internet. Indeed, along with rapid development of internet, more and more data gets dramatically produced especially for Internet and on Web [63]. Nowadays, when encountering any problems, people have a habit of searching on internet to find out the solutions. The addresses such as: forum, social networks, and Wikipedia are reliable sources in almost situations. We don't mention in all situations because knowledge that gained from these sources sometimes aren't acceptable by groups as plural subjects [49]. Moreover, there is a lot of knowledge on internet, but knowledge that called collective knowledge is the knowledge that accepted by all or almost people in a collective. Collective knowledge is considered as resulting from social activities in which a converging discussion has been reached [47]. Google, Wikipedia are also considered as interesting examples of knowledge created by collectives of Internet users [44]. In the case of Wikipedia, there are many contributions from a lot of people around the world. However, each article could be called collective knowledge if they reached a certain balance (i.e., it is no longer modified). In other work, author has shown five ways an organization would benefit[2] from collective knowledge and in these ways, the role of social networks is very important. They are used to create, maintain, and provide access to knowledge that has not been available before. Taking advantage of wisdom of crowd with a lot of things posted by experts existing on internet [21], authors have introduced a method for intelligently growing a list of relevant items. Authors used a collective of simple machine learning components to find these experts and aggregate their lists to produce a single complete and meaningful list. Moreover, collective knowledge has also been investigated in emergency response systems. In [62], author has stated that it

---

[2] http://www.stonecobra.com/
five-benefits-of-using-collective-knowledge/

responded more effectively to the situation. In these systems, collective knowledge gains from social networks such as: Twitter, blogs, wikis, and etc. In tagging systems - conceptual knowledge of community [7], authors have shown that people harnessed collective knowledge through navigation tag clouds.

3.2   Well-Structured Knowledge

With this kind of knowledge structure, the problem determining collective knowledge is related to knowledge integration problem. In general, knowledge integration is understood as determining a representation of a set of opinions from different sources for an issue. In this section, we present an overview of the problems - knowledge integration in which consensus choice methodology has been investigated. The general meaning of consensus method is determining a representation of set of opinions that should be maximally near (or similar) to the given opinions. This method has been proved to be useful in solving conflicts and should be also effective for knowledge inconsistency resolution and knowledge integration [36]. In what follows, we consider the knowledge integration problems that have been solved and deeply investigated for logical, relational structures and for ontology. However, first of all, we consider a general model for knowledge integration investigated in [41,42,44]. In these works, a formal mathematical model for knowledge integration was presented. Besides, the consensus-based knowledge functions for generating integration of knowledge have been defined.

By U we denote a set of objects representing the potential elements of knowledge referring to a concrete real world. The elements of U can represent, for example, logic expressions, tuples etc. Symbol 2^U denotes the powerset of U that is the set of all subsets of U. By $\Pi_k(U)$ we denote the set of all k-element subsets (with repetitions) of set U for $k \in N$ (N is the set of natural numbers), and let

$$\Pi(U) = \bigcup_{k=1}^{\infty} \Pi_k(U)$$

Thus $\Pi(U)$ is the set of all non-empty finite subsets with repetitions of set U. A set $X \in \Pi(U)$ can represent the knowledge of a collective where each element $x \in X$ represents knowledge of a collective member. Note that X is a multi-set. We also call X a collective knowledge profile, or a profile in short. Set U can contain elements which are inconsistent with each other. Two elements $x, y \in U$ are inconsistent if they represent two states of knowledge, which cannot take place simultaneously in the real world to which U refers. A set $Z \subseteq U$ is called inconsistent if all the knowledge states represented by its elements cannot take place in the real world to which U refers, simultaneously, and Z is minimal in the sense that any proper subset of Z does not have this property. Set $Z \subseteq U$ is called consistent if any its subset is not inconsistent.

Each X which belongs to $\Pi(U)$ is called a collective. Elements of U have 2 structures. They are: *macrostructure* and *microstructure*. Microstructure is considered as the knowledge structure of elements in set U whereas macrostructure is understood as relationship between elements. In general, macrostructure is assumed as distance functions for measuring difference between elements of a collective. The definition of macrostructure of the set U is given as follows:

Definition 1. *The macrostructure of the set U is some distance functions*

$$d: U \times U \rightarrow [0,1]$$

which is:

- *Nonnegative*, i.e. $\forall x, y \in U: d(x,y) \geq 0$,
- *Reflexive*, i.e. $\forall x, y \in U: d(x,y) = 0$ if $x = y$, and
- *Symmetrical*, i.e. $\forall x, y \in U: d(x,y) = d(y,x)$,

where [0,1] is the closed interval of real numbers between 0 and 1. Pair (U, d) is called a distance space. The definition of a distance function is independent on the structure of elements of U. Some distance functions for specified structures such as: logic expressions [41,43], relational tuples [36], complex trees [26-27]have been investigated. In Definition 1, function d is a half-metric. Metric conditions, including transitivity, in many cases are too strong.

A collective represents the states of knowledge of its members about the same real world. Now we would like to define the knowledge of the collective as a whole. The knowledge of collectives is determined by so-called integration functions defined as follows:

Definition 2. *By an integration function in space (U, d) we call a function*

$$I: \Pi(U) \rightarrow 2^U.$$

For a collective $X \in \Pi(U)$ the value I(X) is called the knowledge state of collective X. Let $Int(U)$ denotes the set of all integration functions in space (U, d). For $X \in \Pi(U)$ and $x \in U$ let

$$d(x, X) = \sum_{y \in X} d(x, y)$$

and

$$d^2(x, X) = \sum_{y \in X} d^2(x, y)$$

Below we define a set of postulates for knowledge functions.

Definition 3. *An integration function I $\in$ **Int**(U) satisfies the postulate of*

- Unanimity (Un):

$$I(\{n \times x\}) = x$$

for each $n \in N$ and $x \in S$.

- Simplification (Si):

(Collective X is a multiple of Collective Y $\Rightarrow I(X) = I(Y)$

- Quasi-unanimity (Qu):

$$x \notin I(X) \Rightarrow \exists n \in N : x \in I(X \uplus \{n \cdot x\})$$

for each $x \in U$ where symbol $\uplus$ denotes the sum operation for multi-sets.

- Consistency (Co):

$$x \in I(X) \Rightarrow x \in I(X \uplus \{x\})$$

for each $x \in U$.

- Proportion (Pr):

$$(X_1 \subseteq X_2) \wedge (x \in I(X_1)) \wedge (y \in I(X_2)) \Rightarrow d(x, X_1) \leq d(y, X_2)$$

for any $X_1, X_2 \in \Pi(U)$.

- 1-Optimality (O_1): If X is inconsistent then

$$x \in I(X) \Leftrightarrow d(x, X) = min_{y \in U} d(y, X)$$

for any $X \in \Pi(U)$.

- 2-Optimality (O_2): If X is inconsistent then

$$x \in I(X) \Leftrightarrow d^2(x, X) = min_{y \in U} d^2(y, X)$$

for any $X \in \Pi(U)$.

- Complexity (Cp): If X is consistent then $X \subseteq I(X)$.

In [36,41], authors have developed a general model for consensus. The general notion of consensus and postulates for consensus choice functions (in this work also called integration function) were defined. Also in these works, authors have proved that postulates O_1 and O_2 play an important role in determining collective knowledge. These postulates are well-known in Consensus Theory [33,36,41,42]. They are very important criteria because of satisfying these postulates; it implies satisfying the majority of other postulates. Concretely, when a knowledge function satisfying criterion O_1, then it also satisfies postulates Un, Si, Qu, Co, and Pr. Whereas a knowledge function satisfying criterion O_2, it also satisfies postulates Un, Si, Qu, and Co. In addition, classes of consensus functions were defined and analyzed referring to these postulates. However, with general model we couldn't propose concrete distance functions as well as algorithms serve for integration process. It is dependent on specified structure of knowledge. Therefore, in the next section we consider some microstructures of knowledge of collective. These structures mostly have been mentioned by Nguyen in [36,41].

Firstly, Danilowicz and Nguyen [8] have proposed a method of determining a representation of ordered partitions. Knowledge structure is elementary. Two metrics are proposed for measuring difference between ordered partitions as well as criteria for determining a representation of given ordered partitions have been presented.

Then, in [36,37,41], authors defined a relational structure for representing the knowledge being integrated. The conditions of integration results have been specified postulates for *closure of knowledge, consistency of knowledge, superiority of knowledge, and maximal similarity*. All postulates have been analyzed and it has been pointed out that all postulates are not consistent in the sense that not for each profile an integration satisfying all postulates exists. Additionally, some integration algorithms have been worked out. The general idea of the integration method is based on determining subprofiles for attributes and next for each subprofile determining its integration. For determining integration for a subprofile it is possible to use different algorithms for different structures of attribute values. For number intervals several algorithms have been worked out in [34-35]; interval numbers with probability structures have been investigated in [45-46].

Nguyen [31-33,41] also considered two logical structures representing knowledge as a conjunction and disjunction. A knowledge state was represented by a standard logic expression in the form of a conjunction or a disjunction of literals [39]. For conjunction and disjunction the distance functions were defined and the consensus problem was formulated. In [41], author has also presented postulates for consensus function and analyzed them. Some heuristic algorithms for consensus determination were developed. Similar works, for a fuzzy structure of knowledge [41], and for a hierarchical structure [26-27] have been investigated.

The knowledge structure that can be treated as background of knowledge-based system is ontology. In general, problem of ontology integration can be formulated as follows: *For given ontologies $O_1, \ldots, O_n$ one should determine one ontology which best represents them* [40-41]. In these works, authors have considered three levels of ontology conflicts: *instance level*, *concept level*, and *relation level*. For each level the integration problem was formulated and algorithms for its solving were proposed. The advantages of the proposed algorithms were based on the fact that they are not complex and can work for different structures of ontologies. They do not require well-valuation of the parameters in ontologies. In addition, in [59], the authors have presented a heuristic method for propagation of matchable concepts and using consensus techniques for conflict resolution for fuzzy ontology integration fuzzy ontology. All worked out methods were analyzed and it was pointed out that all algorithms are not-complex and in case of heuristic methods they gave quite good results.

Another aspect of knowledge integration is whether a conflict profile is susceptible to consensus. In a real application, the determination of consistent knowledge for a collective is often a complicated and a complex task. By analyzing the elements of a profile we would like to know if it is susceptible to the consensus that is if it is possible to get a compromise among collective members if their opinions are not consistent. Let us consider a situation where the profile consists of only two opinions: yes and no. It is obvious that neither yes nor no is a reliable state of knowledge and could not be acceptable as a common opinion of the collective. For checking the susceptibility to a consensus the following definition has been proposed [41]:

Definition 4. *Collective X is susceptible to consensus iff*

$$\hat{\delta}(X) \geq \hat{\delta}_{min}(X)$$

where: $\quad \hat{\delta}(X) = \frac{\sum_{x,y \in X} \delta^i(x,y)}{n(n+1)} \quad , \quad \hat{\delta}_x(X) = \frac{\sum_{y \in X} \delta^i(x,y)}{n} \quad , \quad \hat{\delta}_{min}(X) = \min_{x \in U} \hat{\delta}_x(X) \quad ,$ $n=card(X), i=\{1,2\}.$

The validation of susceptibility to consensus following from Definition 3 is not easy and therefore we need a criterion (or criteria) that enables stating if a chosen consensus is good enough and can be acceptable as a compromise. It's not easy to define criteria for general cases because of the dependence on structures of knowledge state. Therefore, the mentioned problem will be investigated for concrete structures of the knowledge state. Our preliminary researches pointed out that in many cases the sufficient condition is the odd number of the cardinality of a collective [36,38,41].

The problem of analysis could also be investigated by means inconsistency functions if a chosen consensus was good enough and can be acceptable as the solution for a given collective. It has been proved that, in many cases, the dependency between the criteria for consensus susceptibility and the values of inconsistency functions exist [41]. Owing to this kind of functions one can get to know to what degree the opinions given by the collective members are consistent with each other. Also in that work, author has defined postulates representing the intuitive conditions for consistency degree which should be satisfied by knowledge functions.

In summary, the results of determining collective knowledge: collective knowledge is not worse than the worst state among members of collective. A collective is more intelligent than a single [54]. If all elements of collective have identical distances to real state then, collective knowledge is better than all the members. The larger inconsistency degree, the smaller is the distance between collective knowledge and the real state. If the number of members of a collective is odd, then opinions of this collective are most susceptible to consensus.

4   Conclusions and Future Works

In this paper, we have presented a literature review on research progress related to collective knowledge that is considered as a subfield of collective intelligence. For this aim, firstly, we have introduced some basic concepts about collective intelligence on human and animal environments. Then, the related works to collective knowledge have been analyzed. We classified them in two categories based on knowledge structure. Because research on collective knowledge is multidisciplinary, we couldn't cover all of related works to all of its fields. However, we hope that this paper draws out a general view about research progress of collective knowledge and it is useful for future research on collective knowledge.

The following problems could be considered as future works: collective knowledge is not normal sum of members' knowledge [44] and it is also a "*surplus*" [13]. It can be more or less than the sum of the individuals' knowledge [20]. For some subjects, the number of autonomous members (experts) need to achieve reliable opinions; the relationship between increasing collective knowledge and number of autonomous individuals. For this aim paraconsistent logics can be useful [28].

References

1. Barrick, M.R., Stewart, G.L., Neubert, M.J., Mount, M.K.: Relating Member Ability and Personality to Work-Team Processes and Team Effectiveness. Journal of Applied Psychology 83(3), 377–391 (1998)
2. Beni, G., Wang, J.: Swarm Intelligence. In: Proceeding of the Seventh Annual Meeting of the Robotics Society of Japan, Tokyo, Japan, pp. 425–428 (1988)
3. Bonabeau, E., Theraulaz, G., Deneubourgc, J.L., Aronb, S., Camazined, S.: Self-organization in social insects. Trends in Ecology & Evolution 12(5), 188–193 (1997)
4. Bonabeau, E., Meyer, C.: Swarm Intelligence: A Whole New Way to Think About Business. Harvard Business Review 79(5), 106–114 (2001)
5. Bosse, T., Jonker, C.M., Schut, M.C., Treur, J.: Collective Representational Content for Shared Extended Mind. Cognitive Systems Research 7, 151–174 (2006)
6. Conradt, L.: When it pays to share decisions. Nature 471, 40–41 (2011)
7. Cress, U., Held, C.: Harnessing collective knowledge inherent in tag clouds. Journal of Computer Assisted Learning 29(3), 235–247 (2013)
8. Daniłowicz, C., Nguyen, N.T.: Consensus-based partition in the space of ordered partitions. Pattern Recognition 21, 269–273 (1988)
9. DeLellis, P., Polverino, G., Ustuner, G., Abaid, N., Macrì, S., Bollt, E.M., Porfiri, M.: Collective behaviour across animal species. Scientific Reports 4, Article number: 3723 (2014)
10. Gilbert, M.: Collective Epistemology. Episteme 1, 95–107 (2004)
11. Gruber, T.: Collective Knowledge Systems: Where the Social Web Meets the Semantic Web. Journal of Web Semantics 6, 4–13 (2008)
12. Hackman, R.J.: The interaction of task design and group performance strategies in determining group effectiveness. Organizational Behavior and Human Performance 16(2), 350–365 (1976)
13. Hecker, A.: Knowledge Beyond the Individual? Making Sense of a Notion of Collective Knowledge in Organization Theory. Organization Studies 33(3), 423–445 (2012)
14. Herbert-Read, J.E., Perna, A., Mann, R.P., Schaerf, T.M., Sumpter, D.J.T., Ward, A.J.W.: Inferring the rules of interaction of shoaling fish. Proceedings of the National Academy of Sciences 108(46), 18726–18731 (2011)
15. Hue-Wen, C., Yu-Hsu, N.L., Shyan-Bin, C.H.: Team Cognition, Collective Efficacy, And Performance In Strategic Decision-Making Teams. Social Behavior and Personality 40(3), 381–394 (2012)
16. Hutchins, E.: The social organization of distributed cognition. In: Resnick, L.B., Levine, J.M., Teasley, S.D. (eds.) Socially Shared Cognition, pp. 283–301. American Psychological Association, Washington, D.C. (1991)
17. Kimmerle, J., Moskaliuk, J., Andreas, H., Ulrike, C.: Visualizing Co-Evolution of Individual and Collective Knowledge. Information, Communication & Society 13(8), 1099–1121 (2010)
18. Klimoski, R., Mohammed, S.: Team mental model: Construct or metaphor? Journal of Management 20, 403–437 (1994)
19. Krause, J., Ruxton, G.D., Krause, S.: Swarm intelligence in animals and humans. Trends in Ecology & Evolution 25(1), 28–34 (2010)
20. Lam, A.: Tacit Knowledge, Organizational Learning and Societal Institutions: An Integrated Framework. Organization Studies 21(3), 487–513 (2000)
21. Letham, B., Rudin, C., Heller, K.A.: Growing a list. Data Mining and Knowledge Discovery 27(3), 372–395 (2013)

22. Levy, P.: Collective Intelligence: Mankind's Emerging World in Cyberspace. Perseus Books, Cambridge (1997)
23. Levy, P.: From Social Computing to Reflexive Collective Intelligence: The IEML Research Program. Information Sciences 180, 71–94 (2010)
24. Littlejohn, A., Margaryan, A., Milligan, C.: Charting Collective Knowledge: Supporting Self-Regulated Learning in the Workplace. In: Ninth IEEE International Conference on Advanced Learning Technologies, ICALT 2009 (2009)
25. Luo, S., Xia, H., Yoshida, T., Wang, Z.: Toward Collective Intelligence of Online Communities: A Primitive Conceptual Model. Journal of Systems Science and Systems Engineering 18(2), 203–221 (2009)
26. Maleszka, M., Nguyen, N.T.: A Method for Complex Tree Integration. Cybernetics and Systems 42(5), 358–378 (2011)
27. Maleszka, M., Mianowska, B., Nguyen, N.T.: A method for collaborative recommendation using knowledge integration tools and hierarchical structure of user profiles. Knowledge-Based Systems 47, 1–13 (2013)
28. Nakamatsu, K., Abe, J.M.: The paraconsistent process order control method. Vietnam Journal of Computer Science 1(1), 29–37 (2014)
29. Neuman, G.A.: The Relationship between Work-Team Personality Composition and the Job Performance of Teams. Group & Organization Management 24(1), 28–45 (1999)
30. Newell, A.: Unified theories of cognition. Harvard University Press, Cambridge (1990)
31. Nguyen, N.T.: Using consensus methods for determining the representation of expert information in distributed systems. In: Cerri, S.A., Dochev, D. (eds.) AIMSA 2000. LNCS (LNAI), vol. 1904, pp. 11–20. Springer, Heidelberg (2000)
32. Nguyen, N.T.: Using consensus methods for solving conflicts of data in distributed systems. In: Jeffery, K., Hlaváč, V., Wiedermann, J. (eds.) SOFSEM 2000. LNCS, vol. 1963, pp. 411–419. Springer, Heidelberg (2000)
33. Nguyen, N.T.: Using consensus for solving conflict situations in fault tolerant distributed systems. In: Proceedings of First IEEE/ACM Symposium on Cluster Computing and the Grid 2001, pp. 379–385. IEEE Computer Press (2001)
34. Nguyen, N.T.: Representation choice methods as the tool for solving uncertainty in distributed temporal database systems with indeterminate valid time. In: Monostori, L., Váncza, J., Ali, M. (eds.) IEA/AIE 2001. LNCS (LNAI), vol. 2070, pp. 445–454. Springer, Heidelberg (2001)
35. Nguyen, N.T.: Consensus-based timestamps in distributed temporal databases. The Computer Journal 44, 398–409 (2001)
36. Nguyen, N.T.: Consensus Choice Methods and their Application to Solving Conflicts in Distributed Systems. Wroclaw University of Technology Press (2002) (in Polish)
37. Nguyen, N.T.: Consensus system for solving conflicts in distributed systems. Journal of Information Sciences 147, 91–122 (2002)
38. Nguyen, N.T.: Criteria for consensus susceptibility in conflicts resolving. In: Inuiguchi, M., Tsumoto, S., Hirano, S. (eds.) Rough Set Theory and Granular Computing. STUDFUZZ, vol. 125, pp. 323–333. Springer, Heidelberg (2003)
39. Nguyen, N.T.: Processing inconsistency of knowledge on semantic level. Journal of Universal Computer Science 11, 285–302 (2005)
40. Nguyen, N.T.: Conflicts of ontologies – classification and consensus-based methods for resolving. In: Gabrys, B., Howlett, R.J., Jain, L.C. (eds.) KES 2006, Part II. LNCS (LNAI), vol. 4252, pp. 267–274. Springer, Heidelberg (2006)
41. Nguyen, N.T.: Advanced Methods for Inconsistent Knowledge Management. Springer, London (2008)

42. Nguyen, N.T.: Inconsistency of knowledge and collective intelligence. Cybernetics and Systems: An International Journal 39(6), 542–562 (2008)
43. Nguyen, N.T.: Rough Classification – New Approach and Applications. Journal of Universal Computer Science 15(13), 2622–2628 (2009)
44. Nguyen, N.T.: Processing inconsistency of knowledge in determining knowledge of collective. Cybernetics and Systems: An International Journal 40(8), 670–688 (2009)
45. Nguyen, V.D., Nguyen, N.T.: A Method for temporal knowledge integration using indeterminate model of time. Cybernetics and Systems 44(2-3), 222–244 (2013)
46. Nguyen, V.D., Nguyen, N.T.: A Method for knowledge integration using indeterminate model of time with criterion O_2, KES-AMSTA. In: KES-AMSTA, pp. 325–334 (2013)
47. Padula, M., Reggiori, A., Capetti, G.: Managing Collective Knowledge in the Web 3.0. In: First International Conference on Evolving Internet, INTERNET 2009 (2009)
48. Ridder, J.: Epistemic dependence and collective scientific knowledge. Synthese, 1–17 (2013)
49. Rolin, K.: Science as collective knowledge. Cognitive Systems Research 9(1-2), 115–124 (2008)
50. Russell, P.: The global brain awakens: Our next evolutionary leap, 2nd edn. Global Brain, Inc., USA (1995)
51. Salminen, J.: Collective Intelligence in Humans: A Literature Review. CoRR abs/1204.3401 (2012)
52. Schut, M.C.: On Model Design for Simulation of Collective Intelligence. Information Sciences 180, 132–155 (2010)
53. Shimazu, H., Koike, S.: KM 2.0: Business Knowledge Sharing in the Web 2.0 age. NEC Technical Journal 2(2), 50–54 (2007)
54. Surowiecki, J.: The wisdom of crowds. Random House, New York (2004)
55. Szuba, T.: A Formal Definition of the Phenomenon of Collective Intelligence and Its IQ Measure. Future Generation Computer Systems 17, 489–500 (2001)
56. Taher, N. (ed.): Collective Intelligence: An Introduction. ICEAI University Press (2005)
57. Tovey, M.: What is collective intelligence and what will we do about it? Creating a Prosperous World at Peace. Earth Intelligence Network, Oakton (2008)
58. Trianni, V., Tuci, E., Passino, K.M., Marshall, J.A.R.: Swarm Cognition: an Interdisciplinary Approach to the Study of Self-organizing Biological Collectives. Swarm Intelligence 5, 3–18 (2011)
59. Truong, H.B., Duong, T.H., Nguyen, N.T.: A Hybrid Method for Fuzzy Ontology Integration. Cybernetics and Systems 44(2-3), 133–154 (2013)
60. Tziner, A., Eden, D.: Effects of Crew Composition on Crew Performance: Does the Whole Equal the Sum of Its Parts? Journal of Applied Psychology 70(1), 85–93 (1985)
61. Vanderhaeghen, D., Fettke, P.: Organizational and Technological Options for Business Process Management from the Perspective of Web 2.0: Results of a Design Oriented Research Approach with Particular Consideration of Self-Organization and Collective Intelligence. Business & Information Systems Engineering 2, 15–28 (2010)
62. Vivacqua, A.S., Borges, M.R.S.: Taking advantage of collective knowledge in emergency response systems. Journal of Network and Computer Applications 35(1), 189–198 (2012)
63. Vossen, G.: Big data as the new enabler in business and other intelligence. Vietnam Journal of Computer Science 1(1), 3–13 (2014)
64. Ward, A.J.W., Herbert-Read, J.E., Sumpter, D.J.T., Krause, J.: Fast and accurate decisions through collective vigilance in fish shoals. Proceedings of the National Academy of Sciences 108(6), 2312–2315 (2011)

65. Woolley, A.W., Chabris, C.F., Pentland, A., Hashmi, N., Malone, W.T.: Evidence for a Collective Intelligence Factor in the Performance of Human Groups. Science 330(6004), 686–688 (2010)
66. Woodley, M.A., Bell, E.: Is Collective Intelligence (mostly) the General Factor of Personality? A Comment on Woolley, Chabris, Pentland, Hashmi and Malone. Intelligence 39, 79–81 (2011)
67. Wray, K.B.: Who has scientific knowledge? Social Epistemology 21(3), 335–345 (2007)

Solving Conflicts in Video Semantic Annotation Using Consensus-Based Social Networking in a Smart TV Environment

Trong Hai Duong[1], Tran Hoang Chau Dao[1], Jason J. Jung[2], and Ngoc Thanh Nguyen[3]

[1] International University- Vietnam National University,
Ho Chi Minh, Vietnam
haiduongtrong@gmail.com, dthchau@hcmiu.edu.vn
[2] Yeungnam University,
Dae-Dong, Gyeungsan, Korea
j2jung@gmail.com
[3] Institute of Informatics,
Wroclaw University of Technology, Poland
Ngoc-Thanh.Nguyen@pwr.edu.pl

Abstract. Smart TV media content can be embedded with available information from the Internet and shared via the Social Web. An increasing number of social videos are now available as well as traditional digital videos such as TV programs and video on demand (VOD). However, it is difficult to find relevant content due to a lack of semantic content. Therefore, there is a great need for multimedia content analysis techniques. In this work, we propose a framework of collaborative semantic video annotation using consensus-based social network. The collaborative video annotation process is organized via social networking. The media content is shared with friends of friend who collaboratively annotate it. We use ontologies to semantically describe the media content and share the media content amongst the users. A consensus choice is applied to conciliate conflicts on annotation information among the participants. According to our experiments, a consensus method is an effective approach that can be used to solve conflicts in a collaborative annotation.

Keywords: Video Annotation, Social Network, Ontology, Consensus, Collaboration.

1 Introduction

Ontologies are developed to provide machine-processable information sources. Ontologies allow users to organize information according to taxonomical concepts, with their own attributes. With ontology, we can semantically describe the media content and facilitate the sharing of media content. Each media object can be considered as an instance of a domain concept belonging to the multimedia ontology. It also offers advanced search functionality that includes searches for similar or related information integrated from different social media websites [23,24,25]. Several multimedia ontologies have been developed: LSCOM [7]; and (VERL) [12].

Generally, automatic semantic video annotation (also referred to as video concept detection [28,21,11], video semantic analysis [29], or high-level feature extraction [27])

T.V. Do et al. (eds.), *Advanced Computational Methods for Knowledge Engineering,*   201
Advances in Intelligent Systems and Computing 282,
DOI: 10.1007/978-3-319-06569-4_15, © Springer International Publishing Switzerland 2014

can be performed using machine learning methods. Representative learning-based video annotation methods are presented as follows. The collaborative semantic video annotation methods focus on a web-based annotation interface. The participants collaborate to create and share video annotations via social media networking. Several approaches are presented as follows: The IBM Efficient Video Annotation (EVA) system [30],a server-based tool for semantic concept annotation of large video and image collections, which allows user access using web browsers; Video scene annotation based on a social web activities system [31,32], which is a collaborative tagging for video annotation with two features.

The aforementioned automatic semantic video annotation systems rely heavily on the quality and availability of large training video collections. The annotation of large collections, however, is a time-consuming and error-prone task, since it is mainly performed by users. Therefore, collaborative video annotation is proposed to gain the benefit from sharers' skills and knowledge. However, the above-mentioned collaborative approaches have not yet mentioned any collaborative criteria to evaluate the collaboration process. They also have not explored the following problems: who should share the annotating object in order to create the most opportunities for collaboration and gain the greatest benefit from social user skills and knowledge? Conflicts among the annotated versions of the object are unavoidable. How well can one integrate the conflicting annotated versions of an object with contributions by participants in order to create the best-annotated version of the object?

In this work, a collaborative semantic video annotation using social networking analysis (SNA) with consensus choice is presented. In particular, ontologies are exploited to semantically describe the media content and facilitate the sharing of media content among heterogeneous users or devices (e.g., smart phone, PDA, and so on). Two kinds of ontologies are used. One is used to describe the visual features of the media object such as color, texture, shape, motion, and position (called visual feature media ontology such as that found in MPEG-7 [10]). The other ontology provides a knowledge-base of a specific domain for annotating video content (called domain ontology). The domain ontology used here is LSCOM [7], used for broadcast news video annotation. Each media object in a specific video can be considered as an instance of a domain concept belonging to the LSCOM. The annotator should choose a relevant concept to describe the media object of interest. The collaborative video annotation process is organized via social network. The video content is shared with a large network of friends of friends who are all annotating it based on social networking analysis. Annotators only collaborate with reliable social users. Reliability is expressed through a trust value and other collaborative criteria [9]. Once the annotator has a set of reliable social users, he/she can benefit from their knowledge or skill to fill an annotation. When an annotator is not sure about the annotation information, he/she asks reliable social users for their knowledge/opinions and uses their trust values to decide whether or not the annotating object is interesting for the user. We posit that reliable users provide pertinent opinions while the whole collaboration process is evaluated using collaborative criteria, however they may also give inadequate or conflicting knowledge. A consensus choice is applied to conciliate conflicts regarding annotation information among the participants.

2   Collaborative Semantic Annotation Using Consensus-Based Social Network

2.1   Problem Definition

Collaborative video annotation is a process done amongst a group of participants contributing annotations to a specific video of interest. It has been proven that the use of collaboration in such systems improves performance. Annotation can be improved via collaboration amongst participants, as since each participant can benefit from the others' skills and knowledge. However, how this collaboration is organized and how it is done in order to satisfy collaboration criteria is pertinent. Effective collaboration must be *inclusive, egalitarian, interactive, coordinative*, and *trustworthy*. *Inclusive* in this instance means that there should be enough participants. *Egalitarian* in this sense means that the participants should be afforded as much collaboration as possible. *Interactive* means that there should be an easy way to establish contact with every other collaborator inside the collaboration group. *Coordinated* in this instance means that the annotation information shared within the group should be easy to access. The *trustworthy* in this sense means that the participants should be allowed to edit annotations. Therefore the collaborative video annotation process has a complex workflow. Fortunately, social networking analysis (SNA) has shown many relationships between the collaboration criteria and SNA measure such things as *Density, Degree Centrality, Closeness Centrality, Between Centrality*, and *Flow Centrality* [20]. If the collaborative relationships in video annotation systems can be represented as a social network, then these SNA measures can be used to organize the collaborative annotation and to evaluate the collaboration criteria. For example, the closeness centrality measure can be used to identify the annotators who are in more advantageous positions to obtain support for annotating. The density and centrality measures are useful in evaluating the

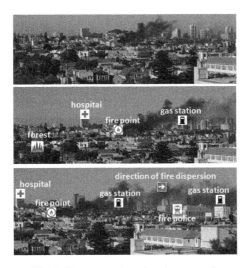

Fig. 1. The annotation situation example

conditions regarding accessing and sharing annotation information among collaborators in a coordinated way.

We consider, as an example of the annotation process, a situation in which an Internet television was live broadcasting a fire in San Francisco on September 27, 2011 at 4:30 pm (see Fig. 1). In order to evacuate the inhabitants of the fire location, a risk map would have to be created to plan the evacuation and guide the rescue and firefighting teams. First, Google Maps would be used to identify the necessary physical objects such as the fire point, nearby hospitals, the nearest firemen, gas stations, and so on. The collaborative annotation of the direction of fire dispersion, the capacity of the hospital, the number of people at the site of the fire, and the number of potentially-affected people are would be essential parts of this task.

The assumption of this work is that the semantic video annotation module uses a domain ontology to semantically describe the media content. Each media object in a specific video can be considered an instance of a domain concept belonging to the ontology. The annotator would choose a relevant concept to describe the media object of interest. For example, an annotator would choose the concept Hospital to describe the hospital nearest to the fire. In addition, the collaborative annotation is organized via social networking. The annotating object would be shared among the heterogeneous users via social networking. Annotators only collaborate with reliable social users. Reliability is expressed through the trust value with which each user labels its relationships. Once the annotator has a set of reliable social users, he/she can benefit from their knowledge or skill in order to fill in the annotations. When an annotator is not sure about the annotation information, that annotator asks his/her social network for knowledge/opinions and uses the trust values of the social network to decide whether or not the annotating object is interesting for the user. In this manner, many social users can participate in the collaboration. We suppose that reliable users provide pertinent opinions, but they may also give inadequate or conflicting knowledge, e.g., some annotators predict that there are around 300 potentially-affected people, but other ones guess different quantities. Therefore, the object would be associated with various and conflicting versions of the annotations created by individuals. The problems we solve in this work are as follows:

– How can annotation information be shared effectively between the users of a social network, and how can the users adequately understand the shared annotation in order to contribute to the collaboration?
– Who should share the annotating object to afford as many opportunities as possible for the collaboration and to benefit as much as possible from the user's social skills and collaboration?
– Conflicts among the annotated versions of the object are unavoidable. How well can the conflict-annotated versions of the object being contributed by participants be integrated in order to obtain the best version of the annotations for the object?

How can annotation information be shared effectively between the users of a social network, and how can the users adequately understand the shared annotation in order to contribute to the collaboration? Ontologies have been developed to provide machine-processable semantic information sources that can be communicated between different agents (software and humans)[26]. The idea of ontologies is to allow users to organize information according to the concept taxonomies, with their own attributes, in order to describe the relationships between the concepts. In this way, a multimedia ontology is exploited to semantically describe the media content and facilitate the sharing of media content among heterogeneous users. Each media object can be considered to be an instance of a domain concept belonging to the multimedia ontology. It also offers advanced search functionality that includes searches for similar or related information integrated from different Social Media websites.

Who should share the annotating object to afford as many opportunities as possible for the collaboration and to benefit as much as possible from the user's social skills and collaboration? To the best of our knowledge, there is no work exploring the afore-mentioned problem. Here, the collaborative video annotation process is organized via social networking. Beyond profiles, friends, comments, and private messaging, social networking sites vary greatly in their features and user base. Some have photo-sharing or video-sharing capabilities; others have built-in blogging and instant messaging tech-nology. There are mobile specific SNSs (e.g., Dodgeball), but some web-based SNSs also support limited mobile interactions (e.g., Facebook, MySpace, and Cyworld). For these social networking advantages, the semantic media content can be shared with a large network of friends of friends who are annotating on it based on their available so-cial profiles. Through this, if a participant does not have enough knowledge/experience to annotate an interesting object in a specific video, he/she will share the object to his/her friends who are also interested in it. Thus, there may be many users participating in the collaborative annotation workflow via social networking. However, the partici-pant does not simply share the object of interest to every friend in his/her network. The shared friends need to satisfy the collaboration criteria. *Egalitarianism*, one of the most important criteria, means that participants need to be afforded as many opportunities for collaboration as possible. To select an egalitarian candidate, we assume that each user in the social network has a profile that describes his/her interests, knowledge, and ex-perience. Therefore, the egalitarian degree of each user is determined by the similarity between his/her profile with the annotating object profile. Another important criterion is *Trust*. How can someone trust a social user's contribution? That is why we do not allow a participant to share the object annotation to everyone in their network. In addition, the degree of trust is determined by the relationship between the annotating participant and his/her friends. If they are more closed, they can better satisfy the *Interactive* criterion.

How well can the conflict-annotated versions of the object being contributed to by participants be integrated in order to obtain the best version of the annotations for the object? There may be many people annotating the same object in a specific video. This means there may be many annotated versions/profiles of the same video being contributed by participants. Even though these participants are trusted via social networking analysis, the conflicts among the annotated versions are unavoidable. Fortunately, one well-known conflict situation is called the conflict profile, and a consensus method is an effective approach that can be used to solve this type of conflict. In a conflict profile, there are different sets of knowledge that explain the same goal or elements in the real world. The consensus aims to determine a reconciled version of knowledge which best represents the given versions. Therefore, the consensus choice can be applied to determine the best annotation of the object from the conflicting participants' points of view.

2.2   Solution Overview

System Architecture. A system architecture of the proposed approach is shown in Fig. 2. There are two kinds of ontologies used to annotate videos. One describes visual feature media objects such as color, texture, shape, motion, and position (i.e. visual feature media ontology such as MPEG-7 [10]). The other one provides a knowledge-base of a specific domain for annotating video content (called a domain ontology). The domain ontology used here is LSCOM [7]. Each media object in a specific video can be considered as an instance of a domain concept belonging to the ontology LSCOM. The annotator needs to choose a relevant concept to describe the media object of interest. In this way, all of the collaborating partners are communicating using the same semantic manner and so it becomes easier for them to understand one another. Using semantics along with ontologies, we can specify, in a structured way, collaboration activities (blog page editions, forum posts, etc.), people and domain knowledge, and arrange them in a meaningful manner. The participant annotates information for the media object; the information is then passed to a matching component in order to generate his/her relevant relationship users who are interested in the object. Annotators only collaborate with reliable social users. Reliability is expressed through the trust value with which each user labels its relationships. Once the annotator has a set of reliable social users, he/she can benefit from their knowledge or skill to fill annotation. When an annotator is not sure about the annotation information, he/she asks his/her reliable social users for their knowledge/opinions and uses their trust values to determine whether or not the annotation object is interesting for the user. Assume that a participant has just finished his/her annotation for the object. This information will be passed to the *Collaboration* component. This component checks the collaboration criteria. If the criteria are satisfied, all of the participants' annotations for the object will be passed to *Consensus Choice* in order to make a consensual version among the participants for the object which is then stored into *Multimedia Ontology*. Otherwise, the annotation information will be matched to user profiles and shared to relevant users who are trusted by the social networking analysis. The information is afterward stored in the user's profile. We suppose that reliable users provide pertinent opinions, but they may also give inadequate knowledge or conflicting opinions. Therefore, an object can be associated with various and conflicting versions of annotation being annotated by the individuals.

The Collaborative Annotation Algorithm. The collaborative annotation algorithm based on consensus choice within social networking is briefly presented in Fig. 3. The algorithm is divided into the following phases:

- Phase 1- Preparatory: The collaboration criteria are given in this phase. The criteria are aimed at guiding and evaluating the collaboration process. In this phase, the media object is also introduced to the participants.
- Phase 2- Contribution: Processes occur in rounds, allowing individuals to contribute or change their opinions of the current version of the annotation information for the object. A participant who is annotating the object can share the object to his/her friends based on the social networking analysis and their profiles. They make a group for collaboration and collaboratively annotate the shared object.

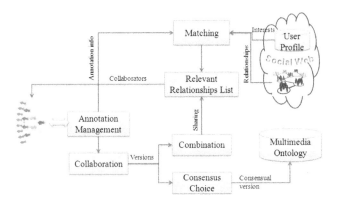

Fig. 2. The collaborative semantic video annotation architecture

– Phase 3- Controlled feedback: If the collaborative criteria are not satisfied, the combination of previous versions will be used as a new version of the object, and the system will show this new version of the group's contribution to each participant in the group. Phase 2 is repeated until the collaborative criteria are satisfied, afterward, the consensus choice is applied to make the final version of the object annotation.

2.3   Consensus Choice for Conflict Resolution

Consensus is a collaborative process allowing an entire group to participate in decision-making in which everyone consents to the decisions of the group. The goal of consensus decision making is to find common ideas and explore these issues until everyone's viewpoint has been recognized and understood by the group. Discussions leading to consensus aim to achieve mutual agreement by addressing all concerns. Consensus does not require unanimity, but people must make a commitment to honest cooperation in order to make final decisions. Consensus is not for individualists or people who want to dominate or coerce others, rather, discussion continues until consensus is achieved. Consensus is not a process for determining whose ideas are best. Rather, it is for searching for the solution that is best for the entire group. Everyone must agree to accept the decision. There are many areas in medicine for which decision-making and practice vary between clinicians. Consensus is used to obtain a majority viewpoint where a range of opinions about specific issues are likely to exist. Two of the methods are the Delphi method and the Nominal Group Technique [38,39]. These have been used in a variety of healthcare settings.

One well-known conflict situation is called a conflict profile, and a consensus method is an effective approach that can be used to solve this conflict. In a conflict profile, there is a set of different versions of knowledge that explains the same goal or elements in the real world. The consensus aims at determining a reconciled version of knowledge which best represents the given versions. In short consensus methods are useful in a process

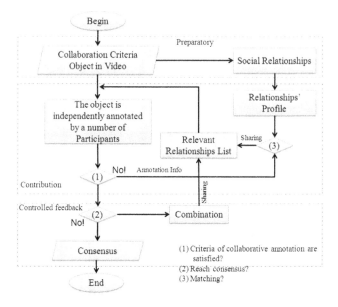

Fig. 3. The collaborative annotation algorithm

of solving conflict or data inconsistency. The scheme of using a consensus method can be presented as follows [17]:

1. Defining the set of potential versions of data
2. Defining the distance function between these versions
3. Selecting a consensus choice criterion
4. Working out an algorithm for consensus choice

Defining the distance functions for potential versions of the data is a very important task and is different for various data structures. For selecting a consensus choice criterion there are two known consensus choice functions. The first is the median function defined by Kemeny [16], which minimizes the sum of the distances between the consensus and given inconsistent versions of the data. The second function minimizes the sum of the squared distances from the consensus to the given elements [18]. As the analysis has shown, the first function gives the closest representative of the given versions whereas the second gives a consensus that is a good compromise of the versions [17]. Let us denote the consensuses chosen by these functions, the O1-consensus and O2-consensus, respectively. The choice of a consensus function should be dependent on the conflict situation. If we assume that the consensus represents an unknown solution of some problem then there are two cases [19]:

  – In the first case the solution is independent on the given versions of the data. Therefore the consensus should be the closest representative to the conflict versions of the data. For this case the criterion for minimizing the sum of distances between the consensus and the conflict versions should be used, that is, the O1-consensus should be determined.

– In the second case the solution is dependent on the given versions of the data. Therefore the consensus should be a compromise which neither harms nor prefers any of the given versions of the data. For this case an O2-consensus should be determined. We will now try to use these consensus functions for collaborative video annotation.

The conflict regarding video annotation can be formulated as follows.

Definition 1. *Let O be $(\mathbf{A}, \mathbf{V})$-based ontology. Let concept (c, A^c, V^c) belong to the ontology O and let the same media object i be annotated by two users u_1 and u_2 based on the concept c in the ontology (the instance i belongs to the concept c), that is $(i, v_1) \in$ Annotation(u_1, c) and $(i, v_2) \in$ Annotation(u_2, c). There is a conflict if $v_1 \neq v_2$.*

For solving conflicts in the aforementioned annotation, consensus methods seem to be very useful. Different criteria, data structures, and algorithms have been determined [Nguyen 2008]. For this kind of conflict, the consensus problem can be defined ny the following:
Given a set of values $X = \{v_1, \ldots, v_n\}$ where v_i is a tuple of type A^c, that is:

$$v_i : A^c \rightarrow V^c \tag{1}$$

for $i = 1, \ldots, n$; $A^c \subseteq \mathbf{A}$ and $V = \bigcup_{a \in A^c} V_a$ we must find the tuple v of type A, such that one or more selected postulates for consensus are satisfied [Nguyen 2008].
    One popular postulate requires minimizing the following sum.

$$\sum_{i=1}^{n} \delta_a(v, v_i) = \min_{v' \in T(A^c)} \sum_{i=1}^{n} \delta_a(v', v_i) \tag{2}$$

where $T(A^c)$ is the set of all tuples of type A^c.
    This postulate is a standard condition for consensus choice and should be useful in reconciling annotation. This criterion is very natural and popular in consensus determining. Its justification is based on the requirement that the annotation of a media object should best represent the given opinions of the participants for the object, thus it should minimally differ from these opinions.
    By using distance functions of type ρ or proportional distance functions of type δ from [19] for a given profile it is possible to determine an integration which satisfies all of the postulates simultaneously; in the general case we would have all of the postulates satisfied except postulate P2. Therefore, it is very important to determine such an integration which satisfies postulates P1 and P6.
    Algorithm 1 presents an algorithm for integration. The idea of this algorithm is based on determining subprofiles for the attributes and then for each subprofile determining its integration.

---

input : Given a set of values $X = \{v_1, \ldots, v_n\}$ where v_i is a tuple of type A^c and
distance functions δ_a for attributes $a \in A^c$.
output: Tuple v^* of of type A^c which is best representation of tuples from X.

1 $A^* = \bigcup A^i$;
2 **foreach** *each* $a \in A^c$ **do**
3 │   Determine a set with repetitions $X_a = \{v_{ia} : v_i \in X \text{ for } i = 1, 2, \ldots, n\}$;
4 **end**
5 **foreach** *each* $a \in A^c$ **do**
6 │   Using distance function δ_a determine a value v_a such that
      $\sum_{i=1}^{n} \delta_a(v_a, v_{ia}) = \min_{v_a' \in V_c} \sum_{i=1}^{n} \delta_a(v_a', v_{ia})$;
7 **end**
8 Create tuple v^* consisting of values v_a for all $a \in A^c$;
9 Return((v^*));

Algorithm 1. An algorithm for reconciling conflicts on annotation

3   Experiments

In the experiment, we evaluate our proposed approach in three respects: technique comparison between traditional collaboration and collaboration organized via Social Networking, collaboration evaluation using Social Networking analysis, and evaluation of consensus choice for solving conflicts on annotation integration.

Data Sets Analysis. We consider three situations in which Internet television is broadcasting regarding a fire at the EVN Twin Towers in Hanoi, Vietnam on December, 15th, 2011 (see *Fig.* 4), the Keangnam Tower in Hanoi, Vietnam on August, 27th, 2011 (see Fig. 5), and a fire in San Francisco on September, 27th, 2011 (see Fig. 1). In order to evacuate the people in the fire, a risk map has to be created in order to plan the evacuation and guide the rescue and firefighting teams.The collaborative annotation of the necessary physical objects such as the nearest hospitals and gas stations, the direction of fire dispersion, the capacity of the hospital, the number of people at the fire, the number of potentially affected people, and so on, are essential pieces of information for this task.

To do so, we created two collaborative groups, A (see Fig. 6) and B (see Fig. 7), for video annotation. Each group includes thirty members. The difference is that the group A had 43 relationships amongst the students whereas group B only had 30 relationships. All members were asked to annotate all of the objects regarding the fires at the aforementioned locations. The members could use any agents, their knowledge, or their own predictions to provide the annotation information.

In both networks, the initial sharer was *Thanh Hien*. Each individual could only share the annotation information to her relationship individuals. They had 50 minutes to complete the assignments. A sample of the annotation information for the fire point-EVN Twin Towers provided by group A is presented in Table 1.

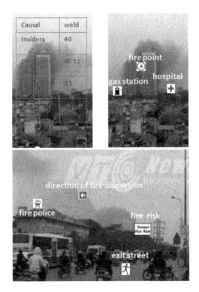

Fig. 4. A fire at EVN Twin Towers in Hanoi

Table 1. The Fire House Annotation Information for EVN Twin Towers

Provider	Causal	Insiders	Dispersion	Floors	Fire at Floor
Thanh Hien	weld	40	W-11	33	-1
Dang Mai	weld	40	W-11	33	1
Thanh Toan	weld	50	W-11	29	5
Le Tinh	weld	30	W-11	33	3
Nguyen Thanh	weld	40	W-11	29	5
Hoai Thuong	weld	30	W-11	33	-1
Duc Huy	weld	25	W-11	33	-1
Tri Dung	weld	40	W-11	33	-1
...	...	...	...	...	...

Evaluation Method. The aforementioned social networking analysis measures were used to analyze how the collaborations is done in the proposed methodology. The annotation results of each collaborative group (called collaborative knowledge) are a consensus among its member's opinions. The real information concerning the fire provided by responsible authorities is considered to be the proper knowledge. To compare the accuracy of annotation results between the two groups, we calculated the similarity between the collaborative knowledge and the proper knowledge.

Fig. 5. A fire at Keangnam Tower in Hanoi

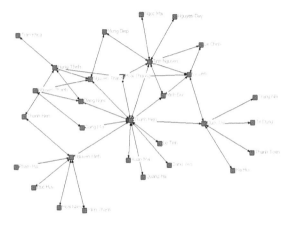

Fig. 6. Collaborative group A

The feature vector of annotation information is provided by user u can be denoted as follows:

$$\vec{F_u} = \{w_1, w_2, w_3, \ldots, w_m\} \tag{3}$$

where, $w_1, w_2, w_3, \ldots, w_m$ are the values of the corresponding attributes of the annotating object $a_1, a_2, a_3, \ldots, a_m$. F_c denotes the collaborative knowledge (consensus) from $u_1, u_2 \ldots, u_n$. F_p denotes the proper knowledge.

The similarity between the annotation information being provided by user u and the proper knowledge can be defined as follows:

$$sim(u, p) = sim(\vec{F_u}, \vec{F_p}) = \frac{\sum_1^5 (F_u^i * F_p^i)}{\sqrt{\sum_1^5 (F_u^i)^2} * \sqrt{\sum_1^5 (F_p^i)^2}} \tag{4}$$

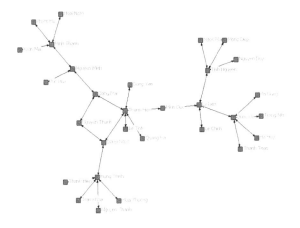

Fig. 7. Collaborative group B

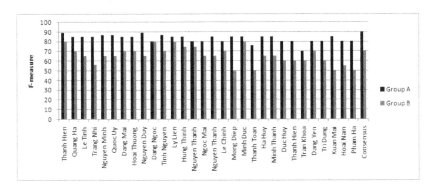

Fig. 8. Consensus among the participants' annotation for the EVN Twin Towers

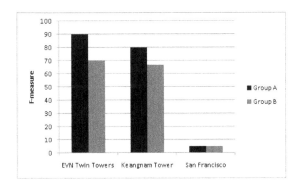

Fig. 9. Consensus among groups for annotation

3.1 The Consensus Choice for Annotation Integration Evaluation

According to the experiment results, we found that there are many people who anno-tate the same object in a specific video. Therefore, many annotated versions/profiles are associated with the same object. Even though these participants are trusted via social networking analysis, conflicts among the annotated versions are unavoidable (see Table 1). This conflict situation is called a conflict profile, and the consensus method [19] is an effective approach that can be used to solve this conflict. The conflict situation among the individuals in Table 1 is considered to be a conflict profile; there is a set of different versions of knowledge that describe the same fire in the EVN Twin Towers. The consensus aims at determining a reconciled version of the knowledge which best represents the given versions. Therefore, consensus choice is applied to determine the best reflect/representative version of annotation for the object from the conflicting par-ticipants' points of view. The consensual result of the annotation is better than most of the individual annotations (see Fig. 8).

The consensual result of group A is more accurate than group B's since the group A satisfies a higher collaborative criteria (*coordination, interaction*) than group B does.

Fig. 9 shows the accuracy of consensual annotation for the fire in San Fransisco is very low in comparison to the two other places. the EVN Twin Towers and Keangnam Tower. This is because the members know the EVN Twin Towers and Keangnam Tower well whereas San Fransisco is not familiar to them. Therefore, the accuracy of the con-sensual result also depends on the collaborative group's knowledge of the annotating object.

Acknowledgment. This research was supported by Vietnam National Foundation for Science and Technology Development (NAFOSTED) under grant number 102.01-2013.12

References

1. Duong, T.H., Jo, G.S., Jung, J.J., Nguyen, N.T.: Complexity Analysis of Ontology Integration Methodologies: A Comparative Study. Journal of Universal Computer Science 15(4), 877–897 (2009)
2. Duong, T.H., Jo, G.S.: Collaborative Ontology Building by Reaching Consensus among Par-ticipants. Information-An International Interdisciplinary Journal 13(5), 1557–1569 (2010)
3. Duong, T.H., Nguyen, N.T., Jo, G.S.: A Hybrid Method for Integrating Multiple Ontologies. Cybernetics and Systems 40(2), 123–145 (2009)
4. Duong, T.H., Nguyen, N.T., Jo, G.S.: Constructing and Mining: A Semantic-Based Academic Social Network. Journal of Intelligent and Fuzzy Systems 21(3), 197–207 (2010)
5. Kim, M.K., Park, J.H.:Demand forecasting and strategies for the successfully development of the smart TV in Korea? In: Proceeding of International Conference on Advanced Com-munication Technology, pp. 1475–1478 (2011)
6. Lo, T.: Overview of smart TV market platforms, business models and solutions, DIGITIMES Research, Taipei (June 20, 2011)
7. Naphade, M., Smith, J.R., Tesic, J., Chang, S.F., Hsu, W., Kennedy, L., Hauptmann, A., Cur-tis, J.: Large-scale concept ontology for multimedia. IEEE Multimedia 13(3), 86–91 (2006)

8. Sikora, T.: The MPEG-7 Visual standard for content description - an overview. IEEE Trans. on Circuits and Systems for Video Technology, special issue on MPEG-7 11(6), 696–702 (2001)

9. Pham, H.: Loglog fault-detection rate and testing coverage software reliability models subject to random environments. Vietnam Journal of Computer Science 1(1), 39–45 (2014)

10. ISO/IEC 15938-5 FDIS Information Technology: MPEG-7 Multimedia Content Description Interface - Part 5: Multimedia Description Schemes (2001)

11. Jeong, J.W., Hong, H.K., Lee, D.H.: Ontology-based Automatic Video Annotation Technique in Smart TV Environment. IEEE Transactions on Consumer Electronics 57(4), 1830–1836 (2011)

12. Francois, A.R.J., Nevatia, R., Hobbs, J., Bolles, R.C., Smith, J.R.: VERL: an ontology framework for representing and annotating video events. IEEE Multimedia 12(4), 76–86 (2005)

13. Auer, S., Bizer, C., Kobilarov, G., Lehmann, J., Cyganiak, R., Ives, Z.G.: DBpedia: A nucleus for a Web of open data. In: Aberer, K., et al. (eds.) ISWC/ASWC 2007. LNCS, vol. 4825, pp. 722–735. Springer, Heidelberg (2007)

14. Bizer, C., Heath, T., Kingsley, I., Berners-Lee, T.: Linked data on the Web. In: Proc. WWW 2008 Workshop: LDOW (2008)

15. Hearst, M., English, J., Sinha, R., Swearingen, K., Yee, P.: Finding the Flow in Web Site Search. Communications of the ACM 45(9), 42–49 (2002)

16. Kemeny, J.G.: Mathematics without Numbers. Daedalus 88(4), 577–591 (1959)

17. Nguyen, N.T.: Using Distance Functions to Solve Representation Choice Problems. Fundamenta Informaticae 48(4), 295–314 (2001)

18. Nguyen, N.T.: Consensus system for solving conflicts in distributed systems. Journal of Information Sciences 147(4), 91–122 (2002)

19. Nguyen, N.T.: Advanced Methods for Inconsistent Knowledge Management. Springer, London (2008)

20. Palau, J., Montaner, M., López, B., de la Rosa, J.L.: Collaboration analysis in recommender systems using social networks. In: Klusch, M., Ossowski, S., Kashyap, V., Unland, R. (eds.) CIA 2004. LNCS (LNAI), vol. 3191, pp. 137–151. Springer, Heidelberg (2004)

21. Park, K.W., Lee, J.H., Moon, Y.S., Park, S.H., Lee, D.H.: OLYVIA: Ontology-based Automatic Video Annotation and Summarization System Using Semantic Inference Rules. In: Proceedings of the Third International Conference on Semantics, Knowledge and Grid (SKG 2007), pp. 170–175. IEEE Computer Society (2007)

22. Renzel, D., Cao, Y., Lottko, M., Klamma, R.: Collaborative Video Annotation for Multimedia Sharing between Experts and Amateurs. In: Proceedings of the 11th International Workshop of the Multimedia Metadata Community, Barcelona, Spain, pp. 7–14 (2010)

23. Wei, S., Zhao, Y., Zhu, Z., Liu, N.: Multimodal fusion for video search reranking. IEEE Transactions on Knowledge and Data Engineering 22(8), 1191–1199 (2010)

24. Hakeem, A., Lee, M.W., Javed, O., Haering, N.: Semantic video search using natural language queries. In: MM 2009: Proceedings of the Seventeen ACM International Conference on Multimedia, pp. 605–608. ACM, New York (2009)

25. Sebastine, S.C., Thuraisingham, B.M., Prabhakaran, B.: Semantic web for content based video retrieval. In: Proceedings of the 3rd IEEE International Conference on Semantic Computing (ICSC 2009), pp. 103–108. IEEE Computer Society, Berkeley (2009)

26. Le, N.T., Nguyen, N.P., Seta, K., Pinkwart, N.: Automatic question generation for supporting argumentation. Vietnam Journal of Computer Science 1(1) (2014)

27. Kraaij, W., Over, P.: TRECVID-2005 high-level feature task: Overview. In: Proc. TRECVID, Gaithersburg, MD (2005)

28. Naphade, M.R., Smith, J.R.: On the detection of semantic concepts at TRECVID. In: Proc. ACM Multimedia, New York, pp. 660–667 (2004)

29. Snoek, C.G., Worring, M., Smeulders, A.W.: Early versus late fusion in semantic video analysis. In: Proc. ACM Multimedia, Singapore, pp. 399–402 (2005)
30. Volkmer, T., Smith, J.R., Natsev, A.: A web-based system for collaborative annotation of large image and video collections: an evaluation and user study. In: Proceedings of the 13th ACM International Conference on Multimedia, Singapore, pp. 892–901 (2005)
31. Yamamoto, D., Masuda, T., Ohira, S., Nagao, K.: Collaborative Video Scene Annotation Based on Tag Cloud. In: Huang, Y.-M.R., Xu, C., Cheng, K.-S., Yang, J.-F.K., Swamy, M.N.S., Li, S., Ding, J.-W. (eds.) PCM 2008. LNCS, vol. 5353, pp. 397–406. Springer, Heidelberg (2008)
32. Yamamoto, D., Masuda, T., Ohira, S., Nagao, K.: Video Scene Annotation Based on Web Social Activities. IEEE MultiMedia 15(3), 22–32 (2008)
33. Dengler, F., Lamparter, S., Hefke, M., Abecker, A.: Collaborative Process Development using Semantic MediaWiki. In: Proceedings of the 5th Conference of Professional Knowledge Management, Solothurn, Switzerland, pp. 97–107 (2009)
34. Gruber, T.R.: Toward principles for the design of ontologies used for knowledge sharing. International Journal of Human Computer Studies 43(5), 907–928 (1995)
35. Barnes, J.: Social Networks. Addison-Wesley, Reading (1972)
36. Scott, J.: Social network analysis. Sage, London (1991)
37. Wasserman, S., Galaskiewicz, J.: Advances in Social Network Analysis. Sage Publications (October 1994)
38. Gallagher, M., Hares, T., Spencer, J., Bradshaw, C., Webb, I.: The nominal group technique: a research tool for general practice? Family Practice 10(1), 76–81 (1993)
39. Pill, J.: The Delphi method: substance, context, a critique and an annotated bibliography. Socioecon. Plann. Sci. 5, 57–71 (1971)
40. Borgatti, S.P., Everett, M.G., Freeman, L.C.: Ucinet for windows: Software for social network analysis. Analytic Technologies, Harvard (2002)
41. Shin, C., Woo, W.: Socially aware tv program recommender for multiple viewers. IEEE Trans. Consumer Electron. 55(2), 927–932 (2009)
42. Matsubara, F.M., Kawamori, M.: Lightweight interactive multimedia environment for TV. IEEE Trans. Consumer Electron. 57(1), 283–287 (2011)
43. Lee, M.H.: The design of a heuristic algorithm for IPTV web page navigation by using remote controller. IEEE Trans. Consumer Electron. 56(3), 1775–1781 (2010)

An Overview of Fuzzy Ontology Integration Methods Based on Consensus Theory[*]

Hai Bang Truong[1] and Xuan Hung Quach[2]

[1] University of Information Technology, VNU-HCM, Ho Chi Minh City, Vietnam
bangth@uit.edu.vn
[2] Institute of Informatics, Wrocław University of Technology, Poland
hung.quach-xuan@pwr.wroc.pl

Abstract. Ontology plays an important role in the organization and management of knowledge in the field of research and various applications. Ontology research has attracted the attention of scientists worldwide. A traditional ontology concept lacks the ability to represent fuzzy information in the field of knowledge uncertainty. It turned out that fuzzy ontology is a good approach for this matter. On the other hand the problem of fuzzy ontology integration is still a problem with many challenges and requires research in both theory and application aspects. This paper presents an overview of selected results of recent research on the methods of resolving conflicts between ontologies in fuzzy ontology integration approaches.

Keywords: Ontology, Fuzzy Ontology, Fuzzy Ontology Integration, Ontology Conflict.

1   Introduction

The integration of heterogeneous information from different data sources will limit ability to share, reuse, and the interaction between distributed systems. Ontology design and ontology integration are the approaches to solving this problem. Recently, there has been much research on ontology management according to the different approaches related to conflict management in the process of knowledge integration, such as ontology matching/alignment [3, 4, 7, 15, 17] ontology merging [1, 6, 15] or ontology mapping [2, 11, 21].

Ontology-based description of traditional logic is not enough to describe fuzzy information and may not fully represent and process uncertain knowledge in different application domains. In 2006, Straccia [16] based on the platform of description logic and fuzzy set theory of Zadeh [22], has represented fuzzy description logic to serve handling uncertain knowledge on the Semantic Web. Since then, which describes fuzzy logic as a basis for knowledge representation and reasoning has been pushed forward. In their studies, Calegari and Ciucci [5] asserted that the method of describ-

---

[*] This research is funded by Vietnam National University Ho Chi Minh City (VNU-HCM) under grant number C2014-26-05.

ing fuzzy logic integrated into the ontology to expand the traditional ontology is more suitable for solving the problem of uncertainty and in which the authors clearly defined the fuzzy description logic, fuzzy ontology, and fuzzy Web Ontology Language (OWL).

Fuzzy ontology integration is important to handle uncertain knowledge on the Semantic Web. However, the existing ontology integration techniques are not sufficient to integrate fuzzy ontologies. Therefore, it is necessary to work out new approaches for fuzzy ontology integration. The results of research on this problem was published in [12], [17] – [20].

Methods for ontology integration can be applied in many practical situations. Particularly, fuzzy ontology enables representing uncertain knowledge, therefore integration aspect of this kind of ontologies should be very important. This paper presents some results of research on fuzzy ontology integration based on consensus theory. This overview can be useful for those who would like use the methods directly to their practical needs.

The rest of this paper is organized as follows. The next section will present the issues related to the methods of reasoning and ontology integration. Section 3 introduces the theory of consensus, section 4 presents the definitions of fuzzy ontology and the methods of resolving conflicts between fuzzy ontologies. Section 5 presents some results of the algorithm settings which were recently published author in [12], [17] - [20]. The last section concludes the paper with information on possible future work in the area.

2   Ontology Integration

The problem of ontology integration had many interested researchers follow different approaches. The recent study has proposed some methods integrate different ontologies: matching, mapping, alignment and merging in the research objectives and specific applications.

Integrated ontology is a process of finding the component similarities between ontologies and create a new ontology to exchange and exploitation of knowledge between systems based on the ontology [20]. To accomplish this, depending on the target application and study, can be done using different tasks of integration can be distinguished such as ontology matching/alignment, ontology mapping, and ontology merging. However, the main task of ontology integration is often finding semantically equivalent elements (objects), the aim of which is to determine an element best representing the given. "Best representation" mentioned here means the following criteria:

(1) All data included in the elements to be integrated should be in the result of integration. This criterion guarantees the completeness that all information included in the component elements will appear in the integrated result.

(2) All conflicts appearing among elements to be integrated should be solved. It often happens that referring to the same subject different elements contain inconsistent information. Such situation is called a conflict. The integrated result should not contain inconsistency, so the conflicts should be solved.

(3) The kind of structure of the integrated result should be the same as of the given elements.

However, the fuzzy ontology contains fuzzy values that are correlations between properties (attribute or relation) and their domain concepts. That means each property belonging to a specific concept is associated with a fuzzy value. Therefore, conflict on fuzzy ontology integration is also expressed by different associated fuzzy values for the same property belonging to the same concept in different ontologies.

3   Introduction to the Theory of Consensus

Problems considered by consensus theory include [12, 20]:

- Problems in which a certain and hidden structure is searched,
- Problems in which inconsistent data related to the same subject are unified.

Basic consensus problem: For a given set (with repetitions) of objects from some universe, it is needed to determine an object which best represents these objects.

Definition 3.1. By a consensus (choice) function in space (U, d) we mean a function
$$C: \Pi(U) \to 2^U$$
For a conflict profile $X \in \Pi(U)$, the set $C(X)$ is called the consensus of X, and an element of $C(X)$ is called a consensus of profile X.

Definition 3.2. A consensus choice function $C \in Con(U)$ satisfies the postulate of:
- *Reliability*

    iff $C(X) \neq \emptyset$,

- *Unanimity*

    iff $C(\{n * x\}) = \{x\}$ $n \in \aleph$ and $x \in U$.

- *Simplification*

    iff (Profile X is a multiple of profile Y) $\Rightarrow C(X) = C(Y)$.

- *Quasi-unanimity*

    iff $(x \notin C(X)) \Rightarrow (\exists n \in \aleph: x \in C(X (n * x)))$ $\forall x \in U$.

- *Consistency*

    iff $(x \in C(X)) \Rightarrow (x \in C(X \{x\}))$ $x \in U$.

- *Condorcet consistency*

    Iff $(C(X_1) \cap C(X_2) \neq \emptyset) \Rightarrow (C(X_1 \cup X_2) = C(X_1) \cap C(X_2))$ $\forall X_1, X_2 \in \Pi(U)$.

- *General consistency*

    iff $C(X_1) \cap C(X_2) \subseteq C(X_1 \cup X_2) \subseteq C(X_1) \cup C(X_2)$, $\forall X_1, X_2 \in \Pi(U_F)$.

- *Proportion*

$$iff\,(X_1 \subseteq X_2 \wedge x \in C(X_1) \wedge y \in C(X_2)) \Rightarrow (d(x,X_1) \le d(y,X_2))$$

- *Optimality* C_1 :

$$iff\,((x \in C(X)) \Rightarrow (d(x, X) = min_{y \in U}\, d(y, X)), for\,any\,X \in \Pi(U)$$

- *Optimality* C_2:

$$iff\,((x \in C(X)) \Rightarrow (d^2(x, X) = min_{y \in U}\, d^2(y, X)), for\,any\,X \in \Pi(U).$$

Optimality C_1: which minimizes the sum of distances between the consensus and the collective members' solutions. Optimality requires the knowledge state to be as near as possible to elements of the collective. The knowledge state generated by a knowledge function satisfying this postulate plays a very important role because it can be understood as the best representative of the collective.

Optimality C_2: which minimizes the sum of squared distances between the consensus and the collective members' solutions. Optimality, also known in the Consensus Theory [12] requires the sum of the squared distances between the knowledge state and the collective elements to be minimal. Notice that the role of the collective knowledge state is not only based on the best representing the knowledge of the collective members, but it should also be "fair," - that is, the distances from the knowledge state of the collective to its elements should be uniform. This requirement is very popular in choice theory and consensus theory [10].

Follow to the consensus theory [8, 13, 14, 25] have proved that postulates C_1 and C_2 play a very important role: If C_1 and C_2 satisfy the remaining postulates satisfied.

4   Fuzzy Ontology Definition and Fuzzy Ontology Integration

4.1   Fuzzy Ontology Definition

We accept the assumption about a real world (A, V) where A is a finite set of attributes and V is the domain of A.

Definition 4.1. (Fuzzy Ontology). A fuzzy (A,V)-based ontology is defined as a 4-tuple: $O=(C, R, I, Z)$, where:

- C is a set of concepts (classes). A fuzzy concept is defined as a 4-tuple: concept $=(c,\, A^c, V^c, f^c)$
  where c is the unique name of the concept, A is a set of attributes describing the concept and V is the attributes domain: where is the domain of the the attribute a and f^c is a fuzzy function: $f^c: A^c \to [0,1]$ representing the degrees to which concept c is described by attributes. Triple (A^c, V^c, f^c) is called the fuzzy structure of concept c.
- R is a set of fuzzy relations between concepts, R $=\{R_1, R_2,\ldots, R_m\}$, $R_i \subseteq C \times C \times (0,1]$, $i = 1,\ldots,m$. A relation is then a set of pairs of concepts with a weight representing the degree to which the relationship should be.
- I is a set of instances.

- Z is a set of axioms, which can be interpreted as integrity constraints or relationship between instances and concepts. It means that Z is a set of restrictions or conditions (necessary & sufficient) to define the concepts in C;

4.2   The Level of Conflict Fuzzy Ontology

Conflicts between fuzzy ontologies may also be considered on the following levels:

- Conflicts on concept level: The same concept has different structures in different ontologies.
- Conflicts on relation level: The relations between the same concepts are different in different ontologies.
- Conflicts on instance level: The same instance has different descriptions in different concepts or ontologies.

Integration on the Concept Level

The same concept has different fuzzy structures in different ontologies. On this level we assume that two ontologies differing from each other in the structures of their concepts. That means these ontologies can contain the same concepts but different structures. The reason of this phenomenon is that these ontologies come from different systems. Therefore, although they refer to the same real world, they can have different structures [12, 20].

Definition 4.2. (Conflict on concept level). For given fuzzy ontology O_1 and O_2. Let concept (c, A^{c_1}, V^{c_1}, f^1) belongs to O_1 and (c, A^{c_2}, V^{c_2}, f^2) belongs to O_2.We say that a conflict takes place on concept level if $A^{c_1} \neq A^{c_2}$ or $V^{c_1} \neq V^{c_2}$ or $f^1 \neq f^2$)._

Problem FOI-1: For given a set of fuzzy structures of the same concept X = $\{(A^i, V^i, f^i)| (A^i, V^i, f^i)$ is the fuzzy structure of concept c in ontology for $i=1,...,n\}$ it is needed to determine a triple c* = $c^* = (A^*, V^*, f^*)$ which best represents the given structures.

Integration on the Relation Revel

The relations for the same concepts are different in different ontologies. In works [13, 14]] the authors have defined conflicts on relation level between non-fuzzy ontologies: Two ontologies are in inconsistency on relation level if referring to a relation a pair of concepts is in this relation in one ontology but in the second ontology it is not. For fuzzy ontologies the definition is similar with taking into account the fuzzy functions of concepts and the weights of relationships between them. Similarly as in [13, 14] we investigate two kinds of ontology inconsistency on relation level. The first kind of inconsistency refers to situations where between the same concepts c and c different ontologies assign different relations. For example, concepts Man and Woman, in one ontology they are in relation Marriage, in another ontology they are in relation Kinship. Notice also that within the same ontology two concepts may be in more than one relation.

Definition 4.3. (Conflict on relation level). Let O_1 and O_2 be (A,V)-based ontology. Let concepts c and c' belong to both ontologies. We say that an inconsistency takes place on relation level if $R_{i1}(c,c') \neq R_{i2}(c,c')$ for some $i \in \{1,...,m\}$.

Problem FOI-2: For given $i \in \{1,...,m\}$ and set $X = \{R_{ij}(c,c'), j = 1,...,n\}$ of relationships between two concepts c and c' in n ontologies it is needed to determine $R_i(c,c')$, of final relationship between c and c' which best represents the given relationships.

Integration on the Instance Level

Definition 4.4. (*Conflict on instance level*). For given fuzzy ontology O_1 and O_2. Let concept $(c, A^{c_1}, V^{c_1}, f^1)$ belongs to O_1 and $(c', A^{c_2}, V^{c_2}, f^2)$ belongs to O_2. Let $(i, v) \in (O_1, c)$ and $(i, v) \in (O_2, c')$. We say that a conflict takes place on instance level if $v(a) \neq v'(a), a \in A^c \cap A^{c'}$

Definition 4.5. (Fuzzy instance). A fuzzy instance of a concept c is described by the attributes from set A^c with values from set 2^{Vx} $(X = A^c)$ and is defined as a pair: instance $= (i, v)$, where:

- i is the unique identifier of the instance in world (A, V)
- v is the value of the instance and is a tuple of type and can be presented as a function: $v: A \to \bar{2}^{A^c}$ such that $v(a) \in 2^{Va}, \forall a \in A^c$

Definition 4.6. (Distance Function). Let $a \in A$ be an attribute, distance function between two values of the same attribute $a \in A$ is defined as follows [13]:
$$d_a : 2^{Va} \times 2^{Va} \to [0, 1]$$
Problem FOI-3: Given a set of instance descriptions $X = \{(i, v_1),..., (i, v_n)\}$, where v_i is a tuple of type $A_i \subseteq A$ that is $v_i: A_i \to v_i, i = 1,..n$ and $v_i = \bigcup_{a \in A_i} V_a$. We need to find a description (i,v) which best agreed set of consensus theory criteria is called integration of the instances described.

5   Conflict Resolution on Fuzzy Integration Using Consensus

5.1   Conflict Resolution on Concept Level

The idea of this algorithm is to determine a best representation for similar concepts with different structure. The redundant and non-consensual attributes are removed. It also recalculates the fuzzy associated values for each remaining attribute based on consensus criterion. The algorithm is briefly presented as follows [12, 20]:

```
Algorithms 1. Determining integration of fuzzy concepts
Input: Given set of fuzzy structures of a concept in n
ontology X = { (A^i, V^i, f^i) | (A^i, V^i, f^i) is the fuzzy structure of
concept c in ontology O_i, i=1,…,n}
Output: c* = (A*, V*, f*) which best represents the elements
from X meets the criteria of the consensus theory.
```

```
Procedure:
  BEGIN
  Set A* =∪ⁿᵢ₌₁Aᵢ;
  Set V* =;∪ⁿᵢ₌₁Vᵢ;
  For each a ∈ A* do
  Begin
    Determine multi-set
      Xₐ ={fⁱ(a): if fⁱ(a) exists and i=1,…,n};
    Calculate f*(a) = ─────── Σᵥ∈ₓₐV;
                      card(Xₐ)
  End.
END.
```

$$Set\ A* =\bigcup_{i=1}^{n} A_i;$$
$$Set\ V* =;\bigcup_{i=1}^{n} V_i;$$
$$X_a =\{f^i(a): if\ f^i(a)\ exists\ and\ i=1,…,n\};$$
$$Calculate\ f*(a) = \frac{1}{card(X_a)}\sum_{v\in X_a} V;$$

Algorithm 1. Algorithm for solving conflict on fuzzy concept

The computation complexity of Algorithm 1 is $O(n^2)$. It is possible to prove that this algorithm determines an integration satisfying - consensus function.

5.2   Conflict Resolution on Relation Level

The inconsistency referring to situations where between the same concepts and different ontologies assign different relations was solved in [12, 20]. The conflict solution mentioned here to solve the problem that the same pair of concepts can belong to the same relation, but with different associated fuzzy value in different ontologies.

```
Algorithm 2. Determining integration of fuzzy relations.
Input:  - Given set X = {Rᵢⱼ ⊆ C × C × (0, 1]: j = 1,..,n}
of relations of the same kind between concepts in ontolo-
gy.
  - These relations are transitive.
Output: Relation Rᵢ ⊆ C × C × (0, 1] which best represents
the elements from X
Procedure:
BEGIN
  Set Rᵢ = ∅;
  For each pair (c,c') ∈ C × C do
  Begin
    Determine multi-set
  X₍c,c'₎ = {v: <c,c', v> ∈ Rᵢⱼ for j = 1,..,n};
    Order set X₍c,c'₎ in increasing order giving
  X = {x₁, x₂, .., xₖ};
    Set interval ⟨X⌈n+1/2⌉, X⌈n+2/2⌉⟩
    Set v as a value belonging to the above defined
  interval;
    Set Rᵢ := Rᵢ ∪ {<c, c', v>}
```

```
End;
For each (c, c', c'') ∈ C × C × C do
Begin
If <c, c', v₁> ∈ Rᵢ, <c, c', v₂> ∈ Rᵢ and <c, c', v₃> ∈ Rᵢ
then
v₃ = min {v₁, v₂};
If only <c, c', v₁> ∈ Rᵢ and <c, c', v₂> ∈ Rᵢ then
Rᵢ := Rᵢ ∪ {<c, c', v>}
where v₃= min{v₁, v₂};
end
END.
```

Algorithm 2. Algorithm for solving conflict on fuzzy relation

The computation complexity of Algorithm 2 is $O(n^2)$. It is possible to prove that in general this algorithm determines an integration satisfying - consensus function. Because of the limited space for the paper, the proof will be included in an extended work.

5.3   Conflict Resolution on Instance Level

In order to integrate different versions of the same instance, we propose some criteria. These criteria are used in order to verify the fulfillment of minimal requirements for the integration process [20]. We define the following criteria for fuzzy instance integration:

- C_1. *Instance closure*: Tuple t^* should be included in the sum of given tuples, that is $t^* \prec \bigcup_{i=1}^{n} t_i$
- C_2. *Instance consistency*: The common part of the given tuples should be included in tuple r*, that is $\bigcap_{i=1}^{n} t_i \prec t^*$
- C_3. *Instance superiority*: If sets of attributes T_i ($i = 1,...,n$) are disjoint with each other then $t^* = [\bigcup_{i=1}^{n} t_i]_{T^*}$, where is the sum restricted to attributes from set T^*
- C_4. *Maximal similarity*: Let d_a be a distance function values of attribute a $\in$ A then the difference between integration t^* and the profile elements should be minimal in the sense that for each $a \in T^*$ the sum $\sum_{r \in Z_a} d(t_a^*, r)$ should be minimal, where $Z_a = \{r_{ia}: r_i \in Z, i = 1, 2,.., n\}$. This is also standard function of consensus theory.

Below we present an algorithm for instance integration which satisfies criteria above. The idea of this algorithm is based on determining consensus for each of attributes and next completing their values for creating the integration.

Algorithm 3. Algorithms for Fuzzy Instance Integration
Input:
- Set of n descriptions of an instance
    $X=\{r_i \in TUPLE(T_i): T_i \subseteq A, i = 1, 2,..,n\}$

– Distance functions $d_a: 2^{V_a} \times 2^{V_a} \rightarrow [0,1]$ for attributes $a \in A$.
Output: Tuple $t*$ of type $T* \subseteq A$ which is the integration of tuples from X being the proper description of the instance.
Procedure:
BEGIN
   1. $A = \bigcup_{i=1}^{n} T_i$;
   2. For each $a \in A$ determine a set with repetitions
$X_a = \{t_{ia}: t_i \in X$ for $i = 1, 2, ..., n\}$;
   3. For each $a \in A$ using distance function d_a determine value $v_a \subset V_a$ such that

$$\sum_{r_{ia} \in X_a} d_a(v_a, r_{ia}) = \min_{v'_a \subseteq V_a} \sum_{r_{ia} \in X_a} d_a(v'_a, r_{ia})$$

   4. Create tuple $t*$ consisting of values v_a for all $a \in A$;
END.

Algorithm 3. Algorithm for solving Conflict on fuzzy instance

The most important step in the above algorithm is step 3: determine value of attributes integration satisfying criterion C_4. In that work a set of postulates for integration has been defined, and algorithms for its determining have been worked out. We can prove that the integration determined by this algorithm satisfies all criteria $C_1, C_2, C_3,$ and C_4. The computational complexity of this algorithm is $O(n^2)$

6   Conclusions

In this paper we present an overview of recent research results on fuzzy ontology integration problem. We have noted that there are still missing clear criteria for this kind of ontology integration. Most often proposed algorithms refer to concrete ontologies and their justification is rather intuitive than formal. The reason is that the semantics of concepts and their relations are not clearly defined, but rather based on default values. We also applied consensus method to solve conflict on fuzzy ontology integration. In this paper, we survey the research on fuzzy ontology integration using consensus theory has been proposed previous [12], [17] – [20], [24].

We can note for working out more effective integration algorithms for fuzzy ontologies one should find out methods for effective concept alignment. Now, these methods are still missing. The difficulty is caused by the fuzzy functions on class attributes and relationships between concepts. The future works should concern working out effective algorithms for this problem, and owing to them new approaches on fuzzy ontology integration can be created. For solving conflicts in fuzzy ontologies one can consider to use paraconsistent logic [23].

References

1. Guzmán-Arenas, A., Cuevas, A.D.: Knowledge accumulation through automatic merging of ontologies. Expert Syst. Appl. 37(3), 1991–2005 (2010)
2. Arenas, A.G., Cuevas, A.C.: Time efficient reconciliation of mappings in dynamic web ontologies. Expert Systems with Applications, 1991–2005 (2010)
3. Belhadef, H.: A new bidirectional method for ontologies matching. Procedia Engineering 23, 558–564 (2011)
4. Bock, J., Hettenhausen, J.: Discrete particle swarm optimisation for ontology alignment. Information Sciences, 152–173 (2012)
5. Calegari, S., Ciucci, D.: Fuzzy Ontology, Fuzzy Description Logics and Fuzzy-OWL. In: Masulli, F., Mitra, S., Pasi, G. (eds.) WILF 2007. LNCS (LNAI), vol. 4578, pp. 118–126. Springer, Heidelberg (2007)
6. Chen, R.C., Bau, C.T., Yeh, C.J.: Merging domain ontologies based on the WordNet system and Fuzzy Formal Concept Analysis techniques. Applied Soft Computing 1, 1908–1923 (2011)
7. Chua, W.W.K., Kim, J.J.: BOAT: Automatic alignment of biomedical ontologies using term informativeness and candidate selection. Journal of Biomedical Informatics 45, 337–349 (2012)
8. Danilowicz, C., Nguyen, N.T.: Consensus - Based Partitions in the Space of Ordered Partitions. Journal of Pattern Recognition 21(3), 269–273 (1998)
9. Duong, T.H., Truong, H.B., Nguyen, N.T.: Local Neighbor Enrichment for Ontology Integration. In: Pan, J.-S., Chen, S.-M., Nguyen, N.T. (eds.) ACIIDS 2012, Part I. LNCS, vol. 7196, pp. 156–166. Springer, Heidelberg (2012)
10. Kemeny, J.G.: Mathematics without Numbers. Daedalus 88, 577–591 (1959)
11. Mao, M., Peng, Y., Spring, M.: An adaptive ontology mapping approach with neural network based constraint satisfaction. Web Semantics: Science, Services and Agents on the World Wide Web, 14–25 (2012)
12. Nguyen, N.T., Truong, H.B.: A Consensus-Based Method for Fuzzy Ontology Integration. In: Pan, J.-S., Chen, S.-M., Nguyen, N.T. (eds.) ICCCI 2010, Part II. LNCS, vol. 6422, pp. 480–489. Springer, Heidelberg (2010)
13. Hernes, M., Nguyen, N.T.: Deriving Consensus for Hierarchical Incomplete Ordered Partitions and Coverings. Journal of Universal Computer Science 13(2), 317–328 (2007)
14. Nguyen, N.T.: Processing Inconsistency of Knowledge in Determining Knowledge of a Collective. Cybernetics and Systems 40(8), 670–688 (2009)
15. Noy, N.F., Musen, M.A.: PROMPT: Algorithm and Tool for Automated Ontology Merging and Alignment. In: AAAI/IAAI 2000, pp. 450–455 (2000)
16. Straccia, U.: A Fuzzy Description Logic for the Semantic Web. In: Sanchez, E. (ed.) Capturing Intelligence: Fuzzy Logic and the Semantic Web, pp. 167–181. Elsevier (2006)
17. Truong, H.B., Nguyen, N.T.: A framework of an effective fuzzy ontology alignment technique. In: International Conference on Systems, Man and Cybernetics, Anchorage, Alaska, USA, pp. 931–935. IEEE (2011) ISBN 978-1-4577-0652-3
18. Truong, H.B., Nguyen, N.T.: A Multi-attribute and Multi-valued Model for Fuzzy Ontology Integration on Instance Level. In: Pan, J.-S., Chen, S.-M., Nguyen, N.T. (eds.) ACIIDS 2012, Part I. LNCS, vol. 7196, pp. 187–197. Springer, Heidelberg (2012)
19. Truong, H.B., Nguyen, N.T., Nguyen, P.K.: Fuzzy Ontology Building and Integration for Fuzzy Inference Systems in Weather Forecast Domain. In: Nguyen, N.T., Kim, C.-G., Janiak, A. (eds.) ACIIDS 2011, Part I. LNCS, vol. 6591, pp. 517–527. Springer, Heidelberg (2011)

20. Truong, H.B., Duong, T.H., Nguyen, N.T.: A Hybrid Method For Fuzzy Ontology Integration. Cybernetics and Systems: An International Journal 44(2-3), 133–154 (2013)
21. Wang, R., Wang, L., Liu, L., Chen, G., Wang, Q.: Combination of the Improved Method for Ontology Mapping. Physics Procedia 25, 2167–2172 (2012)
22. Zadeh, L.A.: Fuzzy sets. Information and Control, 338–358 (2006)
23. Nakamatsu, K., Abe, J.M.: The paraconsistent process order control method. Vietnam Journal of Computer Science 1(1), 29–37 (2014)
24. Nguyen, N.T.: Editorial. Vietnam Journal of Computer Science 1(1), 1–2 (2014)
25. Nguyen, N.T.: Advanced methods for inconsistent knowledge management. Springer, London (2008)

Novel Operations for FP-Tree Data Structure and Their Applications

Tri-Thanh Nguyen and Quang-Thuy Ha

Vietnam National University, Hanoi (VNU),
University of Engineering and Technology (UET)
{ntthanh,thuyhq}@vnu.edu.vn

Abstract. Frequent Pattern Tree (FP-tree) proposed by Han et al. is a data structure that is used for storing frequent patterns (or itemsets) in association rule mining. FP-tree helps to reduce the number of database (DB) scans to only two, and shrink down the number of candidates of frequent patterns. This paper proposes to define some operations on the FP-tree in order to empower its application. With the devised operations, we can: a) incrementally build the FP-tree when only a subset of a DB is ready at a time; b) construct FP-tree in parallel with low cost of communication; c) build local FP-trees independently based on local database and then use them to construct the global FP-tree in a distributed system; d) prune the FP-tree according to different values of minimum support threshold for frequent pattern mining.

Keywords: Association rules, data structure, frequent pattern mining, FP-tree, parallel processing.

1 Introduction

Frequent pattern mining is an important task because its results can be used in a wide range of mining tasks, such as association rule [1], correlation [6], causality [14], sequential pattern [3], etc. In association rule mining [1], it is proved that the step that costs the most is to find frequent patterns (i.e. item sets). The famous algorithm to find these patterns is Apriori [2]. However, this algorithm is slow since it suffers from scanning the database many times and generating several redundant pattern candidates. One effort to improve this algorithm is to parallelize it on a cluster of machines. Another effort is to propose new algorithm [20]. Nevertheless, the above efforts still suffer from scanning the database several times. Han et al. [7] overcame the above mentioned challenges by proposing the *frequent pattern tree* (FP-tree) data structure which is proved to have both completeness and compactness. FP-tree data structure has been studied in lots of literature in both application and modification (or improvement) aspects as summarized in Section 2.

In this paper, we will propose to add one constraint and some operations on the FP-tree data structure so that:

- It is possible to build a FP-tree incrementally. That is, when only a subset of a database is available at a time, we can incrementally update the FP-tree. When the whole database is not ready, we can still get FP-tree corresponding to the snap-short of the data, and get the frequent itemsets.

T.V. Do et al. (eds.), *Advanced Computational Methods for Knowledge Engineering*,
Advances in Intelligent Systems and Computing 282,
DOI: 10.1007/978-3-319-06569-4_17, © Springer International Publishing Switzerland 2014

- it is possible to build a FP-tree in parallel with the minimum cost of communication among the threads of execution. Each thread of execution independently (without communication) works on its part of the database to build the local FP-tree. The only communication among the threads is at the end of the process, when the slave threads send its local FP-tree to the master to construct the final FP-tree.
- it is possible to build a global FP-tree in a distributed system where each server stores only its local transaction database.
- it is possible to mine frequent itemsets with different values of minimum support threshold (*minsup*) without running the algorithm on the database again to rebuild the FP-tree with the new value of *minsup*.

The rest of the paper is organized as follows. Section 2 summarizes some typically related work. Section 3 gives a short description about FP-tree data structure. New operations are given in Section 4 and some case studies to demonstrate the application of the newly proposed operations are discussed in Section 5. Section 6 will conclude the paper and mention about the future directions.

2   Related Work

In this section, we will give a brief survey on typical work related to FP-tree. The first branch of work is to directly apply FP-tree in solving a certain problem, while the second one is to modify or improve this data structure so that it can be used in more complicated problems.

Lin et al. proposed a resource prediction framework (RPF) to predict the resource usage (i.e. the minimum number of virtual machines for a certain application) on a cloud computing platform [11]. For evaluation, the authors parallelized the frequent pattern growth (FP-growth) (i.e. the algorithm to extract frequent patterns on a FP-tree), K-mean, and Particle Swarm Optimization (PSO) as the heavy load algorithms on this platform in order to find out the suitable number of virtual machines. Kumar and Rukmani used both FP-tree and Apriori for web usage mining problem [8]. In the work of Suman et al., the authors claimed that FP-Growth algorithm suffers from pattern extraction since it has to generate conditional tree [16]. Thus, the authors proposed Apriori-Growth algorithm that is based on Apriori and FP-tree structure. In this algorithm, the method of candidate generation step is based on Apriori algorithm, then each candidate is checked against FP-tree to figure out whether is it frequent or not. However, this work did not carry out any implementations to verify the power of the proposed algorithm. Patro et al. proposed to use Huffman coding to compress FP-tree [12]. Yen et al. proposed Search Space Reduced (SSR) algorithm to speed up the pattern extraction from FP-tree [19]. Xu et al. mined associated factors about emotional disease based on FP-tree growing algorithm [18]. Bernecker et al. added probability to FP-tree to mine uncertain databases [5]. Concretely, the authors proposed the first probabilistic FP-Growth (ProFP-Growth) and associated probabilistic FP-Tree (ProFP-Tree), which we use to mine all probabilistic frequent patterns in uncertain transaction databases without candidate generation. Shrivastava et al. mined multiple level association rules based on FP-tree and co-occurrence frequent item tree (CFI) [13]. Lin et al. proposed to add constraints on FP-tree for multi-constraint pattern discovery [10]. When comparing with the multi-constraint Apriori-like algorithm, the performance of the multi-constraint algorithm based on FP-tree data structure prevailed. Wang et al. proposed an algorithm on FP-tree to find k-top frequent closed itemsets [17]. In order to apply the FP-tree for large databases, Li et al. parallelized the FP-tree construction procedure [17]. However, the cost of communication during FP-tree construction is high and the threads of execution need to handshake three times. Singh et al. also carried out experiments on FP-tree construction parallelization, however, they confirmed that it was not successful due to the cost of communication [15].

3   FP-Tree Review

We would like to introduce the FP-tree described in [7]. Let $I = \{a_1, a_2, ..., a_m\}$ be the **set of items** (which can be the list of goods in a supermarket), and a **transaction database** $DB = \langle T_1, T_2, ..., T_n \rangle$, where each T_i $(1 \bullet i \bullet n)$ is a **transaction** which contains a subset of I $(T_i \subseteq I)$. An example of a transaction T_i is the list of goods in a shopping basket. Define the **support** (or the absolute occurrence frequency) of a pattern A $(A \subseteq I)$ is the number of transactions in DB that contains A. Note that, in some other papers, the support is defined as relative occurrence frequency, i.e. the percentage of a pattern in DB. In this paper the absolute occurrence frequency is used to make it easier to follow the examples. Pattern A is called a *frequent pattern* (or *frequent itemset*) if A' support is greater than or equal a predefined *minsup* ξ. The task of finding the *complete set of frequent patterns* in a DB with a *minsup* ξ is called frequent pattern mining problem. This is the most time-consuming task, hence, it needs improving its speed.

Han et al. [7] proposed the FP-tree data structure that can store the *complete set of frequent patterns* using only *two scans* over the *DB*. The biggest contribution to speed up the frequent pattern mining task is the reduction of the number of scans over the *DB* down to only two, since the speed of reading data in the secondary storage is slow. FP-tree is a tree structure as defined below:

1. It consists of one root labeled as "null", a set of **item prefix subtrees** as the children of the root, and a **frequent-item header table**.
2. Each node in the **item prefix subtree** consists of 4 fields: *item-name, count, parent-link*, and *node-link*, where *item-name* registers which item this node represents; *count* registers the number of transactions represented by the portion of the path reaching this node; *node-link* links to the next node in the FP-tree carrying the same item-name (or null if there is none); and the *parent-link* links to the parent node[1].

---

[1] Though this field is not clearly mentioned in [7], it is important in forming the tree, so we list it here for the sake of completeness

3. Each entry in the **frequent-item header table** consists of two fields: (1) *item-name* and (2) *head of node-link*, which points to the first node in the FP-tree carrying the *item-name*.

An example of a FP-tree is given in Figure 1, and now we will study how to construct the FP-tree in this figure. With the *minsup* $\xi=3$, based on the *DB* listed in Table 1. This table shows a simple database of transactions of a supermarket, where the first column is the transaction identity, each row in the second column is the list of items that were bought by a customer. The FP-tree construction is described briefly as follows:

FP-tree construction starts with the first scan over the *DB* to find the list of frequent items (frequent itemsets with the cardinality of 1 having the support not less than ξ). The result of the scan over the *DB* in Table 1 is the list h of $\langle (f: 4), (c:4), (a:3), (b:3), (m:3), (p:3) \rangle$, where the number after the colon ":" is the items' support, and the h is sorted in support descending order denoted as L. h is used to build the *frequent-item header table* (or header table for short), where each entry in the table consists of the *item-name* and a pointer (called *head of node-links*) to the first (appeared) node having the same item-name in the FP-tree as depicted in Figure 1. The second scan will get the list of frequent items of each transaction, sort it according to L, and insert it into the FP-tree. To make it easier to observe, this list of each transaction is showed in the third column of Table 1. The tree construction algorithm is listed in Algorithm 1.

Table 1. Transactions in *DB* and their frequent items

TID	Items Bought	(Ordered) Frequent Items
100	f, a, c, d, g, i, m, p	f, c, a, m, p
200	a, b, c, f, l, m, o	f, c, a, b, m
300	b, f, h, j, o	f, b
400	b, c, k, s, p	c, b, p
500	a, f, c, e, l, p, m, n	f, c, a, m, p

Algorithm 1: FP-tree_construction
Input: A transaction database DB and a *minsup* ξ.
Output: The frequent pattern tree F

```
1. Scan the DB to get the list L of frequent items, and
sort it in support descending order.
 Create a FP-tree F    by
    Create the header table, and set all the
    head-of-node-links to null.
    Create the root node T of the tree having the
    item-name of null.
    Set the parent-link and node-link of T to null.
2. Scan the DB again
    For each transaction Tran in DB do
```

```
  Get the list of frequent items.
  Sort it according to the order L.
  Let this list be [p|P], where p is the first item
  and P is the remaining items.
  Call insert_tree([p|P], T).
EndFor
```

where the *insert_tree*() procedure is defined as:

Algorithm 2: insert_tree
Input: the ordered list [p|P] of frequent items, and a
node T of a FP-tree.
Output: the updated FP-tree.

```
If T has a child node N such that the item name of N and
p is the same, Then
  Increase the count of N by 1 //(*) mark for later
  reference
Else
  Create a new node N.
  Set the item_name of N to p.item_name.
  Set the count of N to 1. //(*)
  Link the parent-link of N to T.
  Set the node-link of N to null.
  If the head-of-node-links of the item h in the header
  table which has the same name as p is null Then
    Set head-of-node-links of h to p;
  Else
    Traverse through the head-of-node-links of h to the
    end of the list, and link the node-link of the
    end-node to p.
  EndIf
EndIf
If P is not empty, Then
  Let P=[p₁|P₁]
Call insert_tree([p₁|P₁], N).
EndIf
```

The *insert_tree*(.) algorithm is used for adding an itemset (having count=1), it is easy to modify the algorithm to add a number of identical item sets (or the item set with the count of $m>1$) by changing the lines marked by (*): we set the support count to (or increase the support count by) m instead of 1.

Han et al. proved the FP-tree has both compactness and completeness [7]. Completeness means it preserves the complete information about transactions in DB for frequent pattern mining, while compactness means it reduces the irrelevant infrequent item sets.

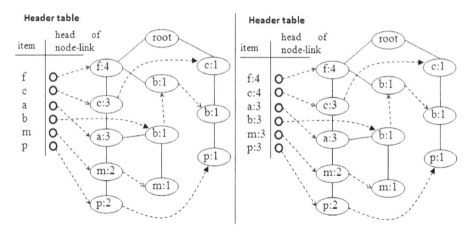

Fig. 1. (a) FP-tree corresponding to the DB in Table 1 (*left*); (b) Header table with support field in a FP-tree (*right*)

4   New Constraint and Operations on FP-Tree

In this section, we will define a new constraint and some operations (or algorithms) on the FP-tree data structure to make it more powerful as mentioned in Section 1. More details of their application will be clearly discussed in Section 5.

4.1   Constraint on FP-Tree Data Structure

Header Table Constraint. the frequent items in the header table must be sorted on support descending order.

Though the FP-tree in Figure 1 (a) is constructed following this constraint, it is not stated clearly in [7]. This constraint is important for later operations, since it helps to make the operations faster. Before defining the operations, let's exploit some more properties of the FP-tree data structure besides those discussed in [7].

4.2   Property on FP-Tree Data Structure

Frequency Retrieval Property. given a FP-tree, we can retrieve the occurrence frequency of all the items in the tree.

Rationale. Since nodes having the same *item-name* are linked via *node-link* structure, we simply follow the *node-link* to calculate the sum of the *count* of these nodes. The sum is the occurrence frequency of the corresponding item.

Since the occurrence frequency of each item is identified after the first scan, if we modify the header table by adding one more field to hold this frequency as depicted in Figure 1(b), then there is no need to traverse the tree to get this information.

Frequent Item List Recovery Property. the frequent item list of all transactions in *DB* can be restored from its corresponding FP-tree.

Rationale. Based on the FP-tree construction steps, the frequent item list of each transaction in *DB* is mapped to one path in the FP-tree. Therefore, we simply traverse the tree to extract these lists. When inserting a list into the FP-tree, the list is sorted in support descending order, so the lowest support item is always the leaves of the tree. Based on this characteristic, we can traverse the FP-tree from the lowest item (which always points to leaf nodes) to extract the frequent item list of each transaction as described in the below *transaction_extraction*() procedure.

```
Algorithm 3: transaction_extraction
Input: a FP-tree
Output: the ordered frequent item list of a transaction
in DB and the updated FP-tree after removing this trans-
action.

If the header table is not empty Then
  Let h be the lowest support item in the table.
  Let p be the node pointed by h.head-of-node-links.
  Traverse the path from p to the root node via its par-
ent-link to get the list l of items.
  For each node q in the path Do
Decrease q.count by 1.
If q.count =0 Then
Remove q out of the path and update   parent-link and
node-link structure to maintain the connectivity of the
remaining path.
    EndIf
  End For
  If p is removed, and it is the last node having the
same item-name as h's Then
    Remove h out of the header table.
  EndIf
  Return l and the updated FP-tree.
 EndIf
```

Note that the returned list *l* is sorted in support ascending order. In addition, the header table constraint helps to make the algorithm faster, since we can directly access the lowest support item without searching the whole header table.

An example of transaction extraction is described in Figure 2. Given the FP-tree in Figure 1 (b), the first call to *transaction_extraction*() will result in the FP-tree in Figure 2 (a) and the extracted list is [*p, m, a, c, f*]. The second call to this procedure will produce the FP-tree in Figure 3-b and the list [*p, m, a, c, f*]. Since the two calls to the procedure return the same list, we can easily modify the algorithm to return the list with a count (i.e. the number of identical lists). For example, a call to the procedure can return [*p, m, a, c, f*] with the count of 2, then we can reduce the number of calls to this procedure. This is very critical (in term of performance) when the count big (e.g. in a FP-tree constructed from dense database which has many identical transactions).

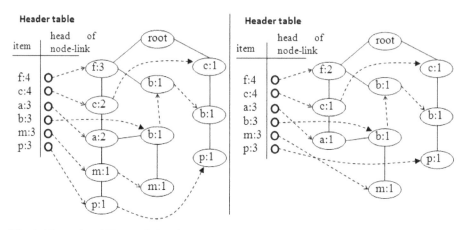

Fig. 2. The updated FP-tree during frequent list extraction: (a) First extracted [p, m, a, c, f] list (*left*); (b) Second extracted [p, m, a, c, f] list (*right*)

4.3   New Operations on FP-Tree Data Structure

Assume the set of items I be fixed in this paper. We propose two operations (algorithms) on FP-tree data structure which make FP-tree more powerful as described in Section 5.

FP-Trees Combination Operation

Given two FP-trees F_1 and F_2 that were constructed independently on two different databases, i.e., DB_1 and DB_2, the problem is how to combine the two trees into one FP-tree F. Since F_1 and F_2 were built based on its local database, the ordered list of frequent items of the F_1 is different from that of F_2. This means we can not update either tree to be the combined one. Our proposal is to construct a new Tree F from the lists extracted from the two given FP-trees. Recall that the construction of a FP-tree needs two scans the database, the first scan is to find the list of frequent items, and the second scan is to get the frequent list of each transaction to insert into the tree. In solving this problem, we strictly follow the original FP-tree construction algorithm, however, we "scan" the two FP-trees F_1 and F_2 instead of a transaction database. Suppose we add the support field in the header table of FP-tree as in Figure 1(b), the FP-tree construction algorithm can be seen as follows:

```
Algorithm 4: FP-tree_combination
Input: Two FP-trees: F₁ and F₂
Output: The combined FP-tree F
1. Go through the header tables of both F₁ and F₂ to get
the list of all items along with their support, and sort
it in support descending order denoted as L.
   Create the output FP-tree F by
   Create the header table, and set all the
   head-of-node-links = null.
   Create the root node T of the tree having
   item-name=null.
   Set the parent-link and node-link of T to null.
```

```
2. For each FP-tree t in [F₁, F₂] do
     Let p=transaction_extraction(t) be a list extracted
     from the FP-tree t
     Sort p according to the order L.
     Call insert_tree(p, T).
   EndFor
```

FP-Tree Pruning Operation

Given a transaction database *DB*, and a *minsup* ξ_1, we can build a FP-tree *F*, what if we want to build another FP-tree from *DB* with another *minsup* ξ_2 (where $\xi_2 > \xi_1$)? We propose to modify (or prune) *F* to yield the expected FP-tree.

Pruning Lemma. Given a FP-tree *F* that was constructed from a transaction database *DB* with the *minsup* ξ_1, if we want to build the FP-tree from *DB* with another *minsup* ξ_2 (where $\xi_2 > \xi_1$), we simply remove nodes that are not frequent (i.e., their support is less than ξ_2) in *F*.

Rationale. Based on the FP-tree construction process, when we build the FP-tree F_1 and F_2 from *DB* with corresponding *minsup* ξ_1 and ξ_2 ($\xi_2 > \xi_1$), the difference between the two trees is that the items whose support is not less than ξ_1, and less than ξ_2, will only appear in F_1. Items, whose support is not less than ξ_2, will appear in both trees with the same structure (i.e., the position in the tree and the link among these nodes). That means we can produce F_2 by removing the different nodes (i.e. nodes of item whose support is less than ξ_2) in F_1.

Since the header table is sorted in support descending order, in order to remove nodes whose support is less than ξ_2, we scan the header table from the lowest support item to remove its linked nodes till we find an item whose support is not less than ξ_2. Note that the lowest support item always links to the leaves of FP-tree. The pruning algorithm is described as follows:

Algorithm 5: FP-tree_Pruning
Input: A FP-tree *F* and *minsup* ξ.
Output: The updated FP-tree *F* after removing nodes whose support is less than ξ

```
Let p be the lowest support item in the header table of F
If p.support >= ξ Then
  Exit
Else
  Follow the node-link structure via
  p. head-of-node-links to remove nodes having the same
  item name as that of p.
  Remove p from header table
  If header table of F is not empty Then
    Call FP-tree_Pruning(F, ξ)
  EndIf
EndIf
```

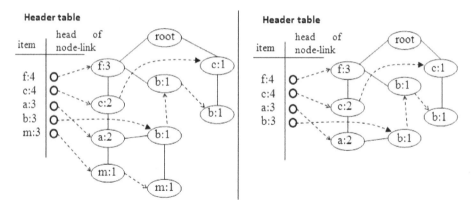

Fig. 3. The updated FP-tree during pruning: (a) Item p is removed (*left*); (b) Item m is removed (*right*)

Given the FP-tree in Figure 2, now we prune the tree with new *minsup* ξ=4, Figure 3 depicted the updated FP-tree when two item p and m are removed.

5   Case Studies for FP-Tree Operations

In this section, we will present some case studies where the proposed operations can be applied.

5.1   Incremental Frequent Pattern Mining

In many real world applications (or systems), e.g. the sale management application of a supermarket, the transactions are added to their database day by day. Thus, we cannot wait until the database stops increasing to start the frequent pattern mining process. It is reasonable to start mining on a *snapshot* of the database, i.e. the database after daily operation. Let the set of transactions of day i be DB_i, then on day N we have the snapshot of the database of $DB = \bigcup_{i=1}^{N} DB_i$.

When we build the FP-tree with a *minsup* ξ on a snapshot of the database DB of day N^{th}, we have to read all the previous DB_i ($i<N$). That means that we do not inherit the mining process of day i^{th}. By exploiting the proposed operations, we can define an *incrementally frequent pattern mining algorithm* that can inherit the scans over the previous snapshot of the database. In other words, in order to build the FP-tree of the today database snapshot, we only need to scan DB_N (i.e. the transaction set added today), other transactions are retrieved from the FP-tree built from yesterday's snapshot. The algorithm is described as bellow:

Algorithm 6: Incrementally_Frequent_Pattern_Mining
Input: A transaction set DB_N, a *minsup* ξ, and another FP-tree F-reuse

Output: The corresponding FP-tree *F*, and the updated F-reuse (for later reuse)

```
If N=1 Then //This is the first day
  F-reuse=FP-tree_construction(DB_N, 1) //Note minsup=1
Else
  F_temp=FP-tree_construction(DB_N, 1) //Note minsup=1
  F-reuse=FP-tree_combination(F-reuse, F_temp)
EndIf
F=FP-tree_pruning(F-reuse, ξ)
```

In this algorithm, we use one more FP-tree (i.e., *F-reuse*) to store the compact database snapshot for later reusing. In order to avoid constructing F_{tmp}, we can easily modify the algorithm to build *F-reuse* directly from *F-reuse* and DB_N with the same idea.

5.2   Frequent Pattern Mining Parallelization

Frequent pattern mining parallelization is a good solution for a large transaction database *DB*, and it is possible to apply FP-tree structure with the above proposed operators to solve this problem as follows:

Algorithm 7: Parallel_Frequent_Pattern_Mining
Input: A large database *DB* and a *minsup* ξ
Output: The corresponding FP-tree *F*
1. Divide *DB* into *N* roughly even subsets DB_i ($i=1..N$) such that $DB_i \cap DB_j = \emptyset$ for $i \cdot j$ and $\bigcup_{i=1}^{N} DB_i = DB$, where *N* is the number of processors, and each DB_i is assigned to a processor.
2. Allow each processor to build the local FP-tree F_i corresponding to its assigned DB_i with the *minsup* of 1.
 F_i =FP-tree_construction(DB_i, 1)
3. Combine the local FP-trees F_i into the final tree *F* in a master processor:
 //Processors have handshake at this point
  $F=F_1$
  For *i*=2 to N Do
    F=FP-tree_combination(F, F_i)
  EndFor
4. Prune *F* with the expected *minsup* ξ:
  F=FP-tree_Pruning(F, ξ)

In this algorithm, the only point the processors need to handshake is Step 3 where the slave processors send its local FP-tree to the master processor to construct the final one.

5.3   Distributed Frequent Pattern Mining

In some situations, we want to mine frequent patterns from multiple databases (sources). For example, a heavy web application that must be load balanced across a number of web servers which even are distributed in different geographical locations. If it is desirable to mine frequent patterns from all the log of each web server, then we can define a distributed mining algorithm as follows:

Algorithm 8: Distributed_Frequent_Pattern_Mining
Input: Transaction database DB_i of N parties (servers), and *minsup* ξ
Output: The corresponding FP-tree F of $DB = \bigcup_{i=1}^{N} DB_i$
1. Each party i constructs the local FP-tree F_i from its DB_i
   F_i =FP-tree_construction(DB_i, 1) //Note *minsup* = 1
2. Each party i sends F_i to all other parties and receives the local FP-trees from these parties
3. Each party constructs the global FP-tree from local FP-trees
   $F=F_1$
   For j=2 to N Do
      F=FP-tree_combination(F, F_i)
   EndFor
   F=FP-tree_pruning(F-reuse, ξ)

Since the frequent patterns may be exploited in each party, the algorithm allows each party to have the global FP-tree. In addition, on each party, new transactions are added to the local database DB_i day by day, so we can use "*Incrementally frequent pattern mining*" algorithm on each party to speed up the performance.

5.4   Frequent Pattern Mining with Different *minsup* Values

Another possible application of the new operations on FP-tree is to support frequent pattern mining with different *minsup* values. In practice, users may be impossible to find a suitable *minsup* value at the beginning, so it is a good idea to try different values of *minsup* to figure out which is the most appropriate one. Thanks to the "*FP-tree pruning*" algorithm, we build the FP-tree with *minsup*=1, then we can use FP-tree pruning algorithm to create appropriate FP-tree with different *minsup* values without scanning the database again.

5.5   Discussion

FP-tree usually has a smaller size than the original transaction database due to the nature that many transactions can have the same prefix, hence, share the same nodes on FP-tree. When the transaction database is dense, its corresponding FP-tree size is much smaller than the database's, and subsequently, our proposed operations will expose its powerfulness.

Though there may be many other applications of the new operations on FP-tree data structure, we just pick up some to make examples. However, the algorithms presented in this Section can only work when the number of items is small enough so that the corresponding FP-tree with *minsup*=1 can fit into the main memory for performing desired operations.

6   Conclusion and Future Work

Though FP-tree has been studied intensively to apply in many problems, in this paper, we proposed some more operations on FP-tree data structure to make it more powerful. With the new operations we can employ FP-tree to solve many problems, such as incremental FP-tree construction, parallel FP-tree construction, distributed FP-tree construction and frequent pattern mining with different *minsup* values. In the next direction, we will apply these new operations in real applications. We also plan to continue studying to add more operations on FP-tree for other applications. One more possible future direction is to explore the space complexity of FP-tree in some domain, and its effect on the proposed operations.

Acknowledgments. This work was partially supported by the VNU Scientist links.

References

1. Agrawal, R., Imielinski, T., Swami, A.N.: Mining Association Rules between Sets of Items in Large Databases. In: Proc. of SIGMOD Conference 1993, pp. 207–216 (1993)
2. Agrawal, R., Srikant, R.: Fast Algorithms for Mining Association Rules. Research Report RJ 9839, IBM Almaden Research Center (1994)
3. Agrawal, R., Srikant, R.: Fast Algorithms for Mining Association Rules in Large Databases. In: VLDB 1994, pp. 487–499 (1994)
4. Agrawal, R., Aggarwal, C., Prasad, V.V.V.: Depth-first generation of large item-sets for association rules. IBM Tech. Report RC21538 (July 1999)
5. Bernecker, T., Kriegel, H.-P., Renz, M., Verhein, F., Züfle, A.: Probabilistic Frequent Pattern Growth for Itemset Mining in Uncertain Databases (Technical Report). CoRR abs/1008.2300 (2010)
6. Brin, S., Motwani, R., Silverstein, C.: Beyond Market Baskets: Generalizing Association Rules to Correlations. In: Proc. of SIGMOD Conference 1997, pp. 265–276 (1997)
7. Han, J., Pei, J., Yin, Y.: Mining Frequent Patterns without Candidate Generation. In: Proc. of SIGMOD Conference 2000, pp. 1–12 (2000)
8. Kumar, B.S., Rukmani, K.V.: Implementation of Web Usage Mining Using APRIORI and FP-Growth Algorithms. Int. J. of Advanced Networking and Applications 1(6), 400–404 (2010)
9. Li, H., Wang, Y., Zhang, D., Zhang, M., Chang, E.Y.: Pfp: parallel fp-growth for query recommendation. In: Proc. of RecSys 2008, pp. 107–114 (2008)

10. Lin, W.-Y., Huang, K.-W., Wu, C.-A.: MCFPTree: An FP-tree-based algo-rithm for multi-constraint patterns discovery. IJBIDM 5(3), 231–246 (2010)
11. Lin, C.-Y., Chen, Y.-A., Tseng, Y.-C., Wang, L.-C.: A flexible analysis and prediction framework on resource usage in public clouds. In: Proc. of CloudCom 2012, pp. 309–316 (2012)
12. Patro, S.N., Mishra, S., Khuntia, P., Bhagabati, C.: Construction of FP Tree using Huffman Coding. International Journal of Computer Science Issues 9(3), 446–469 (2012)
13. Shrivastava, V.K., Kumar, P., Pardasani, K.R.: FP-tree and COFI Based Approach for Mining of Multiple Level Association Rules in Large Data-bases. CoRR abs/1003.1821 (2010)
14. Silverstein, C., Brin, S., Motwani, R., Ullman, J.D.: Scalable Techniques for Mining Caus-al Structures. In: Proc. of VLDB 1998, pp. 594–605 (1998)
15. Singh, M., Ahirwar, R., Kher, N.: FP-Tree Improve Efficiency & Increase Scalability by Applying Parallel Projected. Binary Journal of Data Mining & Networking 1(1), 14–16 (2010)
16. Suman, M., Anuradha, T., Gowtham, K., Ramakrishna, A.: A Frequent Pattern Mining Al-gorithm Based On Fp-Tree Structure Andapriori Algorithm. International Journal of Engi-neering Research and Applications 2(1), 114–116 (2012)
17. Wang, J., Han, J., Lu, Y., Tzvetkov, P.: TFP: An Efficient Algorithm for Mining Top-K Frequent Closed Itemsets. IEEE Trans. Knowl. Data Eng. 17(5), 652–664 (2005)
18. Xu, A.-P., Tang, Y., Wang, Q., Qiao, M.-Q., Zhang, H.-Y., Wei, S.: Mining Associated Factors about Emotional Disease Bases on FP-Tree Growing Algorithm. International Journal of Engineering and Manufacturing 4, 25–31 (2011)
19. Yen, S.-J., Wang, C.-K., Ouyang, L.-Y.: A Search Space Reduced Algorithm for Mining Frequent Patterns. J. Inf. Sci. Eng. 28(1), 177–191 (2012)
20. Zaki, M.J., Parthasarathy, S., Ogihara, M., Li, W.: New Algorithms for Fast Discovery of Association Rules. In: Proc. of KDD 1997, pp. 283–286 (1997)

Policy by Policy Analytical Approach to Develop GAINS-City Data Marts Based on Regional Federated Data Warehousing Framework

Thanh Binh Nguyen

International Institute for Applied Systems Analysis (IIASA),
Schlossplatz 1,
A-2361 Laxenburg, Austria
nguyenb@iiasa.ac.at

Abstract. City-scale data marts are requested to help local policy makers to identify viable and efficient solutions. In this context, the Regional Federated Data warehousing Framework (RFDW), which has been introduced in our previous literatures, is extended and used to develop such city-scale data marts. Afterwards, a case study, namely GAINS-City China, will be presented as a city scale manifested data mart to approve our concepts.

Keywords: Regional Federated Data warehousing Framework (RFDW), policy by policy, Data Mart, Greenhouse Gas - Air Pollution Interactions and Synergies (GAINS).

1   Introduction

Federated Data warehousing approach is a very promising solution for many problems in area of business decision supports [10]. According to [7], there are two types of federation possible in a federated data warehouse, i.e. Regional federation (or more appropriately regional versus corporate/global federation), and Functional federation.

In our current research [12,13], we are focusing on regional federated data warehousing systems, in which data marts are designed for regional analytical requirements and a global data warehouse for global requirements. The data flow from regional data marts to the global data warehouse is defined as upward federation and the data flow from global data warehouse to regional data marts as downward federation. In this context, the RFDW has focused on large (global, regional, national) scale application while the city-scale data marts and their policy options or packages require some additional development, i.e., adding specific source categories, extending multi dimensional schemas down to regional levels of data granularities [11].

In this paper, first we introduce a city scale extension of RFDW [13]. Afterwards, city scale win-win policy by policy requirements will be specified in very formal manner as policy options and packages. By inhering the data structure

T.V. Do et al. (eds.), *Advanced Computational Methods for Knowledge Engineering*,
Advances in Intelligent Systems and Computing 282,
DOI: 10.1007/978-3-319-06569-4_18, © Springer International Publishing Switzerland 2014

and functions from global model, the city-scale data marts will be built based on the RFDW with its just-specified city scale policy options and packages. In the case study context, two major adjustments has been proposed: the source categories of original GAINS model, i.e. regional dimension is extended to city level,i.e. activity pathway data is disaggregated based on city scale; and updating the emission factor database according to local measurements. These improvements would make the GAINS-City model more suitable for city-scale emission estimation and co-benefits assessment.

The rest of this paper is organized as follows: section 2 introduces some research approaches and projects related to our work; in section 3, GAINS-City concepts and the policy by policy analytical approach will be presented. Section 4 will show our development as well as implementation results in term of the GAINS-City China Data Mart. At last, section 5 gives a summary of what have been achieved and future works.

2 Related Work

Our approach is focusing on using of federated data warehousing technologies to specify and build a class of virtual as well as manifest data marts, especially at city scale level. With the amount of data generated increasing continuously, delivering the right and sufficient amount of information at the right time to the right business users has become more complicated and critical [19,20].

According to [10], data warehouse federations are a very promising solution for many problems in area of business decision supports. The authors have moved the idea of federation proposed by [16] to the area of data warehouses.

In our current research within the GAINS model [1,2], we are focusing on regional federated data warehousing systems, in which data marts are designed for regional analytical requirements and a global data warehouse for the corporate requirements. As presented in our previous literature [6,12,13], the painful processes of implementing and maintaining such federated system, including a global data warehouse and its related data marts have been indicated in [18]. In order to improve building and maintaining federated data warehousing systems, we propose a regional federated data warehousing framework (RFDW), including global data warehouse and its virtual data marts in the clouds [13].

In the context of modeling emission at city scale level, according to [11], there have been several attempts to consider scientific advice in the process of development of local air quality strategies where elements of climate change policy would be integrated. However, the previous studies provided a rather general evaluation of policy package at a sector level [8,9], which ignored the individual reduction potential of policy and lacked detailed information to guide policy-by-policy decisions.

3 GAINS-City Concepts and Policy by Policy Analytical Approach

In [13], the cloud-based federated multidimensional data model is formally defined, with the objectives of setting design alternatives in the context of multidimensional data model [4,5,17].

3.1 Regional Federated Multidimensional Data Model(RFDW)

According to [13], the RFDW includes: the global data warehouse, formulated as $G = \langle D, L, V, C \rangle$; a set of dimensions denoted by $D = \{D_1, .., D_n, D_r\}$. The emphasis here is on the regional aspect, presented by regional dimension D_r. The specification presented in [13] also allows us to define a set of *regional* data marts $M^x = \langle D^x, L^x, V^x, C^x \rangle$, which initially are virtual or manifested standalone versions.

3.2 Data Mart Platform-as-a-Services

In the context of the RFDW, there are two main services, namely MSchema and MData are defined follows:

- $MSchema(M^x) =< D^x, L^x, V^x >$ where:
  - $D^x \subset D, L^x \subset L, V^x \subset V$.
  - Furthermore, $\forall D_i^x \in D^x, \exists D_i \in D$, and $DS_i^x = DS_i \cap L^x$.
- $MData(M^x) = C^x \subset C$.

3.3 Policy by Policy Analytical Approach for Modeling GAINS-City Data Marts

By using the RFDW, first the GAINS global data warehouse, according to [14,15,17], has been defined as follows :

- *Region, Fuel activity, Sector, Pollutant, Technology* dimensions are denoted by $D = \{R, F, S, P, T, Y\}$
- *A (Activity Data), T (Technology-specific Activity data), X (Application Rates of technologies), E (Emissions)* are variables or facts.
- *Activity pathway, Emission* are data cubes

Hereafter, the data mart platform-as-a-services are used to generate a classes of city scale data marts, namely GAINS-City, which has also been specified and developed based on the policy by policy analytical approach. In this context, data from GAINS data warehouse have been selected and disaggregated to fulfill GAINS-City data mart requirements, i.e. city level has been added to regional dimension. Afterwards, by simulating the modification of *Fuel activity data R* and *technology T* penetration induced by a specific policy and policy package, the GAINS-City data mart can assess the further reduction potential of emission

[11]. For example, the policy *installation of SCR in 80% of newly-built coal-fired power plants* can be directly converted into the penetration rate of SCR in the GAINS-City model and reduces NOX emissions. However, some simulation needs to involve various parameters and consider potential feedbacks on the energy balance.

Policy Options and Policy Packages. A set of policy options $P = \{p_1, .., p_k\}$ have been defined to fulfill city scale requirements. For example, p_i =*Fuel switch process* is one of policy options [11]. In this context, emission reduction $\triangle E$ of p_i was calculated by:

$$\triangle E = \triangle A_{Bef} e f_{Bef,t,p} x_{Bef,t,p} - (\triangle A_{Bef} e f f_{Bef}/e f f_{Post}) e f_{Post,t,p} x_{Post,tp} \quad (1)$$

Where:

- Bef is baseline (Before) policy activity type, e.g. *coal*,
- Bef is (Post) policy activity type, e.g. *natural gas*
- t is technology, p is pollutant
- $\triangle A$ represents the activity difference inferred from the baseline and the post policies.

Each policy package is a subset of P, which may have significantly contribution to cities.

Ranking Policies or Policy Packages by Using Emission Reduction Index. Having emission under baseline scenario, the priority of single policies or policy packages could be ranked based on emission achieved as follows:

$$index = \sum_i \triangle E_i / E_i \quad (2)$$

where i stands for species, $\triangle E$ presents the emission reduction and E presents the total emission under baseline scenario. The rank is meaningful for picking up high efficiency policy and searching for win-win solutions at city scale level.

4   GAINS-City Data Mart Development

GAINS-City data marts are built on the experience of the regional GAINS model application. It inherits its model structure and functions, but emphasizes the evaluation of co-benefits on GHGs (Greenhouse Gases) and air pollutants of climate-friendly air quality management policies in urban areas [11].

4.1   GAINS-City Requirement Analysis Based on Policy by Policy Analytical Approach

In section 3, policies or policy packages can be ranked by using Emission Reduction Index. Sensitivity analysis is important for such policy making process and helps understanding the reasons for discrepancy in emissions among different policy implementation processes, policy packages, and target species. The priority of policy is an important factor worth consideration can be calculated as shown in equation 2. In the context of GAINS City China project, we have specified top 15 policies which should be considered as system requirements as presented in table 1.

Table 1. 15 policy by policy options and packages

Rank	Remark	Index	Emission reduction in 2030, Gg (Tg for CO2)			
			SO2	NOX	PM25	CO2
1	Proportion of coal-fired boiler replaced by natural gas-fired boiler in fuel consumption: 70% in 2020 and 90% in 2030	0.39	40.67	34.40	5.77	8.47
2	Increasing rate of energy efficiency: 10% in 2020 and 20% in 2030	0.25	9.58	26.33	4.96	16.55
3	Policy impl. 2013, expand cement plant	0.23	4.88	12.82	15.77	3.66
4	Policy impl. 2015, install. FF in cement plant	0.23	0.00	0.00	23.38	0.00
5	Policy impl. 2015, substitution of indust. natural gas-fired kilns for coal-fired ones	0.19	22.21	6.81	6.86	0.00
6	Proportion of coal consumption replaced by natural gas in fuel consumption: 70% in 2020 and 90% in 2030	0.17	15.52	2.77	8.32	1.89
7	Proportion of electricity generated outside city boundary from 2020:85 %	0.17	6.56	18.03	3.40	11.34
8	for newly-built plants and 2015 for old plants	0.16	0.00	52.19	0.00	0.00
9	Proportion of coal-fired boiler replaced by district heating in fuel consumption: 60% in 2020 and 80% in 2030	0.13	12.91	0.08	6.91	0.86
10	Increasing rate of energy efficiency:% 10% in 2020 and 20% in 2030	0.13	4.97	13.65	2.57	8.58
11	Policy impl. 2015 standard, e.g. China VI for heavy duty diesel vehicles	0.06	0.00	17.63	0.82	0.00
12	Policy impl. 2015: Installation of FF in newly-built coal-fired power plants	0.04	0.00	0.00	4.39	0.00
13	Policy impl. 2012 Installation of LNB in power plants	0.03	0.00	9.80	0.00	0.00
14	Proportion of gas vehicles in fuel consumption from 2020:25%; Proportion of alternative energy public bus and taxi in numbers: 30% in 2020 and 40% in 2030;	0.02	0.13	1.96	0.37	2.46
15	Policy impl. 2014 for new standards: e.g. 2014 for China V and 2020	0.01	0.00	1.83	0.08	0.00

4.2   Designing and Deploying GAINS-City Using the RFDW

The following steps show how to design and develop GAINS-City using the
RFDW as illustrated in figure 1.

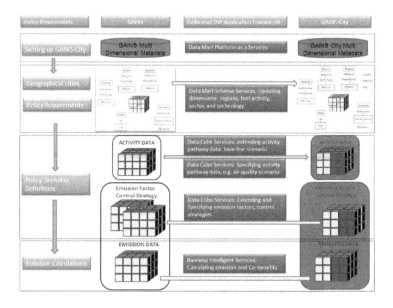

Fig. 1. GAINS City Developing Processes

- Setting up GAINS-City. This step is used to start up a GAINS-City data
  mart M^{city} by using *Data Mart Schema as a Service MSchema(M^{city})*. As
  a result we have $MSchema(M^{city}) =< D^{city}, L^{city}, V^{city} >$, where D^{city}
  are 6 GAINS dimensions, but L^{city} of *Region* dimension is *All Regions $\succ$
  RegionGroups $\succ$ Regions* related to the cities in this step.
- Geographical cities. This step specifies R (regions) dimension of GAINS-City.
  The dimension is defined by three hierarchical levels: *All Country Cities $\succ$
  Province $\succ$ City*. In this case, GAINS RegionGroups and Regions will be
  mapped to the new GAINS-City R dimension, then the *City* level will be
  added to the just-mapped hierarchy.
- Policy Requirements. The specific (15) policies and packages are selected
  in this step based on the local urban requirements as shown in table 1. In
  this context the *T (Technology), F (Fuel activity), S(Sector)* dimensions will
  be justified according to selected policies and packages as shown in table 1.
  For example, according to policy 12 *Policy impl. 2015: Installation of FF
  in newly-built coal-fired power plants*, the sector $INBO$ will be transformed
  to three sub sectors $INBO1, INBO2, INBO3$, and new technologies, e.g.
  IN_ESP1, IN_ESP2 are also assigned to the new *Fuel activity-Sector* combinations as show in figure 2.

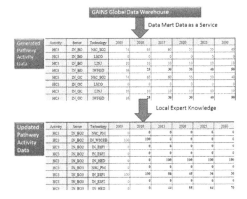

Fig. 2. Applying of policy 12 for justifying *Activity Pathway Data*

- Policy Scenario Definitions. In this context, each GAINS-City data mart contains city scale data, i.e. *Activity Pathway* data cube, which cannot be provided by the GAINS global data warehouse without the help of local expert knowledge. Disaggregation needs to respect the regional pattern present in other data sources. Some of this can be ensured by using simple downscaling algorithms based on population distributions, etc. For example, with policy 12 as shown in figure 2, the first table is *Activity Pathway* data provided by using *Data Mart Data as a Service*, and the second updated table is defined by local expert knowledge. This methodology could be applied to define data for baseline as well as other policy scenarios.
- Emission Calculations. After extending the source according to the city level requirements, emission scenario could be calculated by using GAINS *Business Intelligent Services*.

4.3   Implementation Results: GAINS-City China Case Study

Anthropogenic emissions of air pollutants in China influence not only local and regional environments but also the global atmospheric environment; therefore, it is important to understand how China's air pollutant emissions will change and how they will affect regional air quality in the future [11]. By linking tools developed by Chinese researchers with the GAINS model, the GAINS-City China project adapted the GAINS model to the specific needs of urban planners in order to assess practical policy options for controlling urban air pollution that maximize at the same time co-benefits on greenhouse gas mitigation.

Based on 15 specific policies presented in table 1, a baseline scenario and its two policy scenarios, namely *air quality (AQ)* and *Strict air quality (SAQ)* scenarios have been designed to explore the reduction potential of Beijing. The AQ scenario emphasizes the recently published (after 2010) policy and illustrates the reduction resulting from strict implementation of existing measures. We

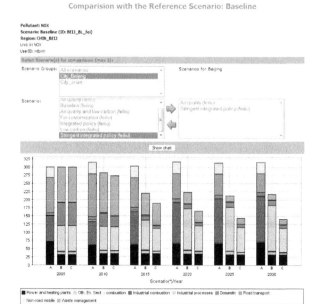

Fig. 3. Comparison with the Reference Scenario: Baseline

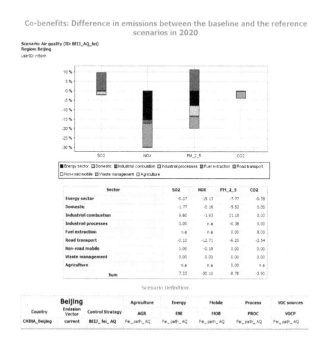

Fig. 4. Co-benefits: Difference in emissions between the baseline and the reference scenarios in 2020

believe that efficient implementation of most recent policies as depicted in the AQ scenario could stimulate further progress in controlling air pollution and resulting in introduction of stricter (more ambitious and more efficient) set of measures, a scenario referred to as Strict Air Quality (SAQ) policy. For example, SAQ additionally introduced *installation of fabric filters in newly built power plants.* Figure 3 presents the comparisons the reduction potential of policies among the three scenarios by sector of Beijing during 2005-2030. In 2030, AQ and SAQ scenarios would reduce 128 and 139 Gg of NOX emission, the results showed great reduction potential: 38-42% of NOX emissions.

Furthermore, figure 4 also shows the emission reductions of multi-pollutants, i.e. SO2,NOX,PM2.5 and CO2, feature of the GAINS-City model.

5   Conclusion

In this paper, the RFDW [13] has been used to develop a class of emission city scale data marts based on urban win-win policy by policy requirements. The policy as well as its policy by policy (policy packages) have been specified in very formal manner. To fulfill such specific policies at city level, two major adjustments has been studied: extending source categories of original GAINS model, i.e. regional dimension is extended to city level, and activity pathway data is disaggregated based on city scale level with the updating of emission factor database according to local measurements. As a result, the GAINS-City data marts are suitable for city-scale emission estimation and co-benefits assessment. In addition, the data mart, namely GAINS City China was presented to demonstrate the city scale model features.

The methodology of GAINS-City could be applied to other large (mega)cities, due to its independent model structure and proposed policy list that are considered essential in most Chinese cities. The large number of source categories included in the model and diverse list of policies focusing on primary sources could support the emission estimation and co-benefits analysis for several cities.

Our future work will be focused on data verification and follow up with the development and implementation of transfer matrices in the GAINS-City model. Finally, estimating costs for the developed measures that would consider specific local factors should be a high priority for future development.

References

1. Amann, M., Bertok, I., Borken, J., Cofala, J., Heyes, C., Hoglund, L., Klimont, Z., Purohit, P., Rafaj, P., Schoepp, W., Toth, G., Wagner, F., Winiwarter, W.: GAINS - potentials and costs for greenhouse gas mitigation in annex i countries. Tech. rep., International Institute for Applied Systems Analysis (IIASA), Laxenburg, Austria (2008), http://gains.iiasa.ac.at/
2. Amann, M., Bertok, I., Borken-Kleefeld, J., Cofala, J., Heyes, C., Hoeglund-Isaksson, L., Klimont, Z., Nguyen, B., Posch, M., Rafaj, P., et al.: Cost-effective control of air quality and greenhouse gases in europe: modeling and policy applications. Environmental Modelling & Software (2011)

3. Aunan, K., Fang, J., Vennemo, H., Oye, K., Seip, H.M.: Co-benefits of climate policy lessons learned from a study in shanxi, china. Energy Policy 32(4), 567–581 (2004), http://www.sciencedirect.com/science/article/pii/S0301421503001563, doi:10.1016/S0301-4215(03)00156-3
4. Nguyen, T.B., Tjoa, A.M., Wagner, R.: Conceptual multidimensional data model based on metaCube. In: Yakhno, T. (ed.) ADVIS 2000. LNCS, vol. 1909, pp. 24–33. Springer, Heidelberg (2000)
5. Hoang, T.A.D., Nguyen, T.B.: State of the art and emerging rule-driven perspectives towards service-based business process interoperability. In: International Conference on Computing and Communication Technologies, RIVF 2009, pp. 1–4 (July 2009)
6. Anh Hoang, D.T., Tran, H., Nguyen, B.T., Tjoa, A.M.: Towards the development of large-scale data warehouse application frameworks. In: Møller, C., Chaudhry, S. (eds.) CONFENIS 2011. LNBIP, vol. 105, pp. 92–104. Springer, Heidelberg (2012)
7. Jindal, R., Acharya, A.: Federated data warehouse architecture. Wipro Technologies white paper (2004)
8. Kennedy, C., Steinberger, J., Gasson, B., Hansen, Y., Hillman, T., Havrainek, M., Pataki, D., Phdungsilp, A., Ramaswami, A., Mendez, G.V.: Greenhouse gas emissions from global cities. Environmental Science & Technology 43(19), 7297–7302 (2009), http://dx.doi.org/10.1021/es900213p, doi:10.1021/es900213p
9. Kennedy, C., Steinberger, J., Gasson, B., Hansen, Y., Hillman, T., Havraenek, M., Pataki, D., Phdungsilp, A., Ramaswami, A., Mendez, G.V.: Methodology for inventorying greenhouse gas emissions from global cities. Energy Policy 38(9), 4828–4837 (2010), http://www.sciencedirect.com/science/article/pii/S0301421509006387, doi:10.1016/j.enpol.2009.08.050
10. Kern, R., Ryk, K., Nguyen, N.T.: A framework for building logical schema and query decomposition in data warehouse federations. In: Jędrzejowicz, P., Nguyen, N.T., Hoang, K. (eds.) ICCCI 2011, Part I. LNCS, vol. 6922, pp. 612–622. Springer, Heidelberg (2011)
11. Liu, F., Klimont, Z., Zhang, Q., Cofala, J., Zhao, L., Huo, H., Nguyen, B., Schoepp, W., Sander, R., Zheng, B., Hong, C., He, K., Amann, M., Heyes, C.: Integrating mitigation of air pollutants and greenhouse gases in chinese cities: development of GAINS-City model for beijing. Journal of Cleaner Production, http://www.sciencedirect.com/science/article/pii/S0959652613001583, doi:10.1016/j.jclepro.2013.03.024
12. Nguyen, T.B., Wagner, F., Schoepp, W.: Cloud business intelligent services to explore the synergies and interactions among climate change, air quality objectives. In: Proceedings of the 1st International Workshop on Cloud Intelligence, Cloud-I 2012, pp. 6:1–6:8. ACM, New York (2012)
13. Nguyen, T.B.: Cloud-based data warehousing application framework for modeling global and regional data management systems. In: Nguyen, N.T., van Do, T., Thi, H.A. (eds.) ICCSAMA 2013. SCI, vol. 479, pp. 319–327. Springer, Heidelberg (2013)
14. Nguyen, T.B., Wagner, F., Schoepp, W.: EC4MACS – an integrated assessment toolbox of well-established modeling tools to explore the synergies and interactions between climate change, air quality and other policy objectives. In: Auweter, A., Kranzlmüller, D., Tahamtan, A., Tjoa, A.M. (eds.) ICT-GLOW 2012. LNCS, vol. 7453, pp. 94–108. Springer, Heidelberg (2012)

15. Nguyen, T.B., Schoepp, W., Wagner, F.: GAINS-BI, p. 332. ACM Press (2008), http://dl.acm.org/citation.cfm?id=1497369, doi:10.1145/1497308.1497369
16. Sheth, A.P., Larson, J.A.: Federated database systems for managing distributed, heterogeneous, and autonomous databases. ACM Comput. Surv. 22(3), 183–236 (1990), http://doi.acm.org/10.1145/96602.96604, doi:10.1145/96602.96604
17. Nguyen, T.B., Tjoa, A.M., Mangisengi, O.: MetaCube XTM: a multidimensional metadata approach for semantic web warehousing systems. In: Kambayashi, Y., Mohania, M., Wöß, W. (eds.) DaWaK 2003. LNCS, vol. 2737, pp. 76–88. Springer, Heidelberg (2003)
18. van Gelder, K.: Elastic data warehousing in the cloud, http://homepages.cwi.nl/~boncz/msc/2011-KeesvanGelder.pdf
19. Watson, H.J., Wixom, B.H.: The current state of business intelligence. Computer 40, 96–99 (2007), http://dl.acm.org/citation.cfm?id=1301970, doi:10.1109/MC.2007.331
20. Wei, X., Xiaofei, X., Lei, S., Quanlong, L., Hao, L.: Business intelligence based group decision support system. In: Proceedings of International Conferences on Info-tech and Info-net, ICII 2001, Beijing, vol. 5, pp. 295–300 (2001)

Next Improvement Towards Linear Named Entity Recognition Using Character Gazetteers

Giang Nguyen, Štefan Dlugolinský, Michal Laclavík, Martin Šeleng, and Viet Tran

Institute of Informatics, Slovak Academy of Sciences
Dúbravská cesta 9, 845 07 Bratislava, Slovakia
{giang.ui,stefan.dlugolinsky,laclavik.ui,
martin.seleng,viet.ui}@savba.sk

Abstract. Natural Language Processing (NLP) is important and interesting area in computer science affecting also other spheres of science; e.g., geographical processing, social statistics, molecular biology. A large amount of textual data is continuously produced in media around us and therefore there is a need of processing it in order to extract required information. One of the most important processing steps in NLP is Named Entity Recognition (NER), which recognizes occurrence of known entities in input texts. Recently, we have already presented our approach for linear NER using gazetteers, namely Hash-map Multiway Tree (HMT) and first-Child next-Sibling binary Tree (CST) with their strong and weak sides. In this paper, we present Patricia Hash-map Tree (PHT) character gazetteer approach, which shows as the best compromise between the both previous versions according to matching time and memory consumption.

Keywords: gazetteer, named entity recognition, natural language processing.

1 Introduction

The task of Named Entity Recognition (NER) is to recognize occurrences of well-known entities in input texts coming from websites, web portals, social media, data dumps, documents, etc. These entities might be of arbitrary type, but the most used types in NER are persons, organizations, locations, times, and quantities. NER is often understood or classified as a subtask of Information Extraction (IE) in Natural Language Processing (NLP Fig. 1). One of the commonly used NER techniques is gazetteer a simple list of well-known entities, which are looked up in input texts. In NER research area, the word "*gazetteer*" is used interchangeably for both the set of entity lists and for the processing resource that uses those lists to find occurrences of named entities in texts.

There are two main approaches of a gazetteer implementation:

- Machine learning techniques: usually known under various rule-based techniques
- Finite State Machines (FSM) techniques.

T.V. Do et al. (eds.), *Advanced Computational Methods for Knowledge Engineering*,
Advances in Intelligent Systems and Computing 282,
DOI: 10.1007/978-3-319-06569-4_19, © Springer International Publishing Switzerland 2014

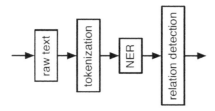

Fig. 1. Natural Language Processing (NLP) and Named Entity Recognition (NER)

An example of a high-level rule-based language for building and customizing NER annotators is described in [5]. Unfortunately, the process of designing the rules themselves is manual and time-consuming, but the rule-based approach performance is comparable to that of machine learning. Pattern-based matching and validation in an unlabeled corpus is described in [6] and [7] respectively. The outcome is that patterns can be quite complex to fulfill different requirements.

Authors in [2] present a gazetteer implemented as an FSM. Gazetteer is built at initialization time starting from the list of phrases that need to be later recognized with contextual dependency. Contextual dependency is also one of the assumptions in [3]. In the work [4], authors deal with automated gazetteer construction as well as with various problems like entity-noun ambiguity, entity-entity ambiguity and entity boundary detection. The overlapping of entities is solved here by selecting and aliasing them.

There are two main approaches that are used to find matches according to the way of handling input text:

- Token-based: input text is split into a sequence of tokens on which a matching is performed
- Character-based: input text is processed character by character. The approach of input handling determines internal representation of the gazetteer list

An example of the token gazetteer is Hash Gazetteer provided by Ontotext[1] in GATE project [1]. It is constructed based on hash tables of words. Authors declare that it takes in average four times less memory and that it works three times faster than GATE's previously optimized character-based FSM implementation. Another example is Large Knowledge Base Gazetteer, a new-generation gazetteer, which provides support for ontology-aware NLP. It allows using of large gazetteer lists and speeds up subsequent loading of data by caching. Linked Data Gazetteer is an experimental gazetteer that uses Linked Open Data for lookups. Besides the Ontotext gazetteers, there is a number of scientific works dealing with NER using comparable approaches. Authors in [2] present an implemented FSM gazetteer, which is built at initialization time by starting from the list of phrases that need to be later recognized with contextual dependency in token level.

---

[1] http://www.ontotext.com/collaborations/gate

2   Character Gazetteers

Despite the large number of listed works in the previous section, existing gazetteers are in various implementation states and availabilities. A lot of them are integrated as a part of other software products. Therefore our work in Information Extraction (IE) area concretely on NER comes from the following requirements:

- Standalone gazetteer independent of 3rd party libraries
- Do not rely on external preprocessing; e.g., tokenization
- Linear complexity lookup algorithm, which provides fast and effective processing of input text as a stream, especially for Big Data
- Editable data structure; i.e., add/remove Named Entities (NEs) between lookups
- Memory efficient data structure; i.e., ability to deal at least with several millions of entities
- Robustness; i.e., input texts of any size and in any language

Gazetteers with token-level tokenization usually faces several problems such as the need of multi-travel search and matching, finding word boundaries, chunking or processing of non-trivial strings and characters. These problems lead to longer running time and low processing performance of this kind of gazetteers. Therefore, we have turned to gazetteers based on character-level tokenization, which provides more precise results and much better performance than token-level equivalent despite that character gazetteer is more memory consuming.

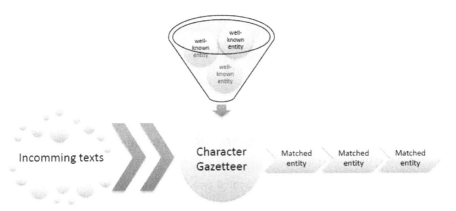

Fig. 2. Character gazetteer

The usage of gazetteers has usually two phases:

- Filling gazetteer structures with datasets of well-known entities: this step is usually performed in the initial phase.
- Matching entity occurrences in continuously incoming input text and reporting these occurrences with their references (URI or MDI, where MID is an ID of an object in Freebase) to the system for next processing steps.

Entity datasets, as the main source of gazetteer lists, are usually large and have tendency to grow; e.g., due to adding of new known entities during the running time. This fact significantly affects memory requirements, running time as well as the tokenization complexity, especially of character gazetteers. Therefore, an efficient data structure is required to effectively fold all the entities at the character level, which also enables a fast linear tokenization of input text and produces required output (references, occurrence quantity and positions in the input text). Tree structure is a powerful way of organizing data hierarchically and suits these considerations very well. Moreover, it provides easy and quick access operations in order to find elements and traverse through the structure.

Possible entity positions in input text are depicted in Fig. 3. The basic situations are case 1 following situation, case 2 embedded situation and case 3 overlapping situation. Case 4, where the entities start at the same position and are embedded in the longest entity, is quite well solved and well-known as a problem of finding the longest or the first matched entity. The case 1, case 2 and case 3 can be freely occupied and they can create various combinations in incoming input texts (e.g. case 5). Therefore, matching process is quite complicated to find all entities, especially when entities are embedded, but not from the start boundary and/or they are also overlapped.

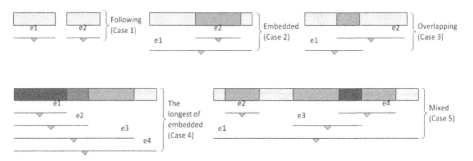

Fig. 3. Possible entity positions in incoming input text (*e* stands for *entity*)

2.1 Hash-Map Multi-way Tree (HTM) and Child-Sibling Tree (CST) Character Gazetteers

In our previous work [9] [10], we presented two representations of the character gazetteer data structure as depicted in Fig. 4.

The HMT gazetteer is based on multi-way tree and implemented using Java HashMap, which guarantees constant-time performance $O(1)$ in average for basic operations. Each node of the HTM tree has a hash map in which it can have arbitrary number of child nodes as its values; i.e. consecutive characters.

The CST gazetteer is based on the first-child next-sibling binary tree and implemented using pure Java simple node object. The CST tree is a binary representation of the HMT tree, where each node refers only to its first child node and its next sibling node. This representation is not as fast as the HMT, but it is efficient in memory consumption the CST uses in average three to four times less memory in comparison to HMT.

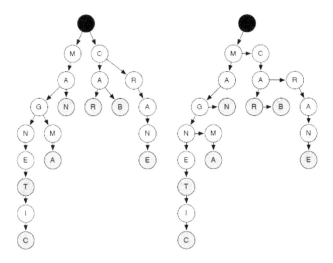

Fig. 4. Multi-way tree structures of HMT gazetteer (left) and first-child next-sibling tree structure of CST gazetteer (right)

The *filling algorithm* [9] loops over all its characters while traversing and building the tree with inserted entities. Finally, a node representing the last character of the input string is marked as a matching node. The difference between two gazetteers is only in the way of how the tree nodes are physically connected.

After the tree gazetteers are filled with the entities during the initialization phase, they are ready to be used for tokenization and finding of well-known entities in the input text. This process is performed by a matching algorithm [9], which can be briefly described as a travelling of the gazetteer tree from the root node (the black circle in Fig. 4) matching entity nodes; i.e., gray circles in Fig. 4.. The matching algorithm for both HMT and CST gazetteers has $O(n)$ complexity, where n is the number of characters in the input text. It means that we need to traverse the input text approximately one time to obtain the results. The matching algorithm is intended to be used on a stream of text of unlimited size.

In paper [10] we present a matching time comparison of existing approaches. In paper [9] we provide an approach of solving situations with embedded and especially with overlapping entities in order to match all the possible well-known entities with negligible increment of matching complexity. This approach of matching algorithm is applied for both HTM and CST gazetteer.

2.2   Patricia Hash-Map Tree (PHT) Character Gazetteer

Although HMT and CST gazetteers work well; i.e., HMT is suitable for machine with more memory and applications that requires very fast processing/matching time; CST is suitable when memory issue is more pushed toward; there is still a gap for a gazetteer, which can work fast and uses less memory. PHT gazetteer (Patricia Hash-map

Tree) seems to be a good candidate to fill the gap. It is named after the Patricia tree data structure and is based on Java HashMap implementation, which has an $O(1)$ complexity for basic operations. Logical representation of the PHT tree is the same as the HMT's. The only difference is that each PHT tree node is collapsible and can contain more characters instead of only one character. In comparison to HMT and CST, nodes in PHT require more memory, but the number of nodes is significantly reduced and therefore the memory usage is more efficient (Fig. 5).

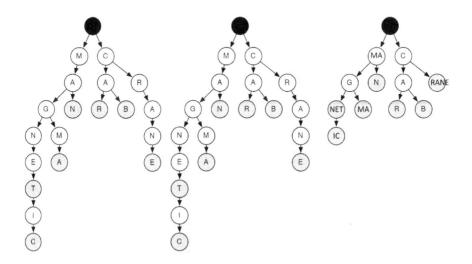

Fig. 5. Tree structure of HMT (left); collapsible nodes (middle) and PHT (right)

Of course the complex collapsible node structure complicates the *filling algorithm*, but except of the usual case, it also deals with cases when:
- The incoming character of the incoming entity splits the existing node into two nodes
- The incoming entity marks new matching/list nodes

In comparison to previous HMT and CST gazetteers, the filling algorithm is modified to deal with the collapsible nodes:

- Stores multiple characters in one node when the node does not have a direct child node
- Breaks the existing node into two nodes according to incoming character; i.e., makes nodes collapsible

The gazetteer tree is created in a flexible way to utilize the memory. Each node can be expanded when necessary and remains as a single node if the incoming character can be stored on it.

```
Algorithm: Insert new entity into PHT tree
 1:  node ← root node
 2:  FOR ALL character in entity DO
 3:      IF character is in the right place in node's list THEN
 4:          IF node is not matching node AND node does not have child THEN
 5:              add character to node's list
 6:          ELSE
 7:              IF node has a child with character THEN
 8:                  node ← child
 9:              ELSE
10:                  child ← new node with character
11:                  add child to node
12:                  node ← child
13:              END IF
14:          END IF
15:      ELSE
16:          first child ← new node with node's sublist from character
17:          add first child to node
18:          second child ← new node with character
19:          add second child to node
20:          node's list ← sublist before character
21:          node ← second child
22:      END IF
23:  END FOR
24:  IF entita is shorter than node's list THEN
25:      child ← new node with node's sublist extends entita
26:      add child to node
27:      node's list ← entita
28:  END IF
29:  mark node as a matching node
```

Fig. 6. PHT filling algorithm: Insert entity into gazetteer's tree

The *matching algorithm* is also modified, and in comparison to HMT case is more complex, but it is compensated with the shorter travel path in the matching algorithm.

```
Algorithm: Matching algorithm of PHT gazetteer
 1:  buf ← empty
 2:  node ← root node
 3:  onNode ← 0
 4:  WHILE character on input DO
 5:      ch ← next character from input
 6:      normalize whitespace or skip multiple
 7:      IF ch is mapped in onNode position of node THEN
 8:          add ch to buf
 9:          onNode++
```

```
10:      IF the last mapped character and node is matching node THEN
11:          found an entity
12:      END IF
13:      ELSE IF node has a child node mapped on ch THEN
14:         add ch to buf
15:         node ← child node mapped on ch
16:         onNode ← 1
17:         IF node has only one character AND node is matching node THEN
18:             found an entity
19:         END IF
20:      ELSE
21:         IF buf contains boundary characters AND entity was found THEN
22:             unread characters from buf back to input until the first
                boundary occurrence
23:         END IF
24:         buf ← empty
25:         node ← root node
26:      END IF
27:  END WHILE
```

Fig. 7. PHT matching algorithm

Although the PHT matching algorithm is a little bit more complicated, the $O(n)$ complexity is guaranteed. We need only one travel though the input text to find all the occurrences of entities. Like the HMT and the CST, the PHT matching algorithm deals with all the entity occurrence cases without significant increment of algorithm complexity; e.g., embedded entities and overlapping entities (Fig. 5).

3   Experiments and Evaluation

There are various sources of named entities available; e.g. Google Data Dumps[2] of FreeBase[3], DBpedia[4]. Therefore, it is not a problem to build up gazetteer lists for miscellaneous domains and testing purposes. Concretely, we use the FreeBase person dataset for the memory consumption testing and for comparisons. The Freebase person dataset has 2,614,401 unique entities of size 163 MB. Matching time was measured in milliseconds by executing the gazetteers over a set of 2,586 documents, 5,107 documents, 7,564 documents and 9,909 documents acquired from CoNLL-2003 datasets [8] with approximately 7 MB, 15 MB, 22 MB and 29 MB of pure text. There is no limitation of the language in the input text for all three versions (HMT, CST and PHT) of our character gazetteers. The test was repeated several times and the measured values were averaged. Results are depicted in Fig. 8.

---

[2] https://developers.google.com/freebase/data
[3] http://www.freebase.com/
[4] http://dbpedia.org/

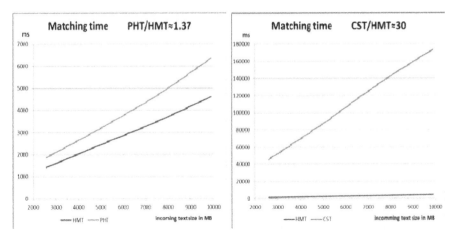

Fig. 8. Matching time comparison: PHT vs. HTM and CST vs. HMT

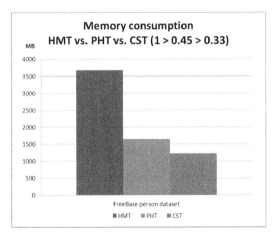

Fig. 9. Matching time comparison between PHT and CST gazetteer

In general, the weakness of the character gazetteer is in its memory consumption, especially for the HMT implementation, where each parent node requires additional memory for its hash map. Also the number of nodes is quite big. Theoretically, each tree node can have so many child nodes as the number of the characters in the charset set, thus the tree structure can be very wide. Fortunately, the tree space is limited by human language and all the combinations will never be covered, thus the management of the gazetteer tree in machine memory is possible also on commodity hardware

The CST version deals the best with the memory issue CST/HMT=0.33 (Fig. 9). That means the CST gazetteer uses only one thirds memory amount in comparison with HMT gazetteer. On the second hand, the matching time of the CST is significantly increased CST/HMT=30 (Fig. 8), that means it is 30 times slower than the HMT gazetteer. This result is not very bad when the matching time is still linear ($O(n)$ matching complexity), but we wanted a better rate.

The PHT gazetteers uses slightly more memory than the CST; i.e., PHT/HMT=0.45 (Fig. 9), that means it uses less than haft of memory amount in comparison with the HMT gazetteer when the CST uses nearly one thirds of memory amount in comparison with the HMT gazetteer. The PHT matching time is linear, a little increased in quite low level PHT/HMT=1.37 (Fig. 8).

4   Conclusion

In this paper, we present a Patricia Hash-map Tree character gazetteer (PHT) and compare it with our two previous Hash-map Multi-way Tree (HMT) and first-Child next-Sibling binary Tree (CST). In comparison to HMT gazetteer, PHT gazetteer benefits from memory saving with slightly acceptable slowdown of the matching time. The PHT fulfils our requirements for improving the tree data structure memory consumption and for more efficient traversing as it is in HTM and in CST respectively.

All the presented implementations of our character gazetteer are available online [5].

Acknowledgment. this work is supported by projects VEGA 2/0185/13, KC INTELINSYS ESF ITMS 26240220072, CLAN APVV-0809-11 and EGI-InSPIRE EU FP7-261323 RI.

References

1. Cunningham, H., Maynard, D., Bontcheva, K., Tablan, V., Aswani, N., Roberts, I., Gorrell, G., Funk, A., Roberts, A., Damljanovic, D., Heitz, T., Greenwood, M.A., Saggion, H., Petrak, J., Li, Y., Peters, W.: Text Processing with GATE, Version 6 (2011), http://tinyurl.com/gatebook
2. Maynard, D., Tablan, V., Ursu, C., Cunningham, H., Wilks, Y.: Named entity recognition from diverse text types. In: Recent Advances in Natural Language Processing 2001 Conference, pp. 257–274 (2001)
3. Liu, X., Zhang, S., Wei, F., Zhou, M.: Recognizing named entities in tweets. In: Proceedings of the 49th Annual Meeting of the Association for Computational Linguistics: Human Language Technologies. HLT 2011, vol. 1, pp. 359–367. Association for Computational Linguistics, Stroudsburg (2002), http://dl.acm.org/citation.cfm?id=2002472.2002519
4. Nadeau, D., Turney, P.D., Matwin, S.: Unsupervised named-entity recognition: Generating gazetteers and resolving ambiguity. In: Lamontagne, L., Marchand, M. (eds.) Canadian AI 2006. LNCS (LNAI), vol. 4013, pp. 266–277. Springer, Heidelberg (2006), http://dx.doi.org/10.1007/11766247_23
5. Chiticariu, L., Krishnamurthy, R., Li, Y., Reiss, F., Vaithyanathan, S.: Domain adaptation of rulebased annotators for named-entity recognition tasks. In: Proceedings of the 2010 Conference on Empirical Methods in Natural Language Processing. EMNLP 2010, pp. 1002–1012. Association for Computational Linguistics, Stroudsburg (1870), http://dl.acm.org/citation.cfm?id=1870658.1870756
6. Laclavik, M., Hluchy, L., Seleng, M., Ciglan, M.: Ontea: Platform for pattern based automated semantic annotation. Computing and Informatics 28(4), 555–579 (2009)

---

[5] http://ikt.ui.sav.sk/gazetteer

7. Kozareva, Z.: Bootstrapping named entity recognition with automatically generated gazetteer lists. In: Proceedings of the Eleventh Conference of the European Chapter of the Association for Computational Linguistics: Student Research Workshop. EACL 2006, pp. 15–21. Association for Computational Linguistics, Stroudsburg (2006), http://dl.acm.org/citation.cfm?id=1609039.1609041

8. Tjong Kim Sang, E.F., De Meulder, F.: Introduction to the conll-2003 shared task: language-independent named entity recognition. In: Proceedings of the Seventh Conference on Natural Language Learning at HLT-NAACL 2003. CONLL 2003, pp. 142–147. Association for Computational Linguistics, Stroudsburg (2003), http://dx.doi.org/10.3115/1119176.1119195

9. Dlugolinský, Š., Nguyen, G., Laclavík, M., Šeleng, M.: Character Gazetteer for Named Entity Recognition with Linear Matching Complexity. In: 3rd World Congress on Information and Communication Technologies, WICT 2013, pp. 364–368 (2013) IEEE Catalog Number: CFP1395H-ART, ISBN: 978-1-4799-3230-6

10. Nguyen, G., Dlugolinský, Š., Laclavík, M., Šeleng, M.: Token gazetteer and character gazetteer for named entity recognition. In: Babič, F., Paralič, J. (eds.) 8th Workshop on Intelligent and Knowledge Oriented Technologies: WIKT 2013 Proceedings. Centre for Information Technologies, Technical University in Košice, pp. 1–6 (2013), http://web.tuke.sk/fei-cit/wikt2013/wikt%202013.pdf

11. Dlugolinský, Š., Krammer, P., Ciglan, M., Laclavík, M.: MSM2013 IE Challenge: Annotowatch. In: Proceedings of the Concept Extraction Challenge at the Workshop on Making Sense of Microposts co-located with the 22nd International World Wide Web Conference (WWW 2013), Rio de Janeiro, Brazil, May 13, vol. 1019, pp. 21–26 (2013)

Part V

Logic Based Methods for Decision Making and Data Mining

Semantic Evaluation of Text Clustering*

Sinh Hoa Nguyen[1], Wojciech Świeboda[2], and Hung Son Nguyen[2]

[1] Polish-Japanese Institute of Inf. Technology,
Koszykowa 86, 02008, Warszawa, Poland
hoa@mimuw.edu.pl
[2] Institute of Mathematics, Warsaw University
Banacha 2, 02-097 Warsaw, Poland
son@mimuw.edu.pl

Abstract. In this paper, we investigate the problem of quality analysis of clustering results using semantic annotations given by experts. We propose a novel approach to construction of evaluation measure, called SEE (Semantic Evaluation by Exploration), which is an improvement of the existing methods such as Rand Index or Normalized Mutual Information. We illustrate the proposed evaluation method on the freely accessible biomedical research articles from Pubmed Central (PMC). Many articles from Pubmed Central are annotated by the experts using Medical Subject Headings (MeSH) thesaurus. We compare different semantic techniques for search result clustering using the proposed measure.

Keywords: Text clustering, semantic evaluation, Pubmed, MeSH.

1 Introduction

Clustering can be understood as an unsupervised data mining task for finding groups of points that are close to each other within the cluster and far from the rest of clusters. Intuitively, the greater the similarity (or homogeneity) within a cluster, and the greater the difference between groups, the "better" the clustering. Clustering is a widely studied data mining problem in the text domains, particularly in segmentation, classification, collaborative filtering, visualization, document organization, and semantic indexing.

It is a fundamental problem of unsupervised learning approaches that there is no generally accepted "ground truth". As clustering searches for previously unknown cluster structures in the data, it is not known a priori which clusters should be identified. This means that experimental evaluation is faced with enormous challenges. While synthetically generated data is very helpful in providing an exact comparison measure, it might not reflect the characteristics of real world data. In recent publications, labeled data, usually used to evaluate

* This work is partially supported by the National Centre for Research and Development (NCBiR) under Grant No. SP/I/1/77065/10 by the Strategic scientific research and experimental development program: "Inter-disciplinary System for Interactive Scientific and Scientific-Technical Information".

T.V. Do et al. (eds.), *Advanced Computational Methods for Knowledge Engineering,*
Advances in Intelligent Systems and Computing 282,
DOI: 10.1007/978-3-319-06569-4_20, © Springer International Publishing Switzerland 2014

the performance of classifiers, i.e. supervised learning algorithms, is used as a substitute [19,6,1]. While this provides the possibility of measuring the performance of clustering algorithms, the base assumption that clusters reflect the class structure is not necessarily valid.

Some approaches therefore resort to the help of domain experts in judging the quality of the result [2,7,6]. When domain experts are available, which is clearly not always the case, they provide very realistic insights into the usefulness of a clustering result. Still, this insight is necessarily subjective and not reproducible by other researchers. Moreover, there is not sufficient basis for comparison, as the clusters that have not been detected are unknown to the domain expert.

There are several suggestions for a measure of similarity between two clusterings. Such a measure can be used to compare how well different data clustering algorithms perform on a data set. These measures are usually tied to the type of criterion being considered in evaluating the quality of a clustering method [10].

The problem becomes even more complicated in evaluation of text clustering with respect to semantic similarity, whose definition is not precise and highly contextual. As the number of results is typically huge, it is not possible to manually analyze the quality of different clustering algorithms.In this paper we extend the results in [11] by the properties of SEE in Section 4.

The remainder of this paper is structured as follows. In Section 2 we present some basic notions and problem statement. This is followed by an overview of external clustering evaluation methods in Section 3. In Section 4 we present the basic semantic evaluation techniques and propose a novel evaluation method based on exploration of expert's tags which is the fundamental contribution of this paper. After this we use the proposed evaluation method to analyze the search result clustering algorithms over the document collection publish by PubMed. An analysis of the methods, the result representation and their interpretability is presented in Section 5, followed by some conclusions in Section 6.

2   Problem Statement

A hard clustering algorithm is any algorithm that assigns a set of objects (e.g. documents) to disjoint groups (called clusters). A soft clustering relaxes the condition on target clusters being disjoint and allows them to overlap. Clustering evaluation measures[16,10] proposed in the literature can be categorized as either *internal criteria of quality* or *external criteria of quality*.

An *internal criterion* is any measure of "goodness" defined in terms of object similarity. These criteria usually encompass two requirements – that of attaining high intracluster similarity of objects and high dissimilarity of objects in different clusters. *External criteria* on the other hand compare a given clustering with information provided by experts. Typically in the literature it is assumed that both the clustering provided by studied algorithm and the clustering provided by experts are hard clusterings. We believe that the requirement that expert knowledge is described in terms of a hard clustering is overly restrictive. In typical applications in text mining, one faces datasets which are manually labeled

Table 1. Illustration of soft clustering found by an algorithm and an expert

Doc.	Soft Cluster			Expert Tag				
	C_1	C_2	C_3	Cosmonaut	astronaut	moon	car	truck
d_1	1			1			1	1
d_2	1				1	1		
d_3	1	1		1				
d_4		1	1				1	1
d_5		1	1			1		
d_6			1					1

by experts, but with each document being assigned a set of tags. We can think of such tags as of soft clusters. In this paper we aim to provide measures of external evaluation criteria that relax both conditions on the clustering and expert clusterings being partitions (hard clusterings).

Typically in the literature it is assumed that the input data to clustering evaluation can be described in a form similar to Table 1, i.e. with exactly one valid cluster C_i and exactly one valid expert cluster E_j for each document. We will relax this condition to allow comparison of soft clustering and a set of expert tags assigned to each document, thus allowing input data as in Table 1.

3 Overview of Clustering Evaluation Methods

In this section we briefly review external evaluation criteria typically used in clustering evaluation. We assume that two partitions (hard clusterings) of objects are given: one by an algorithm, and another one provided by domain experts. Most external evaluation criteria can be naturally grouped in two groups:

Pair-Counting Measures. These measures are defined on a 2×2 contingency matrix Table 2 (left) that summarizes similarity of pairs of objects w.r.t. both clusterings: If there are k objects in the data set, then $a + b + c + d = \binom{k}{2}$. A typical measure that can be expressed in terms of these numbers is

$$Rand\ Index = \frac{a+d}{a+b+c+d}.$$

However, there exit different variants of other similar measures. Pfitzner et al.[16] provide an overview of 43 measures that all fit into this scheme.

Information-Theoretic Measures. on the other hand compare distributions of $c(D)$ and $e(D)$, which denote respectively the cluster and the expert label assigned to a document D drawn randomly from the dataset. These measures can be expressed in terms of joint distribution of $\langle c(D), e(D) \rangle$, i.e. simply by counting objects belonging to each pair $\langle C_i, E_j \rangle$ as shown in Table 2 (right). Numbers in brackets denote expected values of counts assuming independence of $c(D)$ and $e(D)$. Information-theoretic measures thus aim to to measure the

Table 2. Left: All pair-counting measures can be summarized in terms of numbers a, b, c, d. Right: Information-theoretic measures are defined in terms of contingency table (following example in Table 1).

Pairs of documents		Same cluster?	
		Yes	No
Same	Yes	a	b
expert tag?	No	c	d

	C_1	C_2	C_3	Total
E_1	1	0	0	1
E_2	0	0	1	1
E_3	0	1	1	2
E_4	1	1	0	2
Total	2	2	2	6

degree of dependence between these two. An example such measure is *mutual information MMI* between $c(D)$ and $e(D)$, where

$$MMI(X, Y) = \sum_x \sum_y p(x, y) \log \left(\frac{p(x, y)}{p(x)p(y)} \right).$$

A measure typically used in clustering evaluation is *Normalized Mutual Information* [10], though [16] reviews 13 different measures, all defined quite similarly. *Purity* is a measure occasionally used as an external evaluation criterion.

Semantic Evaluation Methods for Soft Clustering. We stress two limitations of measures proposed in the literature and briefly reviewed thus far. The first limitation, already mentioned in the previous section, is the typical assumption that both the clustering algorithm and the experts provide partitions. A more important limitation, though, is that neither of these measures described so far resemble the thought process that the expert himself would undergo if he was faced with the task of manually evaluating a clustering.

Previous works by other authors on this problem include [3] (Fuzzy Mutual Information) and [8] (comparing set covers). In this section we we will describe a novel method of semantic evaluation to deal with the mentioned limitations.

First we describe how to extend a pair-counting measure of similarity of two partitions to a measure of similarity of set covers. In order to fully characterize a pair of documents $\langle d_i, d_j \rangle$, we proposed in [13] to define notions of cluster-similarity and expert-similarity for documents and base pair-counting measures on Table 2. This approach naturally extends any pair-counting measure, with the focus of our prior experiments on Rand Index [17]. We defined very simple notions of similarity: we considered two documents d_i, d_j θ-expert-similar, if

$$\frac{|e(d_i) \cap e(d_j)|}{|e(d_i) \cup e(d_j)|} \geq \theta$$

and we defined θ-cluster-similarity in the same way. This approach allows us to effortlessly apply each of the 43 pair-counting measures reviewed by Pfitzner[16].

Information-theoretic measures can be extended by counting a given document in multiple cells of Table 2 whenever the document is in multiple clusters

Table 3. Information-theoretic measures can be defined if we can describe the joint distribution of clusters and expert labels (see example in Table 1)

	C_1	C_2	C_3
Cosmonaut	0.139	0.083	0
astronaut	0.083	0	0
moon	0.139	0	0
car	0.056	0.125	0.125
truck	0	0.042	0.208

and/or multiple tags are assigned to the document. If we wish to assign an overall equal weight to each document, instead of raw counts, one may further assume that the contribution of a document is inversely proportional to the number of cells that it contributes to. This has a straightforward probabilistic interpretation. The original measures, like $MMI(c(D), e(D))$ are defined for deterministic functions c and e and a random document D. In the proposed extension, c and e are also random variables, with $c(d)$ uniformly distributed across clusters containing document d, and $e(d)$ uniformly distributed across tags assigned to document d. Original formulas themselves, like $MMI(c(D), e(D))$ remain unchanged. This approach is illustrated in Table 3.

4   Semantic Explorative Evaluation

We have mentioned that the calculation of neither of the measures reviewed so far resembles human reasoning. We propose a different approach to the problem of semantic evaluation.

If an expert faced the problem of manual inspection of clustering results, he would try to explain the contents of clusters in terms of expert tags. In essence, a cluster should be valid for an expert if the expert can briefly explain its contents. The expert would find a set of clusters valid if he could provide a short explanation for each cluster. In order to define a measure of semantic validity which is based on this reasoning, we need to specify: (1) the description of clusters in terms of expert concepts (i.e. a model family), (2) the length of such an explanation so that we know if it is short, (3) a penalty incurred if a cluster is indescribable in terms of expert concepts, and (4) the aggregate measure, so that we can evaluate a set of clusters. We specify these three ingredients as follows:

- The explanation of a cluster is in essence a classification model for this cluster in terms of expert tags. The exact choice of the classifier is of secondary importance as long as the same procedure is consistently used to evaluate different clusterings. In our experiments, the classifier of choice is a decision tree with no pruning, with splits defined greedily using Gini index.
- By appealing to Minimum Description Length principle, one may then define a measure of validity of a fixed cluster as the complexity of the model

describing the cluster. The measure of model complexity that we use is the average depth of the resulting decision tree.

- For simplicity we omit a penalty for indescribable clusters at this point, although we guarantee during tree construction that resulting leaves in decision trees contain either objects from the same decision class or objects that are indiscernible given the information about expert tags alone.

Finally, we define the measure of validity of a clustering as the average validity of clusters. Algorithm 1 presents the pseudo-code of this method.

Algorithm 1. SEE – Semantic Explorative Evaluation.

Input:
- $\mathbf{C} = \{C[i,j]\}$: the $k \times n$ document–cluster assignment matrix.
- $\mathbf{E} = \{E[i,j]\}$: the $k \times m$ expert–cluster assignment matrix.
- $\mathcal{L}$: a decision tree construction algorithm.

Output: m: the average mean depth of decision trees describing clusters.

1 **for** $j = 1, \ldots, n$ **do**
2   Construct a decision table $H_j := \left[E; [C_{1,j}, \ldots, C_{k,j}]^T\right]$;
3   // the decision table constructed from $\mathbf{E}$ augmented with the j-th column of matrix $\mathbf{C}$ at the end as the decision variable.
4     $T_j := \mathcal{L}(H_j)$; // Construct the decision tree T_j by applying algorithm L on decision table H_j.
5     $m_j = \text{MeanDepth}(T_j)$
6 **end**
7 Return $m = \frac{m_1 + \ldots + m_n}{n}$;

Example of Semantic Explorative Evaluation. In this Section we demonstrate the proposed evaluation method on the example introduced in Table 1.

This table consists of just 6 documents. Half of these documents, forming cluster C_2, concern vehicles: cars and trucks, whereas the other half concerns cosmonauts and moon: these documents form cluster C_1. Cluster C_1 is the easiest one to explain for the expert: he associates either the concept 'cosmonaut' or 'astronaut' with each document from this cluster. Document d_1 concerns a lunar rover and is an interesting "outlier" that needs to be explicitly excluded from cluster C_3 by the expert: the branch on attribute "moon" in decision tree describing cluster C_3 explicitly addresses this case.

The constructed decision trees T_1, T_2, T_3 for clusters C_1, C_2, C_3 are presented in Fig. 1, and Fig. 2, respectively. According to those trees, cluster C_3 seems to be "hardest" to explain by the expert. Hence the average depth or weighted average depth of the decision tree T_3 are also higher than for T_1.

Depths of decision trees describing clusters C_1, C_2, C_3 are $1\frac{2}{3}$, 2 and $2\frac{1}{4}$, respectively. Thus SEE of the clustering equals approximately 1.97.

| Doc. | Expert Tag | | | | | decision |
	Cosm.	astron.	moon	car	truck	C_1
d_1	1		1	1		1
d_2		1	1			1
d_3	1					1
d_4				1	1	
d_5				1		
d_6					1	

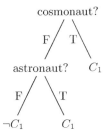

Fig. 1. The decision table H_1 (above) and the decision tree describing cluster C_1 constructed from H_1. Cluster C_1 is the easiest for the expert to explain.

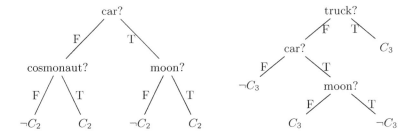

Fig. 2. The decision trees that describe C_2, C_3 in terms of expert knowledge

Properties of Semantic Explorative Evaluation Measures. The fundamental contribution of this paper is related to the semantic evaluation measure that simulates the process that the expert himself would undergo if he have to manually evaluate a clustering result. Let us outline the main properties of the proposed evaluation measure:

1. For hard clustering and hard expert tagging the Mutual Information (MMI) and Semantic Explorative Evaluation (SEE) measures have the similar range;
2. $MMI = 0$ if and only if $SEE = 0$;
3. An arithmetic mean of MMI is an arithmetic mean of SEE for all possible distributions of documents into clusters and expert tags.

Due to the space limitation we omit the formal proof of these properties.

Randomization. The last problem we aim to address is that of comparing different clusterings. With all evaluation methods, either reviewed or introduced in this article, we face the same issue when we aim to compare different clusterings: we lack an explanation why one clustering may be better than the other one. In this section, we introduce a trick which allows us to isolate a specific sub-problem solved by a clustering algorithm, to which we can assign a measure of quality that is easily interpretable.

In what follows, we will think of a clustering algorithm as of a procedure that solves two sub-problems. For hard clustering these are:

– Determining the structure of clusters, i.e. the number of clusters and the number of documents belonging to each cluster.
– The assignment of documents to clusters, while preserving constraints on the structure.

For soft clustering, these two sub-problems are:

– Determining the structure of clusters is actually determining the number of clusters K as well as the joint (rather than the marginal) distribution of the number of documents in each cluster.
– Instead of assigning documents to clusters, an algorithm assigns documents to each of the 2^K possible partitions.

In what follows, we will focus on measuring the quality of a clustering algorithm w.r.t. the second sub-problem, while ignoring the first sub-problem. The idea is to randomize the assignment of documents to clusters while keeping the structure of clusters fixed and calculate the value of m for such randomized assignments so as to determine a meaningful "basis" or benchmark for comparison. Each measure m can thus be transformed into an m-quantile measure, which basically says how often a clustering algorithm outperforms a random assignment, while solving the second sub-problem.

5   The Results of Experiments

The following experiments are the continuation of our previous studies in [14,12,13], although in this work they merely serve as an illustration of the discussed and introduced measures.

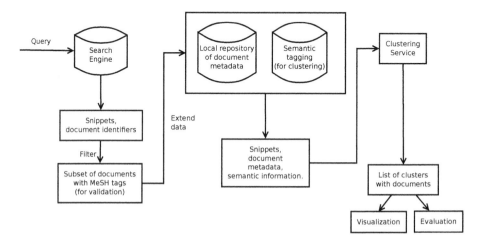

Fig. 3. Experiment diagram

5.1   Experiment Set-Up

We have applied the model-based semantic evaluation measure introduced in this paper to study clusterings induced by different document representations (lexical, semantic and structural) and using different algorithms. The document repository in our study is a subset of PubMed Central Open Access Subset[18].

The majority of documents in PubMed Central are tagged by human experts using headings and (optionally) accompanying subheadings (qualifiers) from a MeSH controlled vocabulary [21]. A single document is typically tagged by 10 to 18 heading-subheading pairs.

There are approximately 25000 subject headings and 83 subheadings. There is a rich structural information accompanying headings: each heading is a node of (at least one) tree and is accompanied by further information (e.g. allowable qualifiers, annotation, etc.). Currently we do not use this information, but in some experiments we use a hierarchy of qualifiers[1] by exchanging a given qualifier by its (at most two) topmost ancestors or roots. The numbers of possible tags in PubMed Central Open Access Subset[18] are summarized as follows:

source	possible tags	expert tags assigned to example document
headings	~ 25000	Internet, MEDLINE, Periodicals as Topic, Publishing
subheadings	83	economics
subheading roots	23	organization & administration

The choice of expert tags determines how precisely we wish to interpret expert opinion. In experiments that we describe in this paper, we interpreted subheadings as the expert tags.

The diagram of our experiments is shown in Figure 3. An experiment path (from querying to search result clustering) consists of three stages:

- Search and filter documents matching to a query. Search result is a list of *snippets* and document identifiers. Usually more than 200 documents are returned for a single query. The result set is then truncated to the top 200 most relevant (in terms of TF-IDF) documents.
- Extend representations of snippets and documents by *citations* and/or *semantically similar concepts* from MeSH ontology (these MeSH terms were automatically assigned by an algorithm[20], whereas MeSH subheadings used for evaluation were manually assigned by human experts).
- Cluster document search results.

In our experiments, we worked with three clustering algorithms: K-Means[9], Lingo[15] and Hierarchical Clustering[4].

In order to perform evaluation (and choose parameters of clustering algorithms) one needs a set of search queries that reflect actual user usage patterns.

---

[1] http://www.nlm.nih.gov/mesh/subhierarchy.html

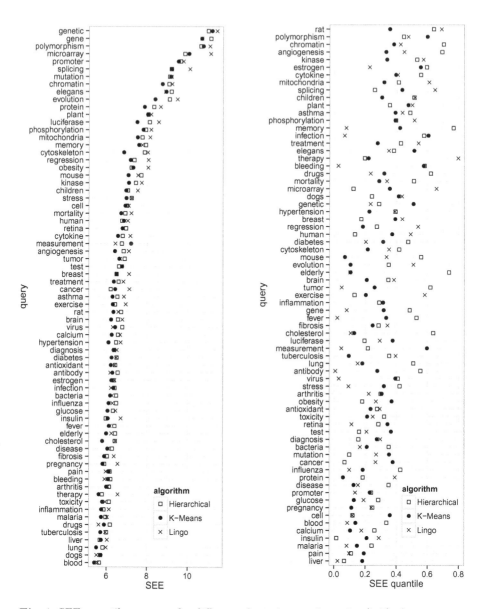

Fig. 4. SEE-quantile measure for different clusterings and queries (left). Average tree depth for different result sets and clustering algorithms (right).

We extracted a subset of most frequent one-term queries from the daily log previously investigated by Herskovic et al. in [5] and retrieved relevant documents from PubMed Central Open Access Subset.

Roughly one fourth of these result sets was used for initial fine-tuning of parameters, whereas the remaining 71 queries were further used in evaluation.

5.2   Experiment Results

We need to stress that we used subheadings as the source of expert tags used for semantic evaluation. There are only 83 possible subheadings in MeSH vocabulary, hence the granularity of information provided for evaluation is very limited. We have not applied pruning to resulting decision trees (the goal of algorithm Algorithm 1 is merely to provide a description, not a model for inference), and the resulting decision trees are somewhat deep, as can be seen from Figure 4.

Nevertheless, as we can see from Figure 4, the m-quantile measure is usually below 0.5 (SEE-quantile value 0.5 corresponds to a random document-to-cluster assignment). Furthermore, result sets for different queries visibly differ in how "hard" they are to cluster: m-quantile measures of different algorithms are significantly correlated. Figure 4 should not be directly interpreted in this way due to different structure of result sets corresponding to different queries (e.g. different number of documents).

6   Conclusions and Future Plans

In this paper we have introduced a novel paradigm of semantic evaluation. Unlike traditional approaches, which are either measures counting pairs of objects or are variations of information theoretic approaches, our proposed procedure resembles the process of human perception, as it is based on a model describing the clustering in terms of expert knowledge. We proposed a specific implementation of this evaluation measure (i.e. a choice of the underlying model structure and optimization framework) and further demonstrated its application to online results clustering evaluation problem. We have observed that even if we only used information about MeSH subheadings assigned to documents as the source of information for evaluation, for most result sets in our experiments clusterings performed better than random assignments of documents to clusters.

References

1. Assent, I., Krieger, R., Müller, E., Seidl, T.: Visa: visual subspace clustering analysis. SIGKDD Explor. Newsl. 9(2), 5–12 (2007)
2. Böhm, C., Kailing, K., Kriegel, H.-P., Kröger, P.: Density connected clustering with local subspace preferences. In: ICDM, pp. 27–34. IEEE Computer Society (2004)
3. Cao, T., Do, H., Hong, D., Quan, T.: Fuzzy named entity-based document clustering. In: Proc. of the 17th IEEE International Conference on Fuzzy Systems (FUZZ-IEEE 2008), pp. 2028–2034 (2008)
4. Hastie, T., Tibshirani, R., Friedman, J.: The Elements of Statistical Learning. Springer Series in Statistics. Springer New York Inc. (2001)
5. Herskovic, J.R., Tanaka, L.Y., Hersh, W., Bernstam, E.V.: A day in the life of pubmed: analysis of a typical day's query log. Journal of the American Medical Informatics Association, 212–220 (2007)
6. Kriegel, H.-P., Kroger, P., Renz, M., Wurst, S.: A generic framework for efficient subspace clustering of high-dimensional data. In: Proceedings of the Fifth IEEE International Conference on Data Mining, ICDM 2005, pp. 250–257. IEEE Computer Society, Washington, DC (2005)

7. Kröger, P., Kriegel, H.-P., Kailing, K.: Density-connected subspace clustering for high-dimensional data. In: Berry, M.W., Dayal, U., Kamath, C., Skillicorn, D.B. (eds.) SDM. SIAM (2004)
8. Lancichinetti, A., Fortunato, S., Kertész, J.: Detecting the overlapping and hierarchical community structure of complex networks. New Journal of Physics 11, 033015 (2009)
9. MacQueen, J.B.: Some methods for classification and analysis of multivariate observations. In: Cam, L.M.L., Neyman, J. (eds.) Proc. of the Fifth Berkeley Symposium on Mathematical Statistics and Probability, vol. 1, pp. 281–297. University of California Press (1967)
10. Manning, C.D., Raghavan, P., Schütze, H.: Introduction to Information Retrieval (2007)
11. Nguyen, H.S., Nguyen, S.H., Swieboda, W.: Semantic explorative evaluation of document clustering algorithms. In: Ganzha, M., Maciaszek, L.A., Paprzycki, M. (eds.) FedCSIS, pp. 115–122 (2013)
12. Nguyen, S.H., Świeboda, W., Jaśkiewicz, G.: Extended document representation for search result clustering. In: Bembenik, R., Skonieczny, Ł., Rybiński, H., Niezgódka, M. (eds.) Intelligent Tools for Building a Scient. Info. Plat. SCI, vol. 390, pp. 77–95. Springer, Heidelberg (2012)
13. Nguyen, S.H., Świeboda, W., Jaśkiewicz, G.: Semantic evaluation of search result clustering methods. In: Bembenik, R., Skonieczny, Ł., Rybiński, H., Kryszkiewicz, M., Niezgódka, M. (eds.) Intell. Tools for Building a Scientific Information. SCI, vol. 467, pp. 393–414. Springer, Heidelberg (2013)
14. Nguyen, S.H., Świeboda, W., Jaśkiewicz, G., Nguyen, H.S.: Enhancing search results clustering with semantic indexing. In: SoICT 2012, pp. 71–80 (2012)
15. Osinski, S., Stefanowski, J., Weiss, D.: Lingo: Search results clustering algorithm based on singular value decomposition. In: Intelligent Information Systems, pp. 359–368 (2004)
16. Pfitzner, D., Leibbrandt, R., Powers, D.: Characterization and evaluation of similarity measures for pairs of clusterings. Knowl. Inf. Syst. 19(3), 361–394 (2009)
17. Rand, W.M.: Objective criteria for the evaluation of clustering methods. J. Amer. Stat. Assoc. 66(336), 846–850 (1971)
18. Roberts, R.J.: PubMed Central: The GenBank of the published literature. Proceedings of the National Academy of Sciences of the United States of America 98(2), 381–382 (2001)
19. Sequeira, K., Zaki, M.: SCHISM: A new approach for interesting subspace mining. In: Proceedings of the Fourth IEEE Conference on Data Mining, pp. 186–193. IEEE Computer Society (2004)
20. Szczuka, M., Janusz, A., Herba, K.: Semantic clustering of scientific articles with use of DBpedia knowledge base. In: Bembenik, R., Skonieczny, Ł., Rybiński, H., Niezgódka, M. (eds.) Intelligent Tools for Building a Scient. Info. Plat. SCI, vol. 390, pp. 61–76. Springer, Heidelberg (2012)
21. United States National Library of Medicine. Introduction to MeSH – 2011 (2011)

An Improved Depth-First Control Strategy for Query-Subquery Nets in Evaluating Queries to Horn Knowledge Bases

Son Thanh Cao[1,2] and Linh Anh Nguyen[2,3]

[1] Faculty of Information Technology, Vinh University
182 Le Duan street, Vinh, Nghe An, Vietnam
sonct@vinhuni.edu.vn
[2] Institute of Informatics, University of Warsaw
Banacha 2, 02-097 Warsaw, Poland
nguyen@mimuw.edu.pl
[3] Faculty of Information Technology
VNU University of Engineering and Technology
144 Xuan Thuy, Hanoi, Vietnam

Abstract. The QSQN evaluation method uses query-subquery nets and allows any control strategy for processing queries to Horn knowledge bases. This paper proposes an improved depth-first control strategy for the QSQN evaluation method to reduce the number of accesses to the intermediate relations and extensional relations. We came up to the improvement by using query-subquery nets to observe which relations are likely to grow or saturate and which ones are not yet affected by the computation and the other relations. Our intention is to accumulate as many as possible tuples or subqueries at each node of the query-subquery net before processing it. The experimental results confirm the outperformance of the improved version.

Keywords: Horn knowledge bases, deductive databases, query processing, Magic-Set transformation, QSQ, QSQR, QSQN.

1 Introduction

In knowledge representation and reasoning, the Horn fragment of first-order logic plays an important role and has received a lot of attention from researchers. Horn knowledge bases are a generalization of Datalog deductive databases as they allow function symbols and do not require the range-restrictedness condition.

Researchers have developed a number of evaluation methods for Datalog deductive databases such as the top-down methods QSQ [16,1], QSQR [16,9,10], QoSaQ [17], QSQN [11] and the bottom-up method Magic-Set [1,2,3,14] (by the Magic-Set method we mean the bottom-up evaluation method that combines the magic-set transformation with the improved semi-naive evaluation method). In [9], Madalińska-Bugaj and Nguyen generalized the QSQR method for Horn

T.V. Do et al. (eds.), *Advanced Computational Methods for Knowledge Engineering*,
Advances in Intelligent Systems and Computing 282,
DOI: 10.1007/978-3-319-06569-4_21, © Springer International Publishing Switzerland 2014

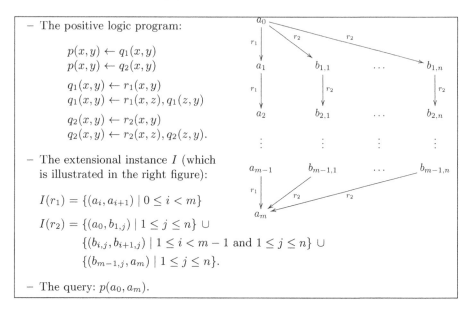

- The positive logic program:

$$p(x,y) \leftarrow q_1(x,y)$$
$$p(x,y) \leftarrow q_2(x,y)$$

$$q_1(x,y) \leftarrow r_1(x,y)$$
$$q_1(x,y) \leftarrow r_1(x,z), q_1(z,y)$$

$$q_2(x,y) \leftarrow r_2(x,y)$$
$$q_2(x,y) \leftarrow r_2(x,z), q_2(z,y).$$

- The extensional instance I (which is illustrated in the right figure):

$$I(r_1) = \{(a_i, a_{i+1}) \mid 0 \le i < m\}$$

$$I(r_2) = \{(a_0, b_{1,j}) \mid 1 \le j \le n\} \cup$$
$$\{(b_{i,j}, b_{i+1,j}) \mid 1 \le i < m-1 \text{ and } 1 \le j \le n\} \cup$$
$$\{(b_{m-1,j}, a_m) \mid 1 \le j \le n\}.$$

- The query: $p(a_0, a_m)$.

Fig. 1. The Horn knowledge base and the query used for Example 1

knowledge bases. Some authors also extended the magic-set technique together with the breadth-first approach for Horn knowledge bases [13,8].

There are several efficient computational procedures for logic programming that use tabled SLD-resolution without redundant recomputations like OLDT [15], SLD-AL [17], linear tabulated resolution [18]. However, they are not directly applicable to Horn knowledge bases to obtain efficient evaluation engines because they are not set-oriented.

In [11,12], we formulated the Query-Subquery Nets (QSQN) framework for evaluating queries to Horn knowledge bases. The aim was to increase efficiency of query processing by eliminating redundant computation, increasing flexibility and reducing the number of accesses to the secondary storage. The framework satisfies: (i) the approach is goal-directed; (ii) each subquery is processed only once and each supplement tuple, if desired, is transferred only once; (iii) operations are done set-at-a-time; (iv) any control strategy can be used.

In [11,12,4], we showed that the QSQN method is better than the QSQR method by reducing redundant recomputations and is more flexible than the Magic-Set method. Inflexibility of the Magic-Set method is illustrated by the following example [12].

Example 1. The orders of program clauses and atoms in their bodies may be specified using the Prolog programming style. In such cases, the top-down depth-first approach may be much more efficient than the breadth-first approach. Figure 1 presents a Horn knowledge base and a query that justify this claim, where p, q_1 and q_2 are intensional predicates, r_1 and r_2 are extensional predicates, x, y and z are variables, a_i and $b_{i,j}$ are constant symbols.   ◁

The QSQN method allows various control strategies with different effects on the number of accesses to the relations. As discussed in [4], QSQN with the depth-first control strategy is better than the other methods w.r.t. the number of accesses to the relations for the cases as in Example 1. The depth-first control strategy proposed in [12,4] is, however, just one of possible realizations of the depth-first approach. The problem is: which successor of the current node should be chosen to explore first?

In this paper, we propose an improved version of the depth-first control strategy to reduce the number of accesses to the intermediate relations and extensional relations. We came up to the improvement by using query-subquery nets to observe which relations are likely to grow or saturate and which ones are not yet affected by the computation and the other relations. Our intention is to accumulate as many as possible tuples or subqueries at each node of the query-subquery net before processing it. The experimental results confirm the outperformance of the improved version.

The rest of this paper is organized as follows. Section 2 recalls the most important notation and definitions of first-order-logic. Section 3 provides a short overview of QSQN. Section 4 presents our improved depth-first control strategy for QSQN and a detailed example for showing the usefulness of this strategy. Section 5 provides our experimental results and Section 6 concludes this work.

2   Preliminaries

A *term* is either a constant, a variable or an expression of the form $f(t_1, \ldots, t_n)$, where $n \geq 1$ and f is an n-ary function symbol and each t_i is a term. An *atom* is an expression of the form $p(t_1, t_2, \ldots, t_n)$, where $n \geq 0$, p is an n-ary predicate and $t_1, t_2, \ldots, t_n$ are terms. An atom is *ground* if it does not use variables. We classify each predicate either as *intensional* or as *extensional*. A *generalized tuple* is a tuple of terms, which may contain function symbols and variables. A *generalized relation* is a set of generalized tuples of the same arity.

A *program clause* is a formula of the form $(A \vee \neg B_1 \vee \ldots \vee \neg B_n)$ with $n \geq 0$, written as $A \leftarrow B_1, \ldots, B_n$, where $A, B_1, \ldots, B_n$ are atoms. A is called the *head*, and $B_1, \ldots, B_n$ the *body* of the program clause. If p is the predicate of A then the program clause is called a program clause defining p.

A *positive* (or *definite*) *logic program* is a finite set of program clauses. From now on, we use the term "program" to mean a positive logic program.

A *goal* is a formula of the form $(\neg B_1 \vee \ldots \vee \neg B_n)$, written as $\leftarrow B_1, \ldots, B_n$, where $B_1, \ldots, B_n$ are atoms. If $n = 1$ then the goal is called a *unary goal*. If $n = 0$ then the goal is called the *empty goal*.

A *Horn knowledge base* is defined to be a pair consisting of a positive logic program for defining intensional predicates and a *generalized extensional instance*, which is a function mapping each extensional n-ary predicate to an n-ary generalized relation. Note that intensional predicates are defined by a positive logic program which may contain function symbols and not be "range-restricted".

We assume that the reader is familiar with the notions of substitution, unification, mgu (most general unifier) and related ones. Given substitutions θ and

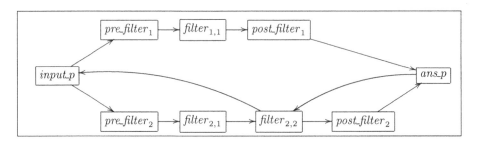

Fig. 2. The QSQ topological structure of the logic program given in Example 2

δ, the composition of θ and δ is denoted by $\theta\delta$, the domain of θ is denoted by $dom(\theta)$, the range of θ is denoted by $range(\theta)$, and the restriction of θ to a set X of variables is denoted by $\theta_{|X}$. The *term-depth* of an expression (resp. a substitution) is the maximal nesting depth of function symbols occurring in that expression (resp. substitution). Given a list/tuple α of terms or atoms, by $Vars(\alpha)$ we denote the set of variables occurring in α.

A *query* to a Horn knowledge base (P, I) is a formula $q(\overline{x})$, where q is an n-ary intensional predicate and $\overline{x}$ is a tuple of n pairwise different variables. An *answer* to the query is an n-ary tuple $\overline{t}$ of terms such that $P \cup I \models q(\overline{t})$, treating I as the corresponding set of atoms of the extensional predicates.

3   An Overview of Query-Subquery Nets

From now on, let P be a positive logic program and $\varphi_1, \ldots, \varphi_m$ be all the program clauses of P, with $\varphi_i = (A_i \leftarrow B_{i,1}, \ldots, B_{i,n_i})$, where $1 \le i \le m$ and $n_i \ge 0$.

Definition 1 (Query-Subquery Net Structure). A *query-subquery net structure* (*QSQ-net structure*) of P is a tuple (V, E, T) such that:

- V is a set of nodes, including
  - *input_p* and *ans_p* for each intensional predicate p of P,
  - *pre_filter$_i$*, *filter$_{i,1}$*, ..., *filter$_{i,n_i}$*, *post_filter$_i$* for each $1 \le i \le m$;
- E is a set of edges, including
  - $(filter_{i,1}, filter_{i,2})$, ..., $(filter_{i,n_i-1}, filter_{i,n_i})$ for each $1 \le i \le m$,
  - $(pre\_filter_i, filter_{i,1})$ and $(filter_{i,n_i}, post\_filter_i)$ for each $1 \le i \le m$ with $n_i \ge 1$,
  - $(pre\_filter_i, post\_filter_i)$ for each $1 \le i \le m$ with $n_i = 0$,
  - $(input\_p, pre\_filter_i)$ and $(post\_filter_i, ans\_p)$ for each $1 \le i \le m$, where p is the predicate of A_i,
  - $(filter_{i,j}, input\_p)$ and $(ans\_p, filter_{i,j})$ for each intensional predicate p and each $1 \le i \le m$ and $1 \le j \le n_i$ such that $B_{i,j}$ is an atom of p;
- T is a function, called the *memorizing type*, mapping each node *filter$_{i,j}$* $\in V$ such that the predicate of $B_{i,j}$ is extensional to *true* or *false*.[1]

---

[1] If $T(filter_{i,j}) = true$ (and the predicate of $B_{i,j}$ is extensional) then subqueries for *filter$_{i,j}$* are always processed immediately, without being accumulated at *filter$_{i,j}$*.

If $(v, w) \in E$ then we call w a *successor* of v, and v a *predecessor* of w. Note that V and E are uniquely specified by P. We call the pair (V, E) the *QSQ topological structure* of P. $\triangleleft$

Example 2. Figure 2 illustrates the QSQ topological structure of the following logic program, where p is an intensional predicate and q an extensional predicate:

$$p(x, y) \leftarrow q(x, y)$$
$$p(x, y) \leftarrow q(x, z), p(z, y).$$

Definition 2 (Query-Subquery Net). A *query-subquery net* (*QSQ-net*) of P is a tuple $N = (V, E, T, C)$ such that (V, E, T) is a QSQ-net structure of P and C is a mapping that associates each node $v \in V$ with a structure called the *contents* of v, satisfying the following conditions:

- If $v = input\_p$ or $v = ans\_p$ for an intensional predicate p, $C(v)$ consists of:
  - $tuples(v)$: a set of generalized tuples of the same arity as p,
  - $unprocessed(v, w)$ for $(v, w) \in E$: a subset of $tuples(v)$;
- If $v = pre\_filter_i$ then $C(v)$ consists of:
  - $atom(v) = A_i$ and $post\_vars(v) = Vars((B_{i,1}, \dots, B_{i,n_i}))$;
- If $v = post\_filter_i$ then $C(v)$ is empty, but we assume $pre\_vars(v) = \emptyset$;
- If $v = filter_{i,j}$ and p is the predicate of $B_{i,j}$ then $C(v)$ consists of:
  - $kind(v) = extensional$ if p is extensional, and $kind(v) = intensional$ otherwise,
  - $pred(v) = p$ (called the predicate of v) and $atom(v) = B_{i,j}$,
  - $pre\_vars(v) = Vars((B_{i,j}, \dots, B_{i,n_i}))$ and $post\_vars(v) = Vars((B_{i,j+1}, \dots, B_{i,n_i}))$,
  - $subqueries(v)$: a set of pairs of the form $(\bar{t}, \delta)$, where $\bar{t}$ is a generalized tuple of the same arity as the predicate of A_i and δ is an idempotent substitution such that $dom(\delta) \subseteq pre\_vars(v)$ and $dom(\delta) \cap Vars(\bar{t}) = \emptyset$,
  - $unprocessed\_subqueries(v) \subseteq subqueries(v)$,
  - in the case p is intensional:
    $unprocessed\_subqueries_2(v) \subseteq subqueries(v)$,
    $unprocessed\_tuples(v)$: a set of generalized tuples of the same arity as p;
- If $v = filter_{i,j}$, $kind(v) = extensional$ and $T(v) = false$ then $subqueries(v) = \emptyset$. $\triangleleft$

A QSQ-net can be viewed as a flow control network for determining which subqueries in which nodes should be processed next, in an efficient way.

If $v \in \{pre\_filter_i, post\_filter_i\}$ or $v = filter_{i,j}$ and $kind(v) = extensional$ then v has exactly one successor: $succ(v)$. If v is $filter_{i,j}$ with $kind(v) = intensional$ and $pred(v) = p$ then v has exactly two successors: $succ(v) = filter_{i,j+1}$ if $n_i > j$; $succ(v) = post\_filter_i$ otherwise; and $succ_2(v) = input\_p$.

By a *subquery* we mean a pair of the form $(\bar{t}, \delta)$, where $\bar{t}$ is a generalized tuple and δ is an idempotent substitution such that $dom(\delta) \cap Vars(\bar{t}) = \emptyset$. For $v = filter_{i,j}$ and p being the predicate of A_i, the meaning of a subquery $(\bar{t}, \delta) \in subqueries(v)$ is that: for processing a goal $\leftarrow p(\bar{s})$ with

$\overline{s} \in tuples(input\_p)$ using the program clause $\varphi_i = (A_i \leftarrow B_{i,1}, \ldots, B_{i,n_i})$, unification of $p(\overline{s})$ and A_i as well as processing of the subgoals $B_{i,1}, \ldots, B_{i,j-1}$ were done, amongst others, by using a sequence of mgu's $\gamma_0, \ldots, \gamma_{j-1}$ with the property that $\overline{t} = \overline{s}\gamma_0 \ldots \gamma_{j-1}$ and $\delta = (\gamma_0 \ldots \gamma_{j-1})_{| Vars((B_{i,j}, \ldots, B_{i,n_i}))}$. Informally, a subquery $(\overline{t}, \delta)$ transferred through an edge to v is processed as follows:

- if $v = filter_{i,j}$, $kind(v) = extensional$ and $pred(v) = p$ then, for each $\overline{t}' \in I(p)$, if $atom(v)\delta = B_{i,j}\delta$ is unifiable with a fresh variant of $p(\overline{t}')$ by an mgu γ then transfer the subquery $(\overline{t}\gamma, (\delta\gamma)_{|post\_vars(v)})$ through $(v, succ(v))$;
- if $v = filter_{i,j}$, $kind(v) = intensional$ and $pred(v) = p$ then
  - transfer the input tuple $\overline{t}'$ such that $p(\overline{t}') = atom(v)\delta = B_{i,j}\delta$ through $(v, input\_p)$ to add a fresh variant of it to $tuples(input\_p)$,
  - for each currently existing $\overline{t}' \in tuples(ans\_p)$, if $atom(v)\delta = B_{i,j}\delta$ is unifiable with a fresh variant of $p(\overline{t}')$ by an mgu γ then transfer the subquery $(\overline{t}\gamma, (\delta\gamma)_{|post\_vars(v)})$ through $(v, succ(v))$,
  - store the subquery $(\overline{t}, \delta)$ in $subqueries(v)$, and later, for each new $\overline{t}'$ added to $tuples(ans\_p)$, if $atom(v)\delta = B_{i,j}\delta$ is unifiable with a fresh variant of $p(\overline{t}')$ by an mgu γ then transfer the subquery $(\overline{t}\gamma, (\delta\gamma)_{|post\_vars(v)})$ through $(v, succ(v))$;
- if $v = post\_filter_i$ and p is the predicate of A_i then transfer the answer tuple $\overline{t}$ through $(post\_filter_i, ans\_p)$ to add it to $tuples(ans\_p)$.

Formally, the processing of a subquery has the following properties:

- every subquery or input/answer tuple that is subsumed by another one or has a term-depth greater than a fixed bound l is ignored;
- the processing is divided into smaller steps which can be delayed to maximize flexibility and allow various control strategies;
- the processing is done set-at-a-time (e.g., for all the unprocessed subqueries accumulated in a given node).

Algorithm 1 in [11] presents our QSQN method for evaluating a query $q(\overline{x})$ to a Horn knowledge base (P, I). It repeatedly selects an "active" edge and "fires" the operation for the edge. The function "active" and the procedure "fire" are defined in [12]. The selection is decided by the adopted control strategy, which can be arbitrary. In [11,12,4] we also proposed two exemplary control strategies: the first one is a depth-first control strategy (DFS, which originally stands for Depth-First Search [4]) and the second one is Disk Access Reduction (DAR), which tries to reduce the number of accesses to the secondary storage. See [12] for details of the algorithm and the two exemplary control strategies.

4   An Improved Depth-First Control Strategy for QSQN

In this section, we present our Improved Depth-First Control Strategy (IDFS) for QSQN. The idea of the improvement is to enter deeper cycles in the considered QSQ-net first and keep looping along the current "local" cycle as long as possible.

This allows to accumulate as many as possible tuples or subqueries at a node before processing it.

We say that a predicate p *directly depends* on a predicate q if the considered program P has a clause defining p that uses q in the body. We define the relation *"depends"* to be the reflexive and transitive closure of "directly depends".

Definition 3. The *priority* of an edge (v, w) in an QSQ-net is a vector defined as follows:

- if $v = input\_p$ and $w = pre\_filter_i$ then $priority(v, w) = (a, b, c)$, where:
  - a is the truth value of $(\varphi_i$ containing at least one intensional predicate r in the body$)$,
  - if $a = false$ then $b = false$,
    else b is the truth value of (one of those predicates r depends on $p)^2$,
  - if $b = false$ then $c = 0$, else c is the modification timestamp of w;
- if $v = ans\_p$ and $w = filter_{i,j}$ then $priority(v, w) = (a, b, c)$, where:
  - a is the truth value of $(p$ depends on the predicate of $A_i)^3$,
  - b is the modification timestamp of w,
  - c is the truth value of $(p$ is the predicate of $A_i)$;
- if v is $filter_{i,j}$ with $kind(v) = intensional$ then $priority(v, w) = (a)$, where $a = 2$ if $w = succ_2(v)$, and $a = 1$ otherwise;
- otherwise, $priority(v, w) = (1)$.

The priorities of two edges (v, w) and (v, w') are compared using the lexico-graphical order, where $false < true$.   ◁

Our IDFS control strategy for QSQN follows the depth-first approach, but adopts a slight modification. It uses a stack of edges of the QSQ-net structure of P. Algorithm 1 in [11] with this control strategy for evaluating a query $q(\bar{x})$ to a Horn knowledge base (P, I) runs as follows:

1. let $\bar{x}'$ be a fresh variant of $\bar{x}$ and set $tuples(input\_q) := \{\bar{x}'\}$;
2. for each edge $(input\_q, v)$ of the QSQ-net do $unprocessed(input\_q, v) := \{\bar{x}'\}$;
3. initialize the stack to the empty one and push all the edges outcoming from $input\_q$ into the stack in the increasing order w.r.t. their priorities (the lower the priority is, the earlier the edge is pushed into the stack);
4. while the stack is not empty and the user wants more answers do:
   (a) pop an edge (u, v) from the stack;
   (b) if this edge is active then "fire" it;
   (c) push all the "active" edges outcoming from v into the stack in the in-creasing order w.r.t. their priorities;
   (d) if v is $filter_{i,j}$ with $kind(v) = intensional$ and the edge $(v, succ_2(v))$ is not "active" then push this edge into the stack (recall that: if $pred(v) = p$ then $succ_2(v) = input\_p$).
5. return $tuples(ans\_q)$.

---

[2] Note that if $b = true$ then p and r mutually depend on each other.
[3] Note that if $a = true$ then p and the predicate of A_i mutually depend on each other.

Example 3. This example shows a comparison between the DFS and the IDFS control strategies. It is a modified version of an example from [9,4]. Consider the following Horn knowledge base (P, I) and the query $s(x)$, where p and s are intensional predicates, q is an extensional predicate, $a - j$ are constant symbols and x, y, z are variables:

- the positive logic program P:

$$s(x) \leftarrow p(a, x)$$
$$p(x, y) \leftarrow q(x, y)$$
$$p(x, y) \leftarrow q(x, z), p(z, y)$$

- the extensional instance I:

$$I(q) = \{(a, b), (a, c), (b, d), (c, d), (c, e), (f, g), (f, i), (g, h), (i, j)\}.$$

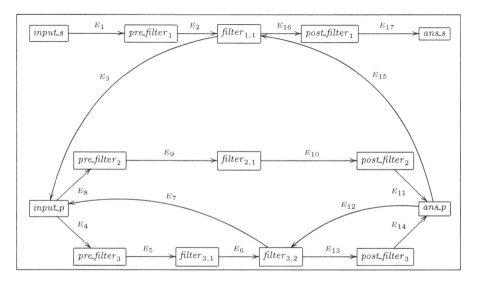

Fig. 3. The QSQ-net structure of the program given in Example 3

The QSQ-net structure of P is presented in Figure 3. For convenience, we name the edges of the net by E_i with $1 \le i \le 17$ as shown in the figure.

We give below a trace of running Algorithm 1 [11] step by step that evaluates the query $s(x)$ to the Horn knowledge base (P, I).

Table 1 shows the steps for the case when using DFS. The "Reading $inp\_/ans\_/sup\_/edb$" column means the number of read operations from *input/answer/supplement/extensional* relations, respectively. Similarly, the "Writing $inp\_/ans\_/sup\_$" column means the number of write operations to *input/answer/supplement*, respectively. The "Intermediate results" column shows the changed values in tuples/subqueries for each step. The list of edges in each row denotes a call of the procedure "fire" for the first edge in the list that triggers transferring data through the subsequent edges of the list. The "active" edges are "fired" in the following order:

Table 1. A trace of running Algorithm 1 by using DFS for the query in Example 3

Id	Intermediate results	Reading				Writing		
		inp_-	ans_-	sup_-	edb	inp_-	ans_-	sup_-
1	**Running Algorithm 1** $tuples(input\_s) = unprocessed(E_1) = \{(x_1)\}$	0	0	0	0	1	0	0
2	$\mathbf{E_1 - E_2}$ $unprocessed(E_1) = \emptyset$ $subqueries(filter_{1,1}) = unprocessed\_subqueries(filter_{1,1}) =$ $\qquad unprocessed\_subqueries_2(filter_{1,1}) = \{((x_1), \{x/x_1\})\}$	1	-	1	-	-	-	1
3	$\mathbf{E_{16} - E_{17}}$ $unprocessed\_subqueries(filter_{1,1}) = \emptyset$	-	1	-	-	-	-	-
4	$\mathbf{E_3}$ $unprocessed\_subqueries_2(filter_{1,1}) = \emptyset$ $tuples(input\_p) = unprocessed(E_4) = unprocessed(E_8) = \{(a, x_2)\}$	2	-	2	-	2	-	-
5	$\mathbf{E_8 - E_9 - E_{10} - E_{11}}$ $unprocessed(E_8) = \emptyset$ $tuples(ans\_p) = unprocessed(E_{12}) = unprocessed(E_{15}) = \{(a, b), (a, c)\}$	3	2	-	1	-	1	-
6	$\mathbf{E_{15}}$ $unprocessed(E_{15}) = \emptyset; \quad unprocessed\_tuples(filter_{1,1}) = \{(a, b), (a, c)\}$	-	3	-	-	-	-	-
7	$\mathbf{E_{16} - E_{17}}$ $unprocessed\_tuples(filter_{1,1}) = \emptyset; \quad tuples(ans\_s) = \{(b), (c)\}$	-	4	3	-	-	2	-
8	$\mathbf{E_{12}}$ $unprocessed(E_{12}) = \emptyset; \quad unprocessed\_tuples(filter_{3,2}) = \{(a, b), (a, c)\}$	-	5	-	-	-	-	-
9	$\mathbf{E_{13} - E_{14}}$ $unprocessed\_tuples(filter_{3,2}) = \emptyset$	-	-	4	-	-	-	-
10	$\mathbf{E_4 - E_5 - E_6}$ $unprocessed(E_4) = \emptyset$ $subqueries(filter_{3,2}) = unprocessed\_subqueries(filter_{3,2}) =$ $\qquad unprocessed\_subqueries_2(filter_{3,2}) = \{((a, x_2), \{y/x_2, z/b\}), ((a, x_2), \{y/x_2, z/c\})\}$	4	-	5	2	-	-	2
11	$\mathbf{E_{13} - E_{14}}$ $unprocessed\_subqueries(filter_{3,2}) = \emptyset$	-	6	6	-	-	-	-
12	$\mathbf{E_7}$ $unprocessed\_subqueries_2(filter_{3,2}) = \emptyset$ $tuples(input\_p) = \{(a, x_2), (b, x_3), (c, x_4)\}$ $unprocessed(E_4) = unprocessed(E_8) = \{(b, x_3), (c, x_4)\}$	5	-	7	-	3	-	-
13	$\mathbf{E_8 - E_9 - E_{10} - E_{11}}$ $unprocessed(E_8) = \emptyset; \quad tuples(ans\_p) = \{(a, b), (a, c), (b, d), (c, d), (c, e)\}$ $unprocessed(E_{12}) = unprocessed(E_{15}) = \{(b, d), (c, d), (c, e)\}$	6	7	-	3	-	3	-
14	$\mathbf{E_{12}}$ $unprocessed(E_{12}) = \emptyset; \quad unprocessed\_tuples(filter_{3,2}) = \{(b, d), (c, d), (c, e)\}$	-	8	-	-	-	-	-
15	$\mathbf{E_{13} - E_{14}}$ $unprocessed\_tuples(filter_{3,2}) = \emptyset$ $tuples(ans\_p) = \{(a, b), (a, c), (b, d), (c, d), (c, e), (a, d), (a, e)\}$ $unprocessed(E_{12}) = \{(a, d), (a, e)\}$ $unprocessed(E_{15}) = \{(b, d), (c, d), (c, e), (a, d), (a, e)\}$	-	9	8	-	-	4	-
16	$\mathbf{E_{12}}$ $unprocessed(E_{12}) = \emptyset; \quad unprocessed\_tuples(filter_{3,2}) = \{(a, d), (a, e)\}$	-	10	-	-	-	-	-

Table 1. *(continued)*

17	$E_{13} - E_{14}$	-	-	9	-	-	-	-
	$unprocessed\_tuples(filter_{3,2}) = \emptyset$							
18	E_{15}	-	11	-	-	-	-	-
	$unprocessed(E_{15}) = \emptyset;\quad unprocessed\_tuples(filter_{1,1}) = \{(b,d),(c,d),(c,e),(a,d),(a,e)\}$							
19	$E_{16} - E_{17}$	-	12	10	-	-	5	-
	$unprocessed\_tuples(filter_{1,1}) = \emptyset;\quad tuples(ans\_s) = \{(b),(c),(e),(d)\}$							
20	$E_4 - E_5 - E_6$	7	-	11	4	-	-	3
	$unprocessed(E_4) = \emptyset$							
	$subqueries(filter_{3,2}) = \{((a,x_2),\{y/x_2,z/b\}),((a,x_2),\{y/x_2,z/c\}),$							
	$\qquad\qquad ((b,x_3),\{y/x_3,z/d\}),((c,x_4),\{y/x_4,z/d\}),((c,x_4),\{y/x_4,z/e\})\}$							
	$unprocessed\_subqueries(filter_{3,2}) = unprocessed\_subqueries_2(filter_{3,2}) =$							
	$\qquad\qquad \{((b,x_3),\{y/x_3,z/d\}),((c,x_4),\{y/x_4,z/d\}),((c,x_4),\{y/x_4,z/e\})\}$							
21	$E_{13} - E_{14}$	-	13	12	-	-	-	-
	$unprocessed\_subqueries(filter_{3,2}) = \emptyset$							
22	E_7	8	-	13	-	4	-	-
	$unprocessed\_subqueries_2(filter_{3,2}) = \emptyset$							
	$tuples(input\_p) = \{(a,x_2),(b,x_3),(c,x_4),(d,x_5),(e,x_6)\}$							
	$unprocessed(E_4) = unprocessed(E_8) = \{(d,x_5),(e,x_6)\}$							
23	$E_8 - E_9 - E_{10} - E_{11}$	9	-	-	5	-	-	-
	$unprocessed(E_8) = \emptyset$							
24	$E_4 - E_5 - E_6$	10	13	13	6	4	5	3
	$unprocessed(E_4) = \emptyset$							

Table 2 shows the steps of running Algorithm 1 [11] using the IDFS control strategy for evaluating the query $s(x)$ to the Horn knowledge base (P, I). The "active" edges for this case are "fired" in the following order:

Table 2. A trace of running Algorithm 1 by using IDFS for the query in Example 3

Id	Intermediate results	Reading				Writing		
		$inp\_$	$ans\_$	$sup\_$	edb	$inp\_$	$ans\_$	$sup\_$
1	Running Algorithm 1	0	0	0	0	1	0	0
	$tuples(input\_s) = unprocessed(E_1) = \{(x_1)\}$							
2	$E_1 - E_2$	1	-	1	-	-	-	1
	$unprocessed(E_1) = \emptyset$							
	$subqueries(filter_{1,1}) = unprocessed\_subqueries(filter_{1,1}) =$							
	$\qquad\qquad unprocessed\_subqueries_2(filter_{1,1}) = \{((x_1),\{x/x_1\})\}$							
3	E_3	2	-	2	-	2	-	-
	$unprocessed\_subqueries_2(filter_{1,1}) = \emptyset$							
	$tuples(input\_p) = unprocessed(E_4) = unprocessed(E_8) = \{(a,x_2)\}$							
4	$E_4 - E_5 - E_6$	3	-	3	1	-	-	2
	$unprocessed(E_4) = \emptyset$							
	$subqueries(filter_{3,2}) = unprocessed\_subqueries(filter_{3,2}) =$							
	$\qquad unprocessed\_subqueries_2(filter_{3,2}) = \{((a,x_2),\{y/x_2,z/b\}),((a,x_2),\{y/x_2,z/c\})\}$							

Table 2. *(continued)*

5	E_7	4	-	4	-	3	-	-
	$unprocessed\_subqueries_2(filter_{3,2}) = \emptyset$							
	$tuples(input\_p) = unprocessed(E_8) = \{(a, x_2), (b, x_3), (c, x_4)\}$							
	$unprocessed(E_4) = \{(b, x_3), (c, x_4)\}$							
6	$\mathbf{E_4 - E_5 - E_6}$	5	-	5	2	-	-	3
	$unprocessed(E_4) = \emptyset$							
	$subqueries(filter_{3,2}) = unprocessed\_subqueries(filter_{3,2}) =$							
	$\quad \{((a, x_2), \{y/x_2, z/b\}), ((a, x_2), \{y/x_2, z/c\}), ((b, x_3), \{y/x_3, z/d\}),$							
	$\quad\quad ((c, x_4), \{y/x_4, z/d\}), ((c, x_4), \{y/x_4, z/e\})\}$							
	$unprocessed\_subqueries_2(filter_{3,2}) =$							
	$\quad \{((b, x_3), \{y/x_3, z/d\}), ((c, x_4), \{y/x_4, z/d\}), ((c, x_4), \{y/x_4, z/e\})\}$							
7	$\mathbf{E_7}$	6	-	6	-	4	-	-
	$unprocessed\_subqueries_2(filter_{3,2}) = \emptyset$							
	$tuples(input\_p) = unprocessed(E_8) = \{(a, x_2), (b, x_3), (c, x_4), (d, x_5), (e, x_6)\}$							
	$unprocessed(E_4) = \{(d, x_5), (e, x_6)\}$							
8	$\mathbf{E_4 - E_5 - E_6}$	7	-	-	3	-	-	-
	$unprocessed(E_4) = \emptyset$							
9	$\mathbf{E_8 - E_9 - E_{10} - E_{11}}$	8	1	-	4	-	1	-
	$unprocessed(E_8) = \emptyset$							
	$tuples(ans\_p) = \{(a, b), (a, c), (b, d), (c, d), (c, e)\}$							
	$unprocessed(E_{12}) = unprocessed(E_{15}) = \{(a, b), (a, c), (b, d), (c, d), (c, e)\}$							
10	$\mathbf{E_{12}}$	-	2	-	-	-	-	-
	$unprocessed(E_{12}) = \emptyset$							
	$unprocessed\_tuples(filter_{3,2}) = \{(a, b), (a, c), (b, d), (c, d), (c, e)\}$							
11	$\mathbf{E_{13} - E_{14}}$	-	3	7	-	-	2	-
	$unprocessed\_tuples(filter_{3,2}) = unprocessed\_subqueries(filter_{3,2}) = \emptyset$							
	$tuples(ans\_p) = unprocessed(E_{15}) = \{(a, b), (a, c), (b, d), (c, d), (c, e), (a, e), (a, d)\}$							
	$unprocessed(E_{12}) = \{(a, e), (a, d)\}$							
12	$\mathbf{E_{12}}$	-	4	-	-	-	-	-
	$unprocessed(E_{12}) = \emptyset$							
	$unprocessed\_tuples(filter_{3,2}) = \{(a, e), (a, d)\}$							
13	$\mathbf{E_{13} - E_{14}}$	-	-	8	-	-	-	-
	$unprocessed\_tuples(filter_{3,2}) = \emptyset$							
14	$\mathbf{E_{15}}$	-	5	-	-	-	-	-
	$unprocessed(E_{15}) = \emptyset$							
	$unprocessed\_tuples(filter_{1,1}) = \{(a, b), (a, c), (b, d), (c, d), (c, e), (a, e), (a, d)\}$							
15	$\mathbf{E_{16} - E_{17}}$	8	6	9	4	4	3	3
	$unprocessed\_tuples(filter_{1,1}) = \emptyset$							
	$tuples(ans\_s) = \{(b), (c), (d), (e)\}$							

As shown in Tables 1 and 2, IDFS performs fewer steps than DFS and reduces the number of accesses to *input/answer/supplement/extensional* relations. ◁

5 Preliminary Experiments

This section presents our experimental results and discussion about the performance of the QSQN evaluation method when used with the DFS and IDFS control strategies. All the experiments have been performed using our codes [5] written in Java and extensional relations stored in a MySQL database. The uploaded package [5] also contains all the experimental results reported below.

In the following tests, we use typical examples that appear in many well-known articles related to deductive databases, including tail/non-tail recursive logic programs as well as logic programs with or without function symbols. Our implementation allows queries of the form $q(\bar{t})$, where $\bar{t}$ is a tuple of terms.

5.1 The Settings

We assume that the computer memory is large enough to load all the involved extensional relations and keep all the intermediate relations in the memory. During execution of the program, for each operation of reading from a relation (resp. writing a set of tuples to a relation), we increase the counter of read (resp. write) operations on this relation by one.

5.2 Experimental Results

We compare the DFS and IDFS control strategies for the QSQN evaluation method using the following tests.

Test 1. This test involves the transitive closure of a binary relation [3,4]. It uses the query $path(a, x)$ and the following logic program P, where x, y, z denote variables, $path$ is an intensional predicate and arc is an extensional predicate:

$$path(x, y) \leftarrow arc(x, y)$$
$$path(x, y) \leftarrow path(x, z), path(z, y).$$

The extensional instance I used for this test is specified as follows, where $a - g$ are constant symbols: $I(arc) = \{(a, b), (b, c), (c, g), (a, d), (d, g), (a, e), (e, f), (f, g)\}$.

Test 2. This test involves a modified version of an example from [9] for showing a case with function symbols. It uses the query $s(x)$ and the following program P:

$$n(x, y) \leftarrow r(x, y)$$
$$n(x, y) \leftarrow q(f(w), y), n(z, w), p(x, f(z))$$
$$s(x) \leftarrow n(c, x)$$

where x, y, z and w are variables, n and s are intensional predicates, p, q and r are extensional predicates, and f is a function symbol. The term-depth bound used for this example is $l = 5$. The extensional instance I for this test is specified

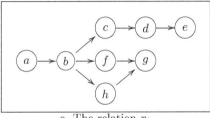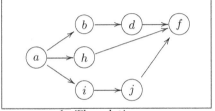

a. The relation r_1 b. The relation r_2

Fig. 4. The extensional relations r_1 and r_2 used for Tests 4 and 5

Table 3. A comparison between the DFS and IDFS control strategies w.r.t. the number of read/write operations. The numbers in the Reading/Writing columns are in the same style as for Tables 1 and 2.

Tests	Strategies	Reading (times) $inp_-/ans_-/sup_-/edb$	Writing (times) $inp_-/ans_-/sup_-$
Test 1	QSQN-DFS	56 (10+21+22+3)	14 (3+5+6)
	QSQN-IDFS	47 (13+14+17+3)	14 (3+4+7)
Test 2	QSQN-DFS	99 (8+42+27+22)	26 (3+20+3)
	QSQN-IDFS	90 (7+38+25+20)	25 (3+19+3)
Test 3	QSQN-DFS	137 (20+59+46+12)	39 (7+25+7)
	QSQN-IDFS	55 (15+13+20+7)	20 (7+6+7)
Test 4	QSQN-DFS	159 (27+61+52+19)	40 (9+22+9)
	QSQN-IDFS	73 (21+16+25+11)	25 (9+7+9)
Test 5	QSQN-DFS	118 (21+42+41+14)	27 (6+14+7)
	QSQN-IDFS	71 (17+19+23+12)	20 (6+7+7)
Test 6	QSQN-DFS	552 (150+152+151+99)	151 (51+50+50)
	QSQN-IDFS	410 (104+102+153+52)	153 (52+50+51)

as follows:

$$I(r) = \{(d, e)\},$$
$$I(p) = \{(c, f(d))\} \cup \{(b_i, f(c)) \mid 0 \le i \le 9\} \cup \{(c, f(b_i)) \mid 0 \le i \le 9\},$$
$$I(q) = \{(f(e), a_0)\} \cup \{(f(a_i), b_i) \mid 0 \le i \le 5\} \cup$$
$$\{(f(b_i), a_{i+1}) \mid 0 \le i \le 4\} \cup$$
$$\{(f^k(a_i), f^k(b_i)) \mid 6 \le i \le 9, \ k = 2 * (i - 5)\} \cup$$
$$\{(f^k(b_i), f^k(a_{i+1})) \mid 5 \le i \le 9, \ k = 2 * (i - 4) - 1\}.$$

Test 3. This test uses the logic program and the query given in Example 3. The extensional instance I for this test consists of the following pairs, where $a - o$ are constant symbols:

$$I(q) = \{(a, b), (b, c), (c, d), (d, e), (e, f), (f, g), (a, h), (h, i), (i, j), (i, d),$$
$$(j, k), (k, f), (a, l), (l, m), (l, i), (m, n), (n, o), (n, k), (o, g)\}.$$

Test 4. This test involves a modified version of an example from [4,11]. It uses the query $p(a, y)$ and the following logic program P, where p, q_1 and q_2 are intensional predicates, r_1 and r_2 are extensional predicates and x, y, z are variables. Figure 4 illustrates the extensional instance I used for this test, where $a - j$ are constant symbols.

$$p(x, y) \leftarrow q_1(x, y)$$
$$q_1(x, y) \leftarrow r_1(x, y)$$
$$q_1(x, y) \leftarrow r_1(x, z), q_1(z, y)$$
$$q_1(x, y) \leftarrow q_2(x, z), r_1(z, y)$$
$$q_2(x, y) \leftarrow r_2(x, y)$$
$$q_2(x, y) \leftarrow r_2(x, z), q_2(z, y).$$

Test 5. This test is a modified version of Test 4. It uses the query $q_1(a, x)$ and the following logic program P, where q_1 and q_2 are intensional predicates, r_1 and r_2 are extensional predicates and x, y, z are variables. In this example, q_1 and q_2 mutually depend on each other. Figure 4 illustrates the extensional instance used for this test, where $a - j$ are constant symbols.

$$q_1(x, y) \leftarrow r_1(x, y)$$
$$q_1(x, y) \leftarrow r_1(x, z), q_1(z, y)$$
$$q_1(x, y) \leftarrow r_1(x, z), q_2(z, y)$$
$$q_2(x, y) \leftarrow r_2(x, y)$$
$$q_2(x, y) \leftarrow r_2(x, z), q_2(z, y)$$
$$q_2(x, y) \leftarrow q_1(x, z), r_2(z, y).$$

Test 6. This test uses the Horn knowledge base and the query given in Example 1, using $n = 50$ and $m = 50$ (r_1 has 50 records and r_2 has 2500 records).

Table 3 shows the comparison between the DFS and IDFS control strategies w.r.t. the numbers of accesses to the intermediate relations and extensional relations. As can be seen in this table, the QSQN method together with the IDFS control strategy is usually better than the DFS control strategy in terms of accesses to the mentioned relations.

6   Conclusions

We have proposed an improved depth-first control strategy for the QSQN evaluation method. The idea is to enter deeper cycles in the considered QSQ-net first and keep looping along the current "local" cycle as long as possible. This allows to accumulate as many as possible tuples or subqueries at a node before processing it. Our experiments confirmed the outperformance of the improved strategy. In the near future, we plan to apply our method to Datalog-like rule languages of the Semantic Web [6,7].

Acknowledgments. This work was supported by Polish National Science Centre (NCN) under Grants No. 2011/02/A/HS1/00395 (for the first author) and 2011/01/B/ST6/02759 (for the second author). The first author would also like to thank the Warsaw Center of Mathematics and Computer Science for support.

References

1. Abiteboul, S., Hull, R., Vianu, V.: Foundations of Databases. Addison Wesley (1995)
2. Bancilhon, F., Maier, D., Sagiv, Y., Ullman, J.D.: Magic sets and other strange ways to implement logic programs. In: Proceedings of PODS 1986, pp. 1–15. ACM (1986)
3. Beeri, C., Ramakrishnan, R.: On the power of magic. J. Log. Program. 10, 255–299 (1991)
4. Cao, S.T.: On the efficiency of Query-Subquery Nets: an experimental point of view. In: Proceedings of SoICT 2013, pp. 148–157. ACM (2013)
5. Cao, S.T.: An implementation of the QSQN evaluation method using the DFS and IDFS control strategies (2014), `http://mimuw.edu.pl/~sonct/QSQN14.zip`
6. Cao, S.T., Nguyen, L.A., Szalas, A.: The Web Ontology Rule Language OWL 2 RL+ and Its Extensions. T. Computational Collective Intelligence 13, 152–175 (2014)
7. Cao, S.T., Nguyen, L.A., Szalas, A.: WORL: a nonmonotonic rule language for the Semantic Web. Vietnam J. Computer Science 1(1), 57–69 (2014)
8. Freire, J., Swift, T., Warren, D.S.: Taking I/O seriously: Resolution reconsidered for disk. In: Naish, L. (ed.) Proc. of ICLP 1997, pp. 198–212. MIT Press (1997)
9. Madalińska-Bugaj, E., Nguyen, L.A.: A generalized QSQR evaluation method for Horn knowledge bases. ACM Trans. on Computational Logic 13(4), 32 (2012)
10. Nejdl, W.: Recursive strategies for answering recursive queries - the RQA/FQI strategy. In: Stocker, P.M., Kent, W., Hammersley, P. (eds.) Proceedings of VLDB 1987, pp. 43–50. Morgan Kaufmann (1987)
11. Nguyen, L.A., Cao, S.T.: Query-Subquery Nets. In: Nguyen, N.-T., Hoang, K., Jędrzejowicz, P. (eds.) ICCCI 2012, Part I. LNCS, vol. 7653, pp. 239–248. Springer, Heidelberg (2012)
12. Nguyen, L.A., Cao, S.T.: Query-Subquery Nets. CoRR, abs/1201.2564 (2012)
13. Ramakrishnan, R., Srivastava, D., Sudarshan, S.: Efficient bottom-up evaluation of logic programs. In: Vandewalle, J. (ed.) The State of the Art in Computer Systems and Software Engineering. Kluwer Academic Publishers (1992)
14. Rohmer, J., Lescouer, R., Kerisit, J.-M.: The Alexander method – a technique for the processing of recursive axioms in deductive databases. New Generation Computing 4(3), 273–285 (1986)
15. Tamaki, H., Sato, T.: OLD resolution with tabulation. In: Shapiro, E. (ed.) ICLP 1986. LNCS, vol. 225, pp. 84–98. Springer, Heidelberg (1986)
16. Vieille, L.: Recursive axioms in deductive databases: The query/subquery approach. In: Proceedings of Expert Database Conf., pp. 253–267 (1986)
17. Vieille, L.: Recursive query processing: The power of logic. Theor. Comput. Sci. 69(1), 1–53 (1989)
18. Zhou, N.-F., Sato, T.: Efficient fixpoint computation in linear tabling. In: Proceedings of PPDP 2003, pp. 275–283. ACM (2003)

A Domain Partitioning Method
for Bisimulation-Based Concept Learning
in Description Logics

Thanh-Luong Tran[1], Linh Anh Nguyen[2,3], and Thi-Lan-Giao Hoang[1]

[1] Faculty of Information Technology, Hue University of Sciences
77 Nguyen Hue, Hue city, Vietnam
{ttluong,hlgiao}@hueuni.edu.vn
[2] Institute of Informatics, University of Warsaw
Banacha 2, 02-097 Warsaw, Poland
nguyen@mimuw.edu.pl
[3] Faculty of Information Technology, VNU University of Engineering and Technology
144 Xuan Thuy, Hanoi, Vietnam

Abstract. We have implemented a bisimulation-based concept learning method for description logics-based information systems using information gain. In this paper, we present our domain partitioning method that was used for the implementation. Apart from basic selectors, we also used a new kind of selectors, called extended selectors. Our evaluation results show that the concept learning method is valuable and extended selectors support it significantly.

Keywords: description logic, concept learning, bisimulation.

1 Introduction

Semantic Technologies have been investigated intensively and applied in many areas such as Bioinformatics, Semantic Web Browser, Knowledge Managements, Software Engineering, etc. One of the pillars of the Semantic Web is ontologies. They are an important aspect in knowledge representation and reasoning for the Semantic Web.

Nowadays, ontologies are usually modeled by using the Web Ontology Language (OWL), which is a standard recommended by W3C for the Semantic Web. In essence, OWL is a language based on description logics (DLs) [8]. Constructing useful ontologies is desirable. In ontology engineering, concept learning is helpful for suggesting important concepts and their definitions.

Concept learning in DLs is similar to binary classification in traditional machine learning. However, the problem in the context of DLs differs from the traditional setting in that objects are described not only by attributes but also by binary relations between objects. Concept learning in DLs has been studied by a number of researchers. Apart from the approaches based on "least common subsumers" proposed by Cohen and Hrish [4], and concept normalization

T.V. Do et al. (eds.), *Advanced Computational Methods for Knowledge Engineering*,
Advances in Intelligent Systems and Computing 282,
DOI: 10.1007/978-3-319-06569-4_22, © Springer International Publishing Switzerland 2014

proposed by Lambrix and Larocchia [10], concept learning was also studied using the approach of inductive logic programming and refinement operators by Badea and Nienhuy-Cheng [2], Iannone et al. [9], Fanizzi et al. [6], and Lehmann et al. [11]. Recently, Nguyen and Szałas [13], Ha et al. [7], and Tran et al. [15,16] used bisimulation for concept learning in DLs.

One of the main settings for concept learning in DLs [7,15] is as follows: given a finite interpretation $\mathcal{I}$ in a DL L, learn a concept C in L such that

$$\mathcal{I} \models C(a) \text{ for all } a \in E^+ \quad \text{and} \quad \mathcal{I} \models \neg C(a) \text{ for all } a \in E^-,$$

where E^+ (resp. E^-) contains positive (resp. negative) examples of C. Note that $\mathcal{I} \not\models C(a)$ is the same as $\mathcal{I} \models \neg C(a)$.

In [13], Nguyen and Szałas divide blocks in partitions of DL-based information systems by using basic selectors. They applied bisimulation in DLs to model indiscernibility of objects. Their work is pioneering in using bisimulation for concept learning in DLs. In [16], Tran et al. generalized and extended the concept learning method of [13] for DL-based information systems. The important problem is: which block from the current partition should be divided first and which selector should be used to divide it? These affect both the "quality" of the final partition and the complexity of the process. Nguyen and Szałas [13] and Tran et al. [16] proposed to use basic selectors and information gain to guide the granulation process, where basic selectors are created from blocks of the partitions appeared in the granulation process. These selectors divide blocks in the partitions and bring good results in favorable cases. However, they are not strong enough for complex cases. To obtain a final partition, the main loop of the granulation process may need to repeat many times. This usually results in too complex concepts, which poorly classify new objects.

In this paper, apart from the so called *basic selectors* that are created from the blocks of partitions, we also propose to use a new kind of selectors, called *extended selectors*, which are created from available selectors (basic and extended selectors) for dividing blocks. We have implemented the bisimulation-based concept learning method [13,16] using the mentioned selectors and information gain measures. The aim is to study effectiveness of basic and extended selectors as well as to provide experimental results for the concept learning method.

The rest of paper is structured as follows. In Section 2, we recall the notation of DLs and the definition of DL-based information systems. Section 3 outlines bisimulation and indiscernibility in DLs. Section 4 presents concept normalization and storage. Basic and extended selectors, information gain measures as well as techniques of concept reduction are presented in Section 5. Our experimental results are reported in Section 6. We conclude this work in Section 7.

2   Notation and DL-Based Information Systems

A *DL-signature* is a finite set $\Sigma = \Sigma_I \cup \Sigma_C \cup \Sigma_R$, where Σ_I is a set of *individuals*, Σ_C is a set of *concept names*, and Σ_R is a set of *role names*. All the sets Σ_I,

$$
\begin{array}{ll}
(r^-)^{\mathcal{I}} = (r^{\mathcal{I}})^{-1} \quad\quad \top^{\mathcal{I}} = \Delta^{\mathcal{I}} \quad\quad \bot^{\mathcal{I}} = \emptyset & (\neg C)^{\mathcal{I}} = \Delta^{\mathcal{I}} \setminus C^{\mathcal{I}} \\
(\forall R.C)^{\mathcal{I}} = \{x \in \Delta^{\mathcal{I}} \mid \forall y\,[R^{\mathcal{I}}(x,y) \Rightarrow C^{\mathcal{I}}(y)]\} & (C \sqcap D)^{\mathcal{I}} = C^{\mathcal{I}} \cap D^{\mathcal{I}} \\
(\exists R.C)^{\mathcal{I}} = \{x \in \Delta^{\mathcal{I}} \mid \exists y\,[R^{\mathcal{I}}(x,y) \wedge C^{\mathcal{I}}(y)]\} & (C \sqcup D)^{\mathcal{I}} = C^{\mathcal{I}} \cup D^{\mathcal{I}} \\
(\geq n\,R.C)^{\mathcal{I}} = \{x \in \Delta^{\mathcal{I}} \mid \#\{y \mid R^{\mathcal{I}}(x,y) \wedge C^{\mathcal{I}}(y)\} \geq n\} & (\geq n\,R)^{\mathcal{I}} = (\geq n\,R.\top)^{\mathcal{I}} \\
(\leq n\,R.C)^{\mathcal{I}} = \{x \in \Delta^{\mathcal{I}} \mid \#\{y \mid R^{\mathcal{I}}(x,y) \wedge C^{\mathcal{I}}(y)\} \leq n\} & (\leq n\,R)^{\mathcal{I}} = (\leq n\,R.\top)^{\mathcal{I}}
\end{array}
$$

Fig. 1. Semantics of complex object roles and complex concepts

Σ_C and Σ_R are pairwise disjoint. We denote individuals by letters like a and b, concept names by letters like A and B, role names by letters like r and s.

We consider some *DL-features* denoted by I (*inverse*), F (*functionality*), N (*unquantified number restriction*), Q (*quantified number restriction*). A *set of DL-features* is a set consisting of some or zero of these names.

Let Σ be a DL-signature and Φ be a set of DL-features. Let $\mathcal{L}$ stand for $\mathcal{ALC}$, which is the name of a basic DL. The DL language $\mathcal{L}_{\Sigma,\Phi}$ allows *roles* and *concepts* defined recursively as follows:

- if $r \in \Sigma_R$ then r is a role of $\mathcal{L}_{\Sigma,\Phi}$,
- if $I \in \Phi$ and $r \in \Sigma_R$ then r^- is a role of $\mathcal{L}_{\Sigma,\Phi}$,
- if $A \in \Sigma_C$ then A is concept of $\mathcal{L}_{\Sigma,\Phi}$,
- if C and D are concepts of $\mathcal{L}_{\Sigma,\Phi}$, R is a role of $\mathcal{L}_{\Sigma,\Phi}$ and n is a natural number then
    - $\top, \bot, \neg C, C \sqcap D, C \sqcup D, \forall R.C$ and $\exists R.C$ are concepts of $\mathcal{L}_{\Sigma,\Phi}$,
    - if $F \in \Phi$ then $\leq 1\,R$ is a concept of $\mathcal{L}_{\Sigma,\Phi}$,
    - if $N \in \Phi$ then $\geq n\,R$ and $\leq n\,R$ are concepts of $\mathcal{L}_{\Sigma,\Phi}$,
    - if $Q \in \Phi$ then $\geq n\,R.C$ and $\leq n\,R.C$ are concepts of $\mathcal{L}_{\Sigma,\Phi}$.

An *interpretation* in $\mathcal{L}_{\Sigma,\Phi}$ is a pair $\mathcal{I} = \langle \Delta^{\mathcal{I}}, \cdot^{\mathcal{I}} \rangle$, where $\Delta^{\mathcal{I}}$ is a non-empty set, called the *domain* of $\mathcal{I}$, and $\cdot^{\mathcal{I}}$ is a mapping, called the *interpretation function* of $\mathcal{I}$, that associates each individual $a \in \Sigma_I$ with an element $a^{\mathcal{I}} \in \Delta^{\mathcal{I}}$, each concept name $A \in \Sigma_C$ with a set $A^{\mathcal{I}} \subseteq \Delta^{\mathcal{I}}$, each role name $r \in \Sigma_R$ with a binary relation $r^{\mathcal{I}} \subseteq \Delta^{\mathcal{I}} \times \Delta^{\mathcal{I}}$.

To simplicity of notation, we write $C^{\mathcal{I}}(x)$ (resp. $R^{\mathcal{I}}(x,y)$) instead of $x \in C^{\mathcal{I}}$ (resp. $\langle x, y \rangle \in R^{\mathcal{I}}$). The interpretation function $\cdot^{\mathcal{I}}$ is extended to complex concepts and complex roles as shown in Figure 1.

An *information system* in $\mathcal{L}_{\Sigma,\Phi}$ is a finite interpretation in $\mathcal{L}_{\Sigma,\Phi}$.

Example 2.1. Let $\Phi = \{I\}$ and $\Sigma = \Sigma_I \cup \Sigma_C \cup \Sigma_R$, where
$\Sigma_I = \{Ava, Britt, Colin, Dave, Ella, Flor, Gigi, Harry\}$,
$\Sigma_C = \{Male, Female, Nephew, Niece\}$ and $\Sigma_R = \{hasChild, hasSibling\}$.
Consider the information system $\mathcal{I}$ specified by:
$\Delta^{\mathcal{I}} = \{a, b, c, d, e, f, g, h\}$, $Ava^{\mathcal{I}} = a$, $Britt^{\mathcal{I}} = b$, ..., $Harry^{\mathcal{I}} = h$,
$Female^{\mathcal{I}} = \{a, b, e, f, g\}$, $Male^{\mathcal{I}} = \Delta^{\mathcal{I}} \setminus Female^{\mathcal{I}} = \{c, d, h\}$,
$hasChild^{\mathcal{I}} = \{\langle a, d \rangle, \langle a, e \rangle, \langle b, f \rangle, \langle c, g \rangle, \langle c, h \rangle\}$,
$hasSibling^{\mathcal{I}} = \{\langle b, c \rangle, \langle c, b \rangle, \langle d, e \rangle, \langle e, d \rangle, \langle g, h \rangle, \langle h, g \rangle\}$,
$Niece^{\mathcal{I}} = (Female \sqcap \exists hasChild^-.(\exists hasSibling.\top))^{\mathcal{I}} = \{f, g\}$,

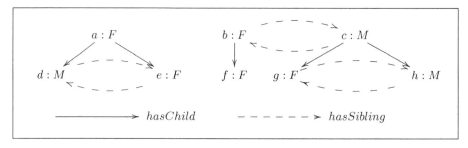

Fig. 2. An illustration for the information system given in Example 2.1

$Nephew^{\mathcal{I}} = (Male \sqcap \exists hasChild^{-}.(\exists hasSibling.\top))^{\mathcal{I}} = \{h\}$.

This interpretation is illustrated in Figure 2. In this figure, each node denotes a person, the letter M stands for $Male$, the letter F stands for $Female$, the solid edges denote assertions of the role $hasChild$, and the dashed edges denote assertions of the role $hasSibling$. ∎

Let C and D be concepts of $\mathcal{L}_{\Sigma,\Phi}$. We say that:

- C is *subsumed* by D, denoted by $C \sqsubseteq D$, if $C^{\mathcal{I}} \subseteq D^{\mathcal{I}}$ for every interpretation $\mathcal{I}$ in $\mathcal{L}_{\Sigma,\Phi}$,
- C and D are *equivalent*, denoted by $C \equiv D$, if $C \sqsubseteq D$ and $D \sqsubseteq C$,
- C is *satisfiable* if there exists an interpretation $\mathcal{I}$ in $\mathcal{L}_{\Sigma,\Phi}$ such that $C^{\mathcal{I}} \neq \emptyset$.

The *modal depth* of a concept C, denoted by $\mathsf{mdepth}(C)$, is defined to be:

- 0 if C is of the form $\top$, $\bot$ or A,
- $\mathsf{mdepth}(D)$ if C is of the form $\neg D$,
- 1 if C is of the form $\geq n\,R$ or $\leq n\,R$,
- $1 + \mathsf{mdepth}(D)$ if C is of the form $\exists R.D$, $\forall R.D$, $\geq n\,R.D$ or $\leq n\,R.D$,
- $\max(\mathsf{mdepth}(D), \mathsf{mdepth}(D'))$ if C is of the form $D \sqcap D'$ or $D \sqcup D'$.

The *length* of a concept C, denoted by $\mathsf{length}(C)$, is defined to be:

- 0 if C is $\top$ or $\bot$,
- 1 if C is a concept name A,
- $\mathsf{length}(D)$ if C is the form $\neg D$,
- 3 if C is of the form $\geq n\,R$ or $\leq n\,R$,
- $2 + \mathsf{length}(D)$ if C is of the form $\exists R.D$ or $\forall R.D$,
- $3 + \mathsf{length}(D)$ if C is of the form $\geq n\,R.D$ or $\leq n\,R.D$,
- $1 + \mathsf{length}(D) + \mathsf{length}(D')$ if C is of the form $D \sqcap D'$ or $D \sqcup D'$.

In this paper, concepts are represented in the negation normal form (i.e., the negation constructor appears only in front of atomic concepts). For this reason, $\mathsf{length}(\neg D)$ is defined to be $\mathsf{length}(D)$.

Example 2.2. Consider the language $\mathcal{L}_{\Sigma,\Phi}$ given in Example 1, we have:

- $\mathsf{mdepth}(Male \sqcap \geq 2\,hasChild.(\exists hasChild.\top)) = 2$,
- $\mathsf{length}(Male \sqcap \geq 2\,hasChild.(\exists hasChild.\top)) = 7$. ∎

We say that a concept C is *simpler* than a concept D if:

- $\mathsf{length}(C) < \mathsf{length}(D)$, or
- $\mathsf{length}(C) = \mathsf{length}(D)$ and $\mathsf{mdepth}(C) \leq \mathsf{mdepth}(D)$.

In fact, the modal depth of a concept is usually restricted by a very small value, while the length of a concept is usually large. This leads to small differences of the modal depth between concepts. In contrast, differences of the length between concepts may be very large. Thus, we choose the length of concepts as the main factor for deciding whether a concept is simpler than another.

Let $\mathbb{C} = \{C_1, C_2, \ldots, C_n\}$ be a set of concepts. A concept $C_i \in \mathbb{C}$ is said to be the *simplest* if it is simpler than any other concept in $\mathbb{C}$.

3   Bisimulation and Indiscernibility

In this section, we recall the notion of bisimulation for the DLs considered in this paper [13,16]. Let Σ and $\Sigma^\dagger$ be DL-signatures such that $\Sigma^\dagger \subseteq \Sigma$, Φ and $\Phi^\dagger$ be DL-features such that $\Phi^\dagger \subseteq \Phi$, $\mathcal{I}$ and $\mathcal{I}'$ be interpretations in $\mathcal{L}_{\Sigma,\Phi}$. An $\mathcal{L}_{\Sigma^\dagger,\Phi^\dagger}$-*bisimulation* between $\mathcal{I}$ and $\mathcal{I}'$ is a binary relation $Z \subseteq \Delta^\mathcal{I} \times \Delta^{\mathcal{I}'}$ satisfies the following conditions for every $a \in \Sigma_I^\dagger$, $A \in \Sigma_C^\dagger$, $r \in \Sigma_R^\dagger$, $x, y \in \Delta^\mathcal{I}$, $x', y' \in \Delta^{\mathcal{I}'}$:

$$Z(a^\mathcal{I}, a^{\mathcal{I}'}) \tag{1}$$

$$Z(x, x') \Rightarrow [A^\mathcal{I}(x) \Leftrightarrow A^{\mathcal{I}'}(x')] \tag{2}$$

$$[Z(x, x') \wedge r^\mathcal{I}(x, y)] \Rightarrow \exists y' \in \Delta^{\mathcal{I}'}[Z(y, y') \wedge r^{\mathcal{I}'}(x', y')] \tag{3}$$

$$[Z(x, x') \wedge r^{\mathcal{I}'}(x', y')] \Rightarrow \exists y \in \Delta^\mathcal{I}[Z(y, y') \wedge r^\mathcal{I}(x, y)], \tag{4}$$

if $I \in \Phi^\dagger$ then

$$[Z(x, x') \wedge r^\mathcal{I}(y, x)] \Rightarrow \exists y' \in \Delta^{\mathcal{I}'}[Z(y, y') \wedge r^{\mathcal{I}'}(y', x')] \tag{5}$$

$$[Z(x, x') \wedge r^{\mathcal{I}'}(y', x')] \Rightarrow \exists y \in \Delta^\mathcal{I}[Z(y, y') \wedge r^\mathcal{I}(y, x)], \tag{6}$$

if $N \in \Phi^\dagger$ then

$$Z(x, x') \Rightarrow \#\{y \mid r^\mathcal{I}(x, y)\} = \#\{y' \mid r^{\mathcal{I}'}(x', y')\}, \tag{7}$$

if $\{N, I\} \subseteq \Phi^\dagger$ then

$$Z(x, x') \Rightarrow \#\{y \mid r^\mathcal{I}(y, x)\} = \#\{y' \mid r^{\mathcal{I}'}(y', x')\}, \tag{8}$$

if $F \in \Phi^\dagger$ then

$$Z(x, x') \Rightarrow [\#\{y \mid r^\mathcal{I}(x, y)\} \leq 1 \Leftrightarrow \#\{y' \mid r^{\mathcal{I}'}(x', y')\} \leq 1], \tag{9}$$

if $\{F, I\} \subseteq \Phi^\dagger$ then

$$Z(x, x') \Rightarrow [\#\{y \mid r^\mathcal{I}(y, x)\} \leq 1 \Leftrightarrow \#\{y' \mid r^{\mathcal{I}'}(y', x')\} \leq 1], \tag{10}$$

if $Q \in \Phi^\dagger$ then

> if $Z(x, x')$ holds then, for every $r \in \Sigma_R^\dagger$, there exists a bijection
> $h : \{y \mid r^{\mathcal{I}}(x, y)\} \to \{y' \mid r^{\mathcal{I}'}(x', y')\}$ such that $h \subseteq Z$, $\qquad$ (11)

if $\{Q, I\} \subseteq \Phi^\dagger$ then

> if $Z(x, x')$ holds then, for every $r \in \Sigma_R^\dagger$, there exists a bijection
> $h : \{y \mid r^{\mathcal{I}}(y, x)\} \to \{y' \mid r^{\mathcal{I}'}(y', x')\}$ such that $h \subseteq Z$. $\qquad$ (12)

An interpretation $\mathcal{I}$ is said to be *finite branching* (or *image-finite*) w.r.t. $\mathcal{L}_{\Sigma^\dagger, \Phi^\dagger}$ if for every $x \in \Delta^{\mathcal{I}}$ and every $r \in \Sigma_R^\dagger$:

- the set $\{y \in \Delta^{\mathcal{I}} \mid r^{\mathcal{I}}(x, y)\}$ is finite, and
- if $I \in \Phi^\dagger$ then the set $\{y \in \Delta^{\mathcal{I}} \mid r^{\mathcal{I}}(y, x)\}$ is finite.

An $\mathcal{L}_{\Sigma^\dagger, \Phi^\dagger}$-bisimulation between $\mathcal{I}$ and itself is called an $\mathcal{L}_{\Sigma^\dagger, \Phi^\dagger}$-*auto-bisimulation* of $\mathcal{I}$. The *largest* $\mathcal{L}_{\Sigma^\dagger, \Phi^\dagger}$-auto-bisimulation of $\mathcal{I}$, denoted by $\sim_{\Sigma^\dagger, \Phi^\dagger, \mathcal{I}}$, is the one that is larger than or equal to ($\supseteq$) any other $\mathcal{L}_{\Sigma^\dagger, \Phi^\dagger}$-auto-bisimulation of $\mathcal{I}$. It always exists [5].

We say that a set Y is *divided* by a set X if $Y \setminus X \neq \emptyset$ and $Y \cap X \neq \emptyset$. Thus, Y is not divided by X if either $Y \subseteq X$ or $Y \cap X = \emptyset$. A partition $\{Y_1, Y_2 \ldots, Y_n\}$ is *consistent* with a set X if Y_i is not divided by X for any $1 \leq i \leq n$.

Theorem 3.1. *Let $\mathcal{I}$ be an interpretation in $\mathcal{L}_{\Sigma, \Phi}$ and let $X \subseteq \Delta^{\mathcal{I}}$, $\Sigma^\dagger \subseteq \Sigma$, $\Phi^\dagger \subseteq \Phi$. Then:*

- *if there exists a concept C of $\mathcal{L}_{\Sigma^\dagger, \Phi^\dagger}$ such that $X = C^{\mathcal{I}}$ then the partition of $\Delta^{\mathcal{I}}$ by $\sim_{\Sigma^\dagger, \Phi^\dagger, \mathcal{I}}$ is consistent with X,*
- *if the partition of $\Delta^{\mathcal{I}}$ by $\sim_{\Sigma^\dagger, \Phi^\dagger, \mathcal{I}}$ is consistent with X then there exists a concept C of $\mathcal{L}_{\Sigma^\dagger, \Phi^\dagger}$ such that $C^{\mathcal{I}} = X$.*

This theorem differs from the ones of [13,16,7] only in the studied class of DLs. It can be proved analogously to [13, Theorem 4].

4   Concept Normalization and Storage

There are different normal forms for concepts [1,12]. By using a normal form, one can represent concepts in a unified way. We propose below a new normal form, which extends the one of [12]. The normal form of a concept is obtained by applying the following normalization rules:

1. concepts are represented in the negation normal form,
2. a conjunction $C_1 \sqcap C_2 \sqcap \cdots \sqcap C_n$ is represented by an "and"-set, denoted by $\sqcap\{C_1, C_2, \ldots, C_n\}$,
3. $\sqcap\{C\}$ is replaced by C,
4. $\sqcap\{\sqcap\{C_1, C_2, \ldots, C_i\}, C_{i+1}, \ldots, C_n\}$ is replaced by $\sqcap\{C_1, C_2, \ldots, C_n\}$,
5. $\sqcap\{\top, C_1, C_2, \ldots, C_n\}$ is replaced by $\sqcap\{C_1, C_2, \ldots, C_n\}$,

6. $\sqcap\{\bot, C_1, C_2, \ldots, C_n\}$ is replaced by $\bot$,
7. if $C_i \sqsubseteq C_j$ and $1 \leq i \neq j \leq n$, then remove C_j from $\sqcap\{C_1, C_2, \ldots, C_n\}$,
8. if $C_i = \overline{C}_j$ and $1 \leq i \neq j \leq n$, then $\sqcap\{C_1, C_2, \ldots, C_n\}$ is replaced by $\bot$, where $\overline{C}$ is the normal form of $\neg C$,
9. $\forall R. \sqcap \{C_1, C_2, \ldots, C_n\}$ is replaced by $\sqcap\{\forall R.C_1, \forall R.C_2, \ldots, \forall R.C_n\}$,
10. $\forall R.\top$ is replaced by $\top$,
11. $\leq n\,R.\bot$ is replaced by $\top$,
12. $\geq 0\,R.C$ is replaced by $\top$,
13. $\geq 1\,R.C$ is replaced by $\exists R.C$,
14. $\geq n\,R.\bot$ is replaced by $\bot$ when $n > 0$,
15. the rules "dual" to the rules 2–10 (for example, dually to the fourth rule, $\sqcup\{\sqcup\{C_1, C_2, \ldots, C_i\}, C_{i+1}, \ldots, C_n\}$ is replaced by $\sqcup\{C_1, C_2, \ldots, C_n\}$).

Each step of the granulation process may generate many concepts. To avoid repetition of the same concepts in the storage, we design the data structure appropriately. If two concepts have the same "normal form" then they are represented only once in the data structure by the normal form. In addition, our program processes only one interpretation (the considered information system) and many concepts may have the same extension (i.e., the same set of objects in the interpretation). Thus, instead of storing all concepts which have the same extension, only the simplest concept is archived in the catalogue. These techniques allow to reduce the memory for representing concepts and their extensions as well as to increase the performance of our program.

5   Granulating Partitions Using Selectors and Information Gain

Let $\mathcal{I}$ be a finite interpretation in $\mathcal{L}_{\Sigma, \Phi}$ given as a training information system. Let $A_d \in \Sigma_C$ be a concept name standing for the "decision attribute". Let $E^+ = \{a \mid a^{\mathcal{I}} \in A_d^{\mathcal{I}}\}$ and $E^- = \{a \mid a^{\mathcal{I}} \in (\neg A_d)^{\mathcal{I}}\}$ be sets of *positive examples* and *negative examples* of A_d in $\mathcal{I}$, respectively. Suppose that A_d can be expressed by a concept C in $\mathcal{L}_{\Sigma^\dagger, \Phi^\dagger}$, where $\Sigma^\dagger \subseteq \Sigma \setminus \{A_d\}$ and $\Phi^\dagger \subseteq \Phi$. How can we learn that concept C on the basis of $\mathcal{I}$, E^+ and E^-? The concept C must satisfy the following condition:

$$\mathcal{I} \models C(a) \text{ for all } a \in E^+ \quad \text{and} \quad \mathcal{I} \models \neg C(a) \text{ for all } a \in E^-.$$

In [13] Nguyen and Szałas proposed a bisimulation-based method for this learning problem, and in [16] Tran et al. extended the method for a large class of DLs. The idea of those works is based on the following observation:

if A_d is definable in $\mathcal{L}_{\Sigma^\dagger, \Phi^\dagger}$ by a concept C then, by the first assertion of Theorem 3.1, $C^{\mathcal{I}}$ must be the union of some equivalence classes of $\Delta^{\mathcal{I}}$ w.r.t. $\sim_{\Sigma^\dagger, \Phi^\dagger, \mathcal{I}}$.

1. A, where $A \in \Sigma_C^\dagger$,
2. $\exists R.\top, \forall R.\bot$,
3. $\exists R.A$ and $\forall R.A$, where $A \in \Sigma_C^\dagger$,
4. $\exists R.C_i$ and $\forall R.C_i$, where $1 \leq i \leq n$,
5. $\leq 1\, R$ if $F \in \Phi^\dagger$,
6. $\leq l\, R$ and $\geq m\, R$, if $N \in \Phi^\dagger$, $0 < l \leq \#\Delta^\mathcal{I}$ and $0 \leq m < \#\Delta^\mathcal{I}$,
7. $\leq l\, R.C_i$ and $\geq m\, R.C_i$, if $Q \in \Phi^\dagger$, $1 \leq i \leq n$, $0 < l \leq \#C_i^\mathcal{I}$ and $0 \leq m < \#C_i^\mathcal{I}$,

where R is a role of $\mathcal{L}_{\Sigma^\dagger,\Phi^\dagger}$.

Fig. 3. Basic selectors

1. $\exists R.D_u$ and $\forall R.D_u$,
2. $\leq l\, R.D_u$ and $\geq m\, R.D_u$, if $Q \in \Phi^\dagger$, $0 < l \leq \#D_u^\mathcal{I}$ and $0 \leq m < \#D_u^\mathcal{I}$,

where R is a role of $\mathcal{L}_{\Sigma^\dagger,\Phi^\dagger}$ and $1 \leq u \leq h$.

Fig. 4. Extended selectors

5.1   Basic Selectors and Extended Selectors

In the granulation process, we denote the blocks created so far in all steps by $Y_1, Y_2, \ldots, Y_n$. We always use a new subscript for each newly created block by increasing n. We take care that, for each $1 \leq i \leq n$, Y_i is characterized by a concept C_i such that $C_i^\mathcal{I} = Y_i$.

Let $\mathbb{Y} = \{Y_{i_1}, Y_{i_2}, \ldots, Y_{i_k}\} \subseteq \{Y_1, Y_2, \ldots, Y_n\}$ be the current partition of $\Delta^\mathcal{I}$ and $\mathbb{D} = \{D_1, D_2, \ldots, D_h\}$ be the current set of selectors, which are concepts of $\mathcal{L}_{\Sigma^\dagger,\Phi^\dagger}$. (At the beginning, $\mathbb{Y} = \{\Delta^\mathcal{I}\}$ and $\mathbb{D} = \emptyset$.) Consider dividing a block $Y_{i_j} \in \mathbb{Y}$. We want to find a selector $D_u \in \mathbb{D}$ to divide Y_{i_j}.

Such a selector D_u should actually divide Y_{i_j} into two non-empty parts (i.e., Y_{i_j} should be divided by $D_u^\mathcal{I}$). We divide Y_{i_j} by D_u as follows:

- $s := n+1, t := n+2, n := n+2$,
- $Y_s := Y_{i_j} \cap D_u^\mathcal{I}, C_s := C_{i_j} \sqcap D_u$,
- $Y_t := Y_{i_j} \cap (\neg D_u)^\mathcal{I}, C_t := C_{i_j} \sqcap \neg D_u$,
- the new partition of $\Delta^\mathcal{I}$ is $\{Y_{i_1}, Y_{i_2}, \ldots, Y_{i_k}\} \cup \{Y_s, Y_t\} \setminus \{Y_{i_j}\}$.

In [16], the used selectors are concepts of $\mathcal{L}_{\Sigma^\dagger,\Phi^\dagger}$ of the forms listed in Figure 3, which were called *basic selectors*. The work [16] showed that, to reach the partition corresponding to the equivalence relation $\sim_{\Sigma^\dagger,\Phi^\dagger,\mathcal{I}}$ it suffices to start from the partition $\{\Delta^\mathcal{I}\}$ and repeatedly granulate it by using only basic selectors.

We have implemented the bisimulation-based concept learning method using basic selectors and information gain [12,16]. Our experiments showed that the resulting concepts obtained by this method have the following characteristics:

- the length of the resulting concept is usually long,
- the accuracy, precision, recall and F1 measures of the resulting classifier are not high for new objects.

The main reason is that basic selectors are not advanced enough. In our implementation, we used a new kind of selectors, called *extended selectors*. They are created from the current set $\mathbb{D}$ of selectors. Each extended selector is a concept of one of the forms listed in Figure 4. Extended selectors play an important role in the granulation process. By using them, we have more selectors that can be used to divide blocks and we can obtain better partitions than by using only basic selectors. Furthermore, by using extended selectors, the number of iterations of the main loop is usually reduced significantly. This leads to simpler resulting concepts with higher accuracy in classifying new objects.

5.2 Information Gain for Granulating Partitions

Let X and Y be subsets of $\Delta^{\mathcal{I}}$, where X plays the role of a set of positive examples for the concept to be learnt. The *entropy* of Y w.r.t. X in $\Delta^{\mathcal{I}}$, denoted by $E_{\Delta^{\mathcal{I}}}(Y/X)$, is defined as follows, where XY stands for $X \cap Y$ and $\overline{X}Y$ stands for $\overline{X} \cap Y$:

$$E_{\Delta^{\mathcal{I}}}(Y/X) = \begin{cases} 0, \text{ if } Y \cap X = \emptyset \text{ or } Y \subseteq X \\ -\dfrac{\#XY}{\#Y} * \log_2 \dfrac{\#XY}{\#Y} - \dfrac{\#\overline{X}Y}{\#Y} * \log_2 \dfrac{\#\overline{X}Y}{\#Y}, \text{otherwise.} \end{cases}$$

The *information gain* of a selector D for dividing Y w.r.t. X in $\Delta^{\mathcal{I}}$, denoted by $IG_{\Delta^{\mathcal{I}}}(Y/X, D)$, is defined as follows, where $D^{\mathcal{I}}Y$ stands for $D^{\mathcal{I}} \cap Y$ and $\overline{D^{\mathcal{I}}}Y$ stands for $\overline{D^{\mathcal{I}}} \cap Y$:

$$IG_{\Delta^{\mathcal{I}}}(Y/X,D) = E_{\Delta^{\mathcal{I}}}(Y/X) - \left(\frac{\#D^{\mathcal{I}}Y}{\#Y} * E_{\Delta^{\mathcal{I}}}(D^{\mathcal{I}}Y/X) + \frac{\#\overline{D^{\mathcal{I}}}Y}{\#Y} * E_{\Delta^{\mathcal{I}}}(\overline{D^{\mathcal{I}}}Y/X) \right)$$

In the case $\Delta^{\mathcal{I}}$ and X are clear from the context, we write $E(Y)$ instead of $E_{\Delta^{\mathcal{I}}}(Y/X)$ and $IG(Y, D)$ instead of $IG_{\Delta^{\mathcal{I}}}(Y/X, D)$.

Suppose that we have the current partition $\mathbb{Y} = \{Y_{i_1}, Y_{i_2}, \ldots, Y_{i_k}\}$ and the current set of selectors $\mathbb{D} = \{D_1, D_2, \ldots, D_h\}$. For each block $Y_{i_j} \in \mathbb{Y}$ (where $1 \leq j \leq k$), let S_{i_j} be the simplest selector from the set $\arg\max_{D_u \in \mathbb{D}}\{IG(Y_{i_j}, D_u)\}$. For the current partition $\mathbb{Y}$, if Y_{i_j} is chosen to be divided then S_{i_j} is the choice for dividing Y_{i_j}. Note that the information gain is used for the selection.

After choosing the selectors for blocks, we decide which block should be divided first. We choose a block Y_{i_j} such that applying the selector S_{i_j} to divide Y_{i_j} maximizes the information gain. That is, we divide a block $Y_{i_j} \in \arg\max_{Y_{i_j} \in \mathbb{Y}}\{IG(Y_{i_j}, S_{i_j})\}$ first.

After dividing a block, we have a new partition. We also add new selectors which are created by the rules in Figures 3 and 4 to the current set of selectors. This set is used to continue granulating the new partition.

Example 5.1. Consider the information system $\mathcal{I}$ given in Example 2.1, the sublanguage $\mathcal{L}_{\Sigma^{\dagger}, \Phi^{\dagger}}$ with $\Sigma^{\dagger} = \{Female, hasChild, hasSibling\}$ and $\Phi^{\dagger} = \{I\}$, and the set $X = \{f, g\}$. One can think of X as the set of instances of the concept $Niece = Female \sqcap \exists hasChild^{-}.(\exists hasSibling.\top)$ in $\mathcal{I}$.

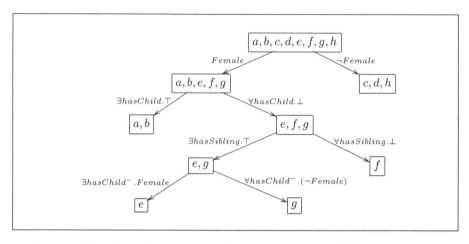

Fig. 5. A tree illustrating the granulation process using only basic selectors

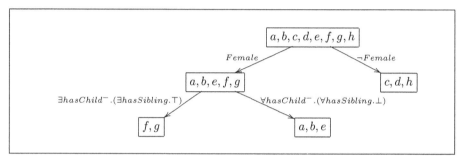

Fig. 6. A tree illustrating the granulation process using basic and extended selectors

1. Learning a definition of X in $\mathcal{L}_{\Sigma^\dagger, \Phi^\dagger}$ using only basic selectors:
   The steps are illustrated by the tree in Figure 5. The resulting concept is

   $$(Female \sqcap \forall hasChild.\bot \sqcap \exists hasSibling.\top \sqcap \forall hasChild^-.(\neg Female)) \sqcup$$
   $$(Female \sqcap \forall hasChild.\bot \sqcap \forall hasSibling.\bot).$$

   This concept can be simplified to

   $$Female \sqcap \forall hasChild.\bot \sqcap (\forall hasChild^-.(\neg Female) \sqcup \forall hasSibling.\bot).$$

2. Learning a definition of X in $\mathcal{L}_{\Sigma^\dagger, \Phi^\dagger}$ using basic and extended selectors:
   The steps are illustrated by the tree in Figure 6. The resulting concept is

   $$Female \sqcap \exists hasChild^-.(\exists hasSibling.\top).$$

In the second case, the concept $\exists hasChild^-.(\exists hasSibling.\top)$ is an extended selector. It is created by applying the rule 1 in Figure 4 to $\exists hasSibling.\top$, which is one of the available selectors.

This example demonstrates that using basic and extended selectors is better than using only basic selectors. The former reduces the number of iterations of the main loop and the length of the resulting concept.   ■

5.3   Reducing the Resulting Concepts

The decision trees generated in the granulation process can be very large. They may give complex concepts which overfit the training datasets and poorly classify new objects. In our implementation, we use some techniques to reduce the size of the decision trees and the length of the resulting concepts. A validating dataset is used to prune the decision tree as well as to reduce the resulting concept. The goal is to increase the accuracy.

- The first technique is pruning. Given a decision tree generated by the granu-lation process, the technique allows to reduce the size of the tree by removing parts that provide little power in classifying objects. Pruning can be done top-down or bottom-up. In our implementation, we use the bottom up fash-ion: we repeatedly choose a node whose successors are leafs and cut the successors if the average accuracy of the resulting concept is not worse on the training and validating datasets.
- The second technique is based on replacement. The resulting concept is usu-ally a disjunction $C_1 \sqcup C_2 \sqcup \cdots \sqcup C_n$. In the case C_i is a too complex concept, we consider replacing it by a simpler one from the set of selectors if they have the same set of objects in the considered information system. The re-placement is done only when the accuracy of the resulting concept is not worse on the validating dataset.
- The third technique is simplification. De Morgan's laws are used to reduce the resulting concept to an equivalent one.

In our experiments, the above three techniques allow to reduce the length of the resulting concepts significantly.

6   Experimental Results

The difficulty we encountered is that there are too few available datasets with linked data that can directly be used for concept learning using our setting. We have to build/get datasets for our setting from some resources on the Internet, including the WebKB [14], PockerHand [3] and Family datasets.

The **WebKB** dataset consists of information about web pages of four depart-ments of computer science (Cornell, Washington, Wisconsin, and Texas univer-sities). It contains information about 877 web pages (objects) and 1608 links between them of one relationship (`cites`). Each object in the dataset is de-scribed by a 0/1-valued word vector indicating the absence/presence of the cor-responding word from the dictionary (1703 words). It is assigned one of five concepts indicating the type of web page: `course`, `faculty`, `student`, `project`, and `staff`. We use data from two of the four departments for training (230

Table 1. Evaluation results on the WebKB, PockerHand and Family datasets using 100 random concepts in the DL $\mathcal{ALCIQ}$

	Avg. Dep. Res./Org.	Avg. Len. Res./Org.	Avg. Acc. [Min;Max]	Avg. Pre. [Min;Max]	Avg. Rec. [Min;Max]	Avg. F1 [Min;Max]
WebKB dataset						
Basic Selectors	0.82/1.02	6.81/4.41	93.84±13.50 [33.69;100.0]	92.09±17.04 [32.08;100.0]	92.82±17.32 [23.08;100.0]	91.59±16.68 [27.69;100.0]
Ba.&Ex. Selectors	0.84/1.02	3.40/4.41	94.60±12.20 [33.69;100.0]	92.81±15.93 [32.08;100.0]	93.14±17.17 [23.08;100.0]	92.33±16.17 [27.69;100.0]
PockerHand dataset						
Basic Selectors	1.41/2.60	37.02/15.97	97.17±08.61 [50.57;100.0]	95.96±14.99 [01.67;100.0]	94.95±14.40 [01.67;100.0]	94.66±14.64 [1.67;100.0]
Ba.&Ex. Selectors	1.23/2.60	3.47/15.97	99.44±02.15 [83.25;100.0]	98.68±09.08 [01.67;100.0]	98.06±09.58 [01.67;100.0]	98.18±09.14 [1.67;100.0]
Family dataset						
Basic Selectors	2.38/3.34	78.50/18.59	88.50±16.65 [27.91;100.0]	90.60±18.57 [04.55;100.0]	85.66±22.36 [07.69;100.0]	86.09±20.10 [08.70;100.0]
Ba.&Ex. Selectors	2.29/3.34	10.20/18.59	92.79±14.35 [27.91;100.0]	91.99±18.40 [04.55;100.0]	91.75 ±19.82 [07.69;100.0]	90.39±19.89 [08.70;100.0]

objects) and validating (195 objects). The two remaining ones (452 objects) are used for testing.

The **Family** dataset consists of information about people from five families (British Royal, Bush, Roberts, Romanov and Stevens families). It contains information about 943 people (objects) and 11062 links between them of seven relationships (hasChild, hasSon, hasDaughter, hasWife, hasHusband, hasBrother, hasSister). Each object is an instance of either the concept Male or Female. The data from two of the five families are used for training (437 objects) and validating (49 objects). The three remaining ones (457 objects) are used for testing.

The **PockerHand** dataset is a subset taken from UCI Machine Learning Repository. It consists of information about 2542 hands, 12710 cards, 119 features of cards (15371 objects in total) and 65220 links between them of six relationships (hasCard, hasRank, hasSuit, sameRank, nextRank, sameSuit). The goal is to predict which among nine classes should be assigned to a hand. These classes are "one pair", "two pairs", "three of a kind", "straight", "flush", "full house", "four of a kind", "straight flush" and "royal flush". Because the number of hands in the classes "four of a kind", "straight flush" and "royal flush" is very small, we remove these classes from our dataset. The dataset is divided into seven subsets. Two subsets are used for training (1343 objects) and validating (1343 objects). The five remaining ones are used for testing (12685 objects).

Given a signature Σ, by $\mathcal{ALCIQ}$ (resp. $\mathcal{ALCI}$ or $\mathcal{ALCQ}$) we denote the logic $\mathcal{L}_{\Sigma,\Phi}$ with $\Phi = \{I, Q\}$ (resp. $\Phi = \{I\}$ or $\Phi = \{Q\}$).

Table 2. Evaluation results on the Family dataset using five popular concepts in the DL $\mathcal{ALCI}$

	Dep. Res.	Len. Res.	Avg. Acc. [Min;Max]	Avg. Pre. [Min;Max]	Avg. Rec. [Min;Max]	Avg. F1 [Min;Max]
Concept: $Grandparent = \exists hasChild.(\exists hasChild.\top)$						
Basic Selectors	2.00	4.00	100.0±00.00 [100.0;100.0]	100.0±00.00 [100.0;100.0]	100.0±00.00 [100.0;100.0]	100.0±00.00 [100.0;100.0]
Ba.&Ex. Selectors	2.00	4.00	100.0±00.00 [100.0;100.0]	100.0±00.00 [100.0;100.0]	100.0±00.00 [100.0;100.0]	100.0±00.00 [100.0;100.0]
Concept: $Grandfather = Male \sqcap \exists hasChild.(\exists hasChild.\top)$						
Basic Selectors	2.00	36.00	95.90±01.39 [94.28;97.67]	87.38±06.81 [80.00;96.43]	79.15±17.38 [57.45;100.0]	81.44±08.35 [72.00;92.31]
Ba.&Ex. Selectors	2.00	07.00	99.46±00.77 [98.37;100.0]	100.0±00.00 [100.0;100.0]	95.74±6.02 [87.23;100.0]	97.73±03.21 [93.18;100.0]
Concept: $Grandmother = Female \sqcap \exists hasChild.(\exists hasChild.\top)$						
Basic Selectors	2.00	18.00	89.74±01.30 [88.37;91.49]	100.0±00.00 [100.0;100.0]	15.32±04.47 [09.30;20.00]	26.31±06.85 [17.02;33.33]
Ba.&Ex. Selectors	2.00	07.00	99.91±00.13 [99.73;100.0]	100.0±00.00 [100.0;100.0]	99.22±01.10 [97.67;100.0]	99.61±00.55 [98.82;100.0]
Concept: $Niece = Female \sqcap \exists hasChild^-.(\exists hasBrother.\top \sqcup \exists hasSister.\top)$						
Basic Selectors	3.00	151.00	85.57±09.47 [72.21;93.02]	57.92±32.09 [12.66;83.33]	64.70±29.35 [23.26;87.50]	60.69±31.33 [16.39;83.33]
Ba.&Ex. Selectors	2.00	11.00	100.0±00.00 [100.0;100.0]	100.0±00.00 [100.0;100.0]	100.0±00.00 [100.0;100.0]	100.0±00.00 [100.0;100.0]
Concept: $Nephew = Male \sqcap \exists hasChild^-.(\exists hasBrother.\top \sqcup \exists hasSister.\top)$						
Basic Selectors	3.00	178.00	91.40±05.74 [83.38;95.74]	77.04±26.30 [40.22;100.0]	88.40±01.99 [86.05;90.91]	79.82±17.72 [54.81;93.75]
Ba.&Ex Selectors	2.00	11.00	100.0±00.00 [100.0;100.0]	100.0±00.00 [100.0;100.0]	100.0±00.00 [100.0;100.0]	100.0±00.00 [100.0;100.0]

We have used Java language (JDK 1.6) to implement the bisimulation-based concept learning method for information systems in the DL $\mathcal{ALCIQ}$ using basic and extended selectors as well as the information gain discussed in Section 5. The reduction techniques mentioned in that section have been integrated into our program. The program and datasets can be downloaded from http://www.mimuw.edu.pl/~ttluong/ConceptLearning.rar.

We tested our method on the above mentioned three datasets using 100 random origin concepts in the DL $\mathcal{ALCIQ}$. For each random origin concept C, we used $E^+ = \{a \mid a^{\mathcal{I}} \in C^{\mathcal{I}}\}$ as the set of positive examples and $E^- = \{a \mid a^{\mathcal{I}} \in \Delta^{\mathcal{I}} \setminus C^{\mathcal{I}}\}$ as the set of negative examples, where $\mathcal{I}$ is the considered interpretation used as the training information system. These concepts have different depths and lengths. We ran the program on each dataset and each concept in two cases: using only basic selectors and using

Table 3. Evaluation results on the Family dataset using six sets of objects and the DL $\mathcal{ALCQ}$

	Dep. Res.	Len. Res.	Avg. Acc. [Min;Max]	Avg. Pre. [Min;Max]	Avg. Rec. [Min;Max]	Avg. F1 [Min;Max]
One pair						
Basic Selectors	4.0	109.00	42.57±01.48 [40.71;45.24]	16.74±00.87 [15.64;18.05]	76.00±4.03 [71.67;81.67]	27.44±01.42 [25.67;29.45]
Ba.&Ex. Selectors	5.00	15.00	100.0±00.00 [100.0;100.0]	100.0±00.00 [100.0;100.0]	100.0±00.00 [100.0;100.0]	100.0±00.00 [100.0;100.0]
Two pairs						
Basic Selectors	4.00	25.00	36.33±00.47 [35.48;36.67]	17.16±0.53 [16.34;17.70]	90.33±4.14 [83.33;95.00]	28.83±00.96 [27.32;29.84]
Ba.&Ex. Selectors	5.00	15.00	100.0±00.00 [100.0;100.0]	100.0±00.00 [100.0;100.0]	100.0±00.00 [100.0;100.0]	100.0±00.00 [100.0;100.0]
Three of a kind						
Basic Selectors	4.00	48.00	52.52±02.16 [50.71;56.67]	20.92±1.01 [19.75;22.77]	83.33±01.83 [80.00;85.00]	33.43±01.39 [31.68;35.92]
Ba.&Ex. Selectors	3.00	11.00	100.0±00.00 [100.0;100.0]	100.0±00.00 [100.0;100.0]	100.0±00.00 [100.0;100.0]	100.0±00.00 [100.0;100.0]
Straight						
Basic Selectors	5.00	97.00	81.24±02.01 [80.00;85.24]	39.65±04.62 [36.36;48.72]	58.33±04.94 [53.33;65.00]	47.13±4.41 [43.24;55.07]
Ba.&Ex. Selectors	5.00	32.00	98.67±00.68 [97.62;99.52]	96.35±03.44 [90.32;100.0]	94.33±02.00 [91.67;96.67]	95.31±02.35 [91.80;98.31]
Flush						
Basic Selectors	2.00	10.00	94.33±00.80 [92.86;95.24]	71.71±02.79 [66.67;75.00]	100.0±00.00 [100.0;100.0]	83.49±01.92 [80.00;85.71]
Ba.&Ex. Selectors	3.00	7.00	100.0±00.00 [100.0;100.0]	100.0±00.00 [100.0;100.0]	100.0±00.00 [100.0;100.0]	100.0±00.00 [100.0;100.0]
Full house						
Basic Selectors	4.00	68.00	60.48±03.05 [57.62;64.76]	25.95±01.45 [24.23;28.00]	94.67±2.45 [91.67;98.33]	40.71±01.73 [38.33;43.08]
Ba.&Ex. Selectors	2.00	6.00	100.0±00.00 [100.0;100.0]	100.0±00.00 [100.0;100.0]	100.0±00.00 [100.0;100.0]	100.0±00.00 [100.0;100.0]

both basic and extended selectors. Table 1 summarizes the evaluation of our experiments in:

- The average (Avg.) modal depth (Dep.) of the origin concepts (Org.),
- The average length (Len.) of the origin concepts,
- The average modal depth of the resulting concepts (Res.),
- The average length of the resulting concepts,
- The average accuracy (Acc.), precision (Pre.), recall (Rec.) and F1 measures,
- The standard variant, minimum (Min) and maximum (Max) values of accuracy, precision, recall and F1 measures.

As can be seen in Table 1, the accuracy, precision, recall and F1 measures of the resulting concepts in classifying new objects are usually very high. This demonstrates that the bisimulation-based concept learning method is valuable.

In addition, we tested the method using specific concepts on the Family and PockerHand datasets. For the former, we use the following five popular concepts in the DL $\mathcal{ALCI}$:

1. $Grandparent = \exists hasChild.(\exists hasChild.\top)$,
2. $Grandfather = Male \sqcap \exists hasChild.(\exists hasChild.\top)$,
3. $Grandmother = Female \sqcap \exists hasChild.(\exists hasChild.\top)$,
4. $Nephew = Male \sqcap \exists hasChild^{-}.(\exists hasBrother.\top \sqcup \exists hasSister.\top)$,
5. $Niece = Female \sqcap \exists hasChild^{-}.(\exists hasBrother.\top \sqcup \exists hasSister.\top)$.

For the PockerHand dataset, we tested the method using six sets of objects corresponding to six concepts (classes) in the DL $\mathcal{ALCQ}$. They are described below:

1. "one pair" - one pair of equal ranks within five cards,
2. "two pairs" - two pairs of equal ranks within five cards,
3. "three of a kind" - three equal ranks within five cards,
4. "straight" - five cards, sequentially ranked with no gaps,
5. "flush" - five cards with the same suit,
6. "full house" - pair + different rank three of a kind.

Table 2 provides the evaluation results on the Family dataset using the mentioned popular concepts. Table 3 provides the evaluation results on the Pocker-Hand dataset using the above six classes.

From Tables 1, 2 and 3, it is clear that extended selectors are highly effective for reducing the length of the resulting concepts and for obtaining better classifiers. This demonstrates that extended selectors efficiently support the bisimulation-based concept learning method.

7   Conclusions

We have implemented the bisimulation-based concept learning method for description logics-based information systems [13,16]. Apart from basic selectors proposed in [13,16], we introduced and used extended selectors for this method. We tested the method using basic and extended selectors for different datasets. Our experimental results show that the method is valuable and extended selectors support it significantly.

Acknowledgments. This paper was written during the first author's visit at Warsaw Center of Mathematics and Computer Sciences (WCMCS). He would like to thank WCMCS for the support. We thank Prof. Hung Son Nguyen for allowing us to use a server to run our program. This work was also supported by Polish National Science Centre (NCN) under Grant No. 2011/01/B/ST6/02759 as well as by Hue University under Grant No. DHH2013-01-41.

References

1. Baader, F., Calvanese, D., McGuinness, D.L., Nardi, D., Patel-Schneider, P.F. (eds.): The Description Logic Handbook: Theory, Implementation, and Applications. Cambridge University Press (2003)
2. Badea, L., Nienhuys-Cheng, S.-H.: A refinement operator for description logics. In: Cussens, J., Frisch, A.M. (eds.) ILP 2000. LNCS (LNAI), vol. 1866, pp. 40–59. Springer, Heidelberg (2000)
3. Cattral, R., Oppacher, F., Deugo, D.: Evolutionary data mining with automatic rule generalization (2002)
4. Cohen, W.W., Hirsh, H.: Learning the CLASSIC description logic: Theoretical and experimental results. In: Proceedings of KR 1994, pp. 121–133 (1994)
5. Divroodi, A., Nguyen, L.: On bisimulations for description logics. In: Proceedings of CS&P 2011, pp. 99–110 (2011) (see also arXiv:1104.1964)
6. Fanizzi, N., d'Amato, C., Esposito, F.: DL-FOIL concept learning in description logics. In: Železný, F., Lavrač, N. (eds.) ILP 2008. LNCS (LNAI), vol. 5194, pp. 107–121. Springer, Heidelberg (2008)
7. Ha, Q.-T., Hoang, T.-L.-G., Nguyen, L.A., Nguyen, H.S., Szałas, A., Tran, T.-L.: A bisimulation-based method of concept learning for knowledge bases in description logics. In: Proceedings of the Third Symposium on Information and Communication Technology, SoICT 2012, pp. 241–249. ACM (2012)
8. Horrocks, I., Kutz, O., Sattler, U.: The even more irresistible $\mathcal{SROIQ}$. In: KR, pp. 57–67. AAAI Press (2006)
9. Iannone, L., Palmisano, I., Fanizzi, N.: An algorithm based on counterfactuals for concept learning in the semantic web. Applied Intelligence 26(2), 139–159 (2007)
10. Lambrix, P., Larocchia, P.: Learning composite concepts. In: Proceedings of DL 1998 (1998)
11. Lehmann, J., Hitzler, P.: Concept learning in description logics using refinement operators. Machine Learning 78(1-2), 203–250 (2010)
12. Nguyen, L.A.: An efficient tableau prover using global caching for the description logic $\mathcal{ALC}$. Fundam. Inform. 93(1-3), 273–288 (2009)
13. Nguyen, L.A., Szałas, A.: Logic-based roughification. In: Skowron, A., Suraj, Z. (eds.) Rough Sets and Intelligent Systems. ISRL, vol. 42, pp. 517–543. Springer, Heidelberg (2013)
14. Sen, P., Namata, G.M., Bilgic, M., Getoor, L., Gallagher, B., Eliassi-Rad, T.: Collective classification in network data. AI Magazine 29(3), 93–106 (2008)
15. Tran, T.-L., Ha, Q.-T., Hoang, T.-L.-G., Nguyen, L.A., Nguyen, H.S.: Bisimulation-based concept learning in description logics. In: Proceedings of CS&P 2013, pp. 421–433. CEUR-WS.org (2013)
16. Tran, T.-L., Ha, Q.-T., Hoang, T.-L.-G., Nguyen, L.A., Nguyen, H.S., Szalas, A.: Concept learning for description logic-based information systems. In: Proceedings of the 2012 Fourth International Conference on Knowledge and Systems Engineering, KSE 2012, pp. 65–73. IEEE Computer Society (2012)

Measuring the Influence of Bloggers in Their Community Based on the H-index Family

Dinh-Luyen Bui, Tri-Thanh Nguyen[*], and Quang-Thuy Ha

Vietnam National University, Hanoi (VNU),
University of Engineering and Technology (UET)
{luyenbd_54,ntthanh,thuyhq}@vnu.edu.vn

Abstract. Nowadays, people in social networks can have impact on the actual society, e.g. a post on a person's space can lead to real actions of other people in many areas of life. This is called social influence and the task of evaluating the influence is called social influence analysis which can be exploited in many fields, such as typical marketing (object oriented advertising), recommender systems, social network analysis, event detection, expert finding, link prediction, ranking, etc. The h-index, proposed by Hirsch in 2005, is now a widely used index for measuring both the productivity and impact of the published work of a scientist or scholar. This paper proposes to use h-index to measure the blogger influence in a social community. We also propose to enhance information for h-index (as well as its variants) calculation, and our experimental results are very promising.

Keywords: social network, influence of blogger, h-index.

1 Introduction

In real life, people usually tend to consult others (e.g. family members, relatives, friends, or experts) before making decisions, especially important ones. As reviewed by [1], 83% of people ask others for experience before trying a restaurant, 71% of people do the same before buying a prescription drug or visiting a place, and 61% of people talk to others before watching a movie. Thanks to the characteristic of social networks that makes the information distribution almost at real-time, it leads to the change of daily behaviors of people who participate in a social network. For example, before buying a certain product (e.g. a mobile phone), people tend to search for others' available comments, experiences or evaluation on the product. As a result, if the content of a user's post is interesting and reliable, it can have a certain impact on other people in that network community. In other words, people have one more source of consultant affecting their daily habits.

A recent typical example that shows the influence of a user on a social network on economy is two tweets of Carl Icahn on Tweeter in August 2013: *"We currently have a large position in APPLE. We believe the company to be extremely undervalued.*

---

[*] Corresponding author.

T.V. Do et al. (eds.), *Advanced Computational Methods for Knowledge Engineering*,
Advances in Intelligent Systems and Computing 282,
DOI: 10.1007/978-3-319-06569-4_23, © Springer International Publishing Switzerland 2014

Spoke to Tim Cook today. More to come", and "*Had a nice conversation with Tim Cook today. Discussed my opinion that a larger buyback should be done now. We plan to speak again shortly.*" The two tweets had a big impact on Apple's stock market. The value of Apple's stocks increased more 12 billion US dollars with about 200.000 stock transactions soon after the appearance of the tweets. Such fact raised a new topic called *social influence analysis* which evaluates the influence capacity of a user (in a social network) on the others. In other words, it evaluates how much an action (described in a user's post) can lead to certain actions of other people in the community as well as real society.

N. Agarwal et al. [1] proposed a model (called iFinder) which attempts to figure out top k influential bloggers having highest scores. The key idea is to score all the posts of bloggers in a community, and select the highest score of one's posts to be his/her influence score (more details of the model will be given in Section 2). Naturally, influence score should be a value that is accumulatively calculated and increased over new posts. Hence, if the influence score relies on only one post, we do not take the contribution of other posts into account, and it does not seem reasonable. In addition, such a score is not reliable in some situations, such as spamming in which spammers simply make some effort to increase the score of only one of his posts. Though the authors claimed that it is possible to use the mean score of all posts as the influence score, this calculation method, again, has a drawback, i.e. it takes into account both influential and non-influential posts. Finally, based on the fact that the life time (time to have attention) of posts in social networks is short, if we rely on a single post score, and when this post is obsolete, it is not reasonable to use its score as blogger's score.

In this paper, we propose to apply the h-index [8] to calculate the influence score of bloggers which will better reflect the reality. The h-index was proposed by Hirst to measure both the productivity and impact of the published papers of a researcher. If a researcher has N published papers in which there are h papers ($h \leq N$) each of which has at least h (inbound) citations, then his h-index is h. It is easily noted that the productivity is the number of papers (h) that have impact (as the number of citations h). When the h-index is applied to rank bloggers, we do not rely on a single post anymore, and also calculate non-influential (or less influential) posts.

However, as we can see, the h-index does not take outbound citations into account. This is not appropriate for social networks where inbound and outbound links and other related information play the role of essential constructs for information navigation and distribution. In this paper, we propose to utilize the post score of iFinder which incorporates several properties (besides inbound links) in the first step of h-index calculation.

The next problem we faced in this work is that the posting score of iFinder is a real number (in the range of [0..1)) which cannot be directly used for the h-index calculation. We use two methods to convert a real number post score to an integer for h-index calculation. Finally, since the h-index was introduced, there have been several proposed variants with improvements. In this work, we also calculated influence score using h-index variants for evaluation.

The rest of the paper is organized as follows. Section 2 briefly introduces related work. Section 3 presents our model to calculate influence scores. Section 4 shows the experimental results and evaluation. Finally, Section 5 concludes the paper and gives some potential future directions.

2   Related Work

2.1   Influential Blogger Identification

The people whose experiences, opinions, and suggestions are sought after are called the influentials [2]. As stated by M. Momma et al. [13], social influence has two forms: the first one is the action (or behavior) (stated in the post) itself, and the second is that this action can lead to the action of other people. The second form is the object of this paper that reflects the impact of influential on other individuals in the community. As reviewed in [1], the identification of the influential bloggers can benefit all in developing innovative business opportunities, forging political agendas, discussing social and societal issues, and lead to many interesting applications [5, 7, 10, 11, 12, 14]. For example, the influentials are often market-movers. Since they can influence buying decisions of the fellow bloggers, identifying them can help companies better understand the key concerns and new trends about products interesting to them, and smartly aspect them with additional information and consultation to turn them into unofficial spokesmen. Approximately 64% advertising companies have acknowledged this phenomenon and are shifting their focus toward blog advertising. As representatives of communities, the influentials could also sway opinions in political campaigns, elections, and aspect reactions to government policies. Tracking the influentials can help understand the changing interests, foresee potential pitfalls and likely gains, and adapt plans timely and pro-actively (not just reactively). The influentials can also help in customer support and troubleshooting since their solutions are trustworthy in the sense of their authority in term of being influentials.

The influential blogger identification can be roughly defined as: *Given a set of M bloggers (in a certain community), find out K (K ≤ M) bloggers who have highest scores (according to a certain estimation).*

Nitin Agarwal et al. [1] proposed a model called iFinder for calculating blogger influence score, which will be introduced in detail in Section 3.

2.2   H-index Family

In this section, we briefly introduce h-index as well as its variants which will be used in our research.

The h-index was proposed by Hirsch in 2005 [8] to be used as an index of a scientist or scholar. It is defined as follows:

A scientist has index h if h of his/her N_p papers have at least h citations each, and the other (N_p h) papers have no more than h citations each.

Let C be the set of top most cited papers of a scientist, U be the set of all the scientist's papers, $cite(p)$ be the function returning the number of citations to paper p, then the h-index h of the scientist is defined as follows:

$$h = arg \max_{C \subseteq U} |C|$$

$$such \ that \ \forall p \in C, cite(p) \geq |C| \wedge \forall p \in U \backslash C, cite(p) < |C| \tag{1}$$

For example, a scientist published 6 papers. Assuming that for two top most cited papers, each has 6 references, while each of the rest has 2 references. Then the h-index of this scientist is 2. The common sense of the h-index is that it increases as the number of papers and citations accumulate, and thus it depends on the 'academic age' of the scientist. It also has quantitative aspect: As reviewed by the author, for physicists, a value for h of about 12 might be typical for advancement to associate professor at major research universities. A value of about 18 could have a full professorship; 15–20 could gain a fellowship in the American Physical Society; and 45 or higher could mean membership in the United States National Academy of Sciences. This indicates the h-index to be a stable and consistent estimator of scientific achievement. Thus, it is currently used to rank objects bigger than a person, such as a department, a university, a country or a journal.

L. Egghe [6], in 2006, argued that h-index has a problem of assigning the same weight to all papers that contribute to h-index, since when a researcher has the index h, and one of his papers has much more citations than h, this paper contributes the same weight as that of the top h papers. Egghe proposed another index called g-index as follows:

Given a set of articles ranked in decreasing order of the number of citations that they received, the g-index is the (unique) largest number g such that the top g articles received a total of at least g^2 citations.

Let C be the set of g top most cited papers ($C = \{p_1, p_2, ..., p_{|C|}\}$) the formula for g-index can be defined as follows:

$$g\text{-}index = arg \max_{C \subseteq U} g \ such \ that \ g^2 \leq \sum_{i=1}^{|C|} cite(p_i) \tag{2}$$

We can notice that total number of top g papers is used in g-index calculation, hence, a paper of higher number of citations contributes more weight to the index than a smaller one. With the same argument as that of Egghe, Jin [9], in 2006, proposed another variant of h-index called A-index. If a researcher has the h-index h constructed from the set C of top most cited papers ($C = \{p_1, p_2, ..., p_h\}$), then A-index is defined as follows:

$$A\text{-}index = \frac{1}{h} \sum_{j=1}^{h} cite(p_j) \tag{3}$$

However, this formula still has a problem as stated in [4]. Consider the following situation: an author X_1 published 20 papers, in which one paper has 10 citations while each of the rest has only one citation; another author X_2 published 30 papers, in which one paper has 30 citations while each of the rest has 2 citations. Naturally, author X_2

should be considered to be better than X_1. Nonetheless, H-indices of X_1 and X_2 are 1 and 2, correspondingly, whereas, the *A-indices* of the two authors X_1 and X_2 are 10 and 6, correspondingly. This drawback comes from the fact that *A-index* formula has a division by h. Suppose an author has h-index h, based on the set of h top most cited papers, J. BiHui et al. [4], in 2007, proposed another one called R-index which is defined as,

$$R\text{-}index = \sqrt{\sum_{j=1}^{h} cite(p_j)} \qquad (4)$$

Peter Vinkler [15], in 2009, proposed *π-index* to improve the h-index. Suppose the total number of papers of a scientist is T that are sorted in the acceding order of number of citations, let the elite set P_π be $\lfloor \sqrt{T} \rfloor$ top most cited papers, $C(P_\pi) = \sum_{p \in P_\pi} cite(p)$, then *π-index* is defined as follows:

$$\pi\text{-}index = 0.01C(P_\pi) \qquad (5)$$

Due to the limitation of *A-index*, we will not use it in our experiments. The h-index is used to measure the productivity as well as impact in the whole academic life of a scientist, so it should increase over time. However, when it is used to rank bloggers, we can calculate h-index of a blogger based on the data in a certain duration (not the whole), so that it can increase or decrease depending on the data. In other words, it is possible to compare the influence of the blogger in different time durations.

3   Using the H-index to Measure Influence

3.1   Rationale

Based on the intuition that when paper A refers to another one B, A tends to borrow information from B. In other words, B is an information source. The more references B has, the more interesting it is. Thus, the h-index bases only on inbound citation information for calculation. The situation is completely changed in World Wide Web or social networks. Let's analyze some important properties other than inbound reference (citation) which should be considered in index calculation: a) *Outbound links* also play important roles in information navigation or distribution. For a website of an organization, the home page has a crucial role, because it stores the links as a map to guide users to navigate to their expected pages. For social network sites, such as Twitter, when a user A follows (or links) to another one B, then B's new tweets will appear in (or be distributed to) A's home page. In this case, the outbound link (from A to B) servers as a clue for information distribution. b) *The content of the post* (or webpage or tweet) is an important property in the context whether it is a hot/contemporary topic in the real world. This may be the most important aspect, however, it is the most difficult aspect to estimate. c) *Response*: a post can attack others to respond in a form of comments/discussions. The more comments a post has, the more interesting it tends to be. d) *Related information* of the user in real life (e.g. the position of job or

expertise): as seen in the example of Icahn's tweets, the position of Icahn has a big effect on the others. However, this information is difficult (even impossible) to obtain.
e) *The number of reads* (or visits): may indicate a certain interesting level of the post.
f) *Activeness*: an active user may usually have new information to post.

From this discussion, we propose to integrate some more properties (information) into h-index calculation. After a review, we noticed that iFinder has exploited and incorporated some additional properties in their model, thus, we reuse the calculation model of iFinder as the first step for h-index calculation. Before introducing our model, we briefly present the iFinder model in the next subsection.

3.2   iFinder Model

Influential Blogger definition: A blogger is influential if s/he has at least one *influential blog post*
For a blogger b_k who has N blog posts $\{p_1, p_2, ..., p_N\}$; denote the influence score of ith post as $I(p_i)$, then b_k influence index (*iIndex*) is defined as follows:

$$iIndex = \arg\max_{i=1..N} I(p_i) \tag{6}$$

A blog post p_i is deemed influential iff $I(pi) \geq \alpha$, where α is a threshold determined at the calculation time based on the number of the most influential bloggers.

Problem Statement: Given a set U of M bloggers $\{b_1, b_2, ..., b_M\}$, the problem of identifying influential bloggers is defined as determining an ordered subset V of K [1] most influential bloggers (with highest *iIndex* values): $V = \{b_{j1}, b_{j2}, ..., b_{jK}\}$ sorted by their *iIndex* in the descending order such that
$V \subseteq U$ and $K \leq M$, i.e. $iIndex(b_{j1}) \geq iIndex(b_{j2}) \geq ... \geq iIndex(b_{jK})$. In this problem, we can see that the threshold α is equal to $iIndex(b_{jK})$.

As stated by K. Apostolos et al. [3], the graphs (based on the links) of blog sites are very sparse, hence, it is not suitable to rank blog posts using Web ranking algorithms (e.g. the PageRank algorithm). N. Agarwal et al. [1] proposed an alternative model to identify influential bloggers called iFinder which is described below.

The initial properties (or parameters) used to calculate the influence score of a blog post are: its set of inbound links (ι); its set of comments (γ); its set of outbound links (θ); and the length of the post (λ).

Let $I(p)$ denote the influence score of a node p (e.g. a blog post) in the graph representing a blog site, then the *InfluenceFlow(.)* across that node is given as follows:

$$InfluenceFlow(p) = w_{in} \sum_{m=1}^{|\iota|} I(p_m) - w_{out} \sum_{n=1}^{|\theta|} I(p_n) \tag{7}$$

where w_{in} and w_{out} are weights used to adjust the contribution of inbound and outbound influence, respectively; p_m ($1 \leq m \leq \theta$) is a post that has a link to p; p_n ($1 \leq n \leq \theta$) is a post that is referred by p;

---

[1] K is a user specified parameter.

InfluenceFlow(.) measures the difference between the total incoming influence of all inbound links and the total outgoing influence by all outbound links of the blog post p. It accounts for the part of influence of a blog post that depends upon inbound and outbound links. The intuitive aspect of this function is that: if a blog post is referred by another one, then it seems to have novelty, and then it gets bonus score; however, when a post links to another post, then its content seems to 'borrow' information from an external source, and it gets penalty score.

In addition, the post's comments also indicate that the post is interesting or has novelty, hence influence $I(p)$ is proportional to the number of comments (γ_p),

$$I(p) = w_c \gamma_p + InfluenceFlow(p) \tag{8}$$

where w_c is the contribution weight of the total number of comments γ_p on the post p.

The last parameter is the length of the post λ_p. It is not simply to use λ_p as a weight, Agarwal proposed to convert λ_p to a weight by a function $w(.)$, and the final formula for $I(p)$ (from Eq. 8) is written as follows:

$$I(p) = w(\lambda_p) \times \left(w_c \gamma_p + InfluenceFlow(p) \right) \tag{9}$$

The influence score of each post $I(p)$ is normalized in the range of $[0..1)$.

Given a set U of M bloggers who have a set P of N blog posts $P = \{p_1, p_2, ..., p_N\}$, denote A as the adjacency matrix, where each entry A_{ij} represents the link between the post p_i and p_j. i.e. if p_i refers to p_j, then $A_{ij}=1$; otherwise $A_{ij}=0$. Matrix A represents the outbound links among posts, consequently, A^T represents the inbound links among the posts. Define the vectors of post length $\vec{\lambda}$, comments $\vec{\gamma}$, influence $\vec{\imath}$, and influence flow $\vec{f}$ as follows:

$$\vec{\lambda} = \left(w(\lambda_{p_1}), w(\lambda_{p_2}), ..., w(\lambda_{p_N}) \right)^T,$$
$$\vec{\gamma} = \left(\gamma_{p_1}, \gamma_{p_2}, ..., \gamma_{p_N} \right)^T,$$
$$\vec{\imath} = \left(I(p_1), I(p_2), ..., I(p_N) \right)^T,$$
$$\vec{f} = (f(p_1), f(p_2), ..., f(p_N))^T$$

Now, Eq. 7 can be rewritten as follows:

$$\vec{f} = w_{in} A^T \vec{\imath} - w_{out} A \vec{\imath} = (w_{in} A^T - w_{out} A) \vec{\imath} \tag{10}$$

and Eq. 9 can be rewritten as follows:

$$\vec{\imath} = diag(\vec{\lambda})\left(w_c \vec{\gamma} + \vec{f}\right) \tag{11}$$

Combine Eq. 10 and Eq. 11, we have

$$\vec{\imath} = diag(\vec{\lambda})(w_c \vec{\gamma} + (w_{in} A^T - w_{out} A)\vec{\imath}) \tag{12}$$

It is possible to solve the iterative Eq. 12 using *power iteration method* as described in Algorithm 1 [1].

Input: A set P of blog posts, the termination parameters: number of iteration *iter*, the similarity threshold τ
Output: The influence vector $\vec{\imath}$ representing the influence score of all the blog posts in P

> Compute the adjacency matrix A
> Compute vectors post length $\vec{\lambda}$, comments $\vec{\gamma}$
> Initialize $\vec{\imath} = \vec{\imath_0}$
> **repeat**
> $\qquad \vec{\imath'} = diag(\vec{\lambda})(w_c\vec{\gamma} + (w_{in}A^T - w_{out}A)\vec{\imath})$
> $\qquad iter \leftarrow iter - 1$
> **until** $(cosine\_similarity(\vec{\imath}, \vec{\imath'}) < \tau)$ or $(iter \leq 0)$

Algorithm 1. Influence calculation (blog posts' score calculation)

After experiments, the author found out the contribution order of the 4 properties used in the iFinder model is: inbound links > comments > outbound links > blog post length, and the combination of the four gives the highest performance indicating that the selection of the four properties is suitable.

3.3   Our Model

In this section, we describe the details of our model for finding top K influential bloggers based on the h-index family. In comparison with scientific articles, the life time of posts (from the time the post appeared to the last time it was referred) in social networks is shorter, thus using the h-index family for measuring the influence is a more meaningful than the measuring method of iFinder which only bases on a single post. Since when the post represented for a blogger's influence score is obsolete, it should not be the representative anymore. Our model to identify influential bloggers is based on the h-index family, which is different from that of iFinder, we redefine an influential blogger as,

A blogger has the influence score of h if h is his/her h-index (or its variant) value.
And the influential blogger identification problem is defined as follows:

Input: A set U of M bloggers who have N blog posts and a parameter K $(K \leq M)$
Output: The set V of K top h-index bloggers.

Our model is described in Fig. 1, which has following steps:
Preprocessing: for each post, we parse each post to extract essential information for next steps, e.g. the post title; the content of the post; the length of the post; the number of inbound links; the number of outbound links; the author (blogger) of the post; the number of comments; the tags of the post; the timestamp (post time).

Post score estimation: as discussed in Section Rationale, we would like to integrate some more properties (besides inbound links). However, due to some limitation (e.g. the availability of data), we finally selected same four properties as those of iFinder, i.e., the number of inbound links; the number of outbound links; the number of comments; the post's content (estimated as the post length). We apply iFinder model to estimate the score of each post. The results of this step are the scores of each post in the range of $[0..1]$.

Post score conversion: since the post score (returned by the previous step) in the range of $[0..1]$ is not compatible for h-index calculation, we propose to use *binning* for transforming a post score into an integer. There are two binning methods:

- *Equal-frequency (or equal-depth) binning*: given m posts, equal-frequency binning method divides them into n bins, so that the bins have an equal number of posts. Formally, let $pos(p)$ denote the position of post p in the sorted list by score in the ascending order, the bin number of p is $bin(p) = \lfloor pos(p)/n \rfloor$.
- *Equal-width binning*: in this method, each bin will have the same interval range of value instead of number of posts. Denote l, r as the lower and upper bounds of the target integer range, correspondingly. The interval range (*irange*) of each bin is $irange = \frac{r-l}{n}$, and the range of ith bin is $[l + (i-1) * irange, l + i * irange)$ where $(1 \leq i \leq n)$. Given a post p then $bin(p) = i$ if $score(p) \in bin_i$

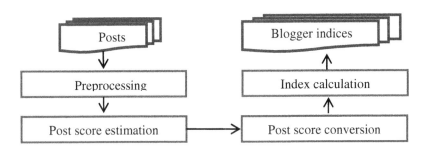

Fig. 1. The ranking model based on the h-index family

Index calculation: for each blogger, we collect the $bin(.)$ values of all his posts to use as the number of citation (i.e., $cite(.)$ function), and then calculate the values of the variant of h-index. After this step, we have the index of all bloggers, hence, we can sort the blogger list by their index and return K top highest index bloggers.

In the real world, the influence of a blogger may increase or decrease (not always increase as h-index for a scientist). However, as discussed in Section 2, it is possible to apply our model to calculate the index of a blogger based on the data subset collected at a certain duration in order to track the influence change of the blogger over time to reflect the real situation.

4   Experiments and Evaluation

4.1   Data Set and Experimental Setup

Thanks to the support of Nitin Agarwal et al. [1], we had the data set "The Unofficial Apple Weblog" (TUAW) which consists of about 10,000 blog posts from 35 bloggers. The dataset was manually investigated to rank bloggers based on their activeness. The parameter settings used in iFinder model (cf. Algorithm 1) are those recommended by the author. In equal-depth binning, we set the number of bins to 100. In equal-width binning, we set $l = 1; r = 1000; n = 1000$ (or $irange = 1$).

4.2   Experimental Results and Evaluation

We ran our model with two binning methods (i.e., equal-depth and equal-width) which both gave the same set of top 5 of most influential bloggers. To evaluate our model, similar to iFinder, we compare top 5 bloggers returned by our model (with the rank of equal-depth binning) with those of iFinder and TUAW as shown in Table 1.

Table 1. Comparison of top 5 bloggers

TUAW	iFinder	Our model
Erica Sadun	*Erica Sadun*	*Scott McNulty*
Scott McNulty	Dan Lurie	C. K. Sample, III
Mat Lu	*David Chartier*	Dave Caolo
David Chartier	*Scott McNulty*	*David Chartier*
Micheal Rose	**Laurie A. Duncan**	**Laurie A. Duncan**

As claimed by Nitin Agarwal, an influential blogger can be, but not necessarily, an active one. Thus the results returned by iFinder are not the same as top 5 active bloggers. Refer to Table 1, iFinder shares three bloggers (in italic) with TUAW, while our model shares two bloggers with TUAW (in italic), and shares 3 bloggers with iFinder. As reviewed by Agarwal, *Dan Lurie* is not active (i.e. not in the top of TUAW) but influential. Because, *Dan* has 4 influential posts and, especially, one of them writing about IPhone attacked a large number of discussion, and iFinder selects this highest post score as the influence score of a blogger resulting in *Dan* appearing in top 5. However, recalling the discussion in Section 1 that this score selection is a drawback of iFinder where spammers simply try to boost one of his posts to have a high score leading them to be influentials.

Our model did not put *Dan Lurie* in the top 5 influentials thanks to the difference in blogger score calculation. Another example is *Erica Sadun* who is marked as the first ranked influential blogger by both TAUW and iFinder. His most influential post is a keynote speech of Apple Inc. CEO Steve Jobs, which fostered a big number of comments and inbound links (two of the most influential properties contributing to the post score) giving him the highest score in iFinder model. Nonetheless, the h-index family does not rely on a single post, and assigns *Erica Sadun* a lower score in comparison with the fifth blogger *Laurie A. Duncan*. That is also the reason why two bloggers: *C. K. Sample* and *C. K. III Dave Caolo* appear in top 5 of our model.

Observation from equal-width and equal-depth binning experiments, the two methods produced the same top 5 influential set with 4 different indexes (i.e. *h-index*, *g-index*, *r-index* and *π-index*), however, the blogger's index values are different. There are 3 different bloggers in top 10 set between the two methods indicating that top influential bloggers seem to be stable in two binning methods and 4 indexes. In addition, equal-depth binning gave higher index values than equal-width binning, though the scale of equal-width binning (in the range of [1..1000]) is larger than that of equal-depth binning (in the range of [1..100]). This is from the fact that the post scores do not distribute equally in the range but group in discrete clusters. At the moment, we haven't found out a suitable method to evaluate which index among the four is the best. This is a potential problem for our future.

Table 2. g-index of top 5 bloggers over time

Blogger	2004	2005	2006	2007
Scott Mcnulty	0	92	98	98
C. K. Sample, III	0	94	95	95
Dave Caolo	0	90	95	85
David Chartier	0	86	96	96
Laurie A. Duncan	43	87	94	94

We also carried out experiments to observe the change of blogger's influence score over time. As discussed in Section 3, we calculated the index (e.g. *g-index*) of a blogger based on a data subset (e.g. in one year duration). From the four year results of top 5 bloggers' *g-index* in Table 2, we can notice that the index can increase or decrease depending on the actual data. This means it is possible to use an index to follow the influential change of a blogger.

5   Conclusion and Future Work

In this paper, we proposed to use the h-index family for ranking bloggers in order to find out the top most influential ones. For enhancing the information used in h-index calculation, we proposed to integrate some more properties (in addition to inbound reference). The experimental results proved our proposed model are comparable to the iFinder model. Moreover, our model may avoid the drawback of iFinder model, i.e. vulnerable to spam. For the future work, we plan to integrate some more properties as discussed in Section 3, and apply our model to other domain than blogosphere, such as Facebook or Twitter.

Since the life time (the time of having attention) of a post is much shorter than that of a scientific paper, we plan to incorporate some information (e.g. the post time) in score estimation.

Another future direction is h-index threshold determination, as estimated by Hirsch in 2005 [8], a certain h-index value a physicist has can be appropriate for a certain academic position or award (e.g. associate/full professor, cf. Section 2). We plan to figure out the threshold to judge a blogger to be influential instead of simply returning the top ranked ones.

The final future stuff is to judge which index (in the h-index family) is the most suitable for measuring influences.

Acknowledgments. This work was partially supported by the VNU Scientist links and Grant No. BB-2012-B42-29.

References

1. Agarwal, N., Liu, H., Tang, L., Yu, P.S.: Modeling blogger influence in a community. Social Netw. Analys. Mining 2(2), 139–162 (2012)
2. Akritidis, L., Katsaros, D., Bozanis, P.: Identifying the Productive and Influential Bloggers in a Community. IEEE Transactions on Systems, Man, and Cybernetics, Part C 41(5), 759–764 (2011)
3. Kritikopoulos, A., Sideri, M., Varlamis, I.: BLOGRANK: Ranking We-blogs based on Connectivity and Similarity Features. CoRR abs/0903.4035 (2009)
4. BiHui, J., LiMing, L., Rousseau, R., Egghe, L.: The R- and AR-indices: complementing the h-index. Chinese Science Bulletin 52(6), 855–963 (2007)
5. Egghe, L.: The Hirsch index and related impact measures. In: ARIST, pp. 65–114 (2010)
6. Egghe, L.: Theory and practise of the g-index. Scientometrics, 131–152 (2006)
7. Goyal, A.: Social Influence and its Applications: An algorithmic and data mining study. PhD Thesis, The University of British Columbia, Vancouver (2013)
8. Hirsch, J.E.: An index to quantify an individual's scientific research output. Proc. of the National Academy of Sciences of the United States of America 102(46), 16569–16572 (2005)
9. Jin, B.: H-index: an evaluation indicator proposed by scientist. Science Focus, 8–9 (2006)
10. Keller, E., Berry, J.: One American in ten tells the other nine how to vote, where to eat and, what to buy. They are The Inuentials. The Free Press (2003)
11. Lee, Y., Jung, H.-Y., Song, W., Lee, J.-H.: Mining the blogosphere for top news stories identification. In: SIGIR 2010, pp. 395–402 (2010)
12. Lin, Y.-R., Sundaram, H., Chi, Y., Tatemura, J., Tseng, B.L.: Splog detection using self-similarity analysis on blog temporal dynamics. In: Proceedings of the 3rd International Workshop on Adversarial Information Retrieval on the Web, AIRWeb (2007)
13. Momma, M., Chi, Y., Lin, Y., Zhu, S., Yang, T.: Influence Analysis in the Blogosphere. CoRR abs/1212.5863 (2012)
14. Romero, D.M., Galuba, W., Asur, S., Huberman, B.A.: Influence and Passivity in Social Media. In: Gunopulos, D., Hofmann, T., Malerba, D., Vazirgiannis, M. (eds.) ECML PKDD 2011, Part III. LNCS, vol. 6913, pp. 18–33. Springer, Heidelberg (2011)
15. Vinkler, P.: The pi-index: a new indicator for assessing scientific impact. Information Science (JIS), 602–612 (2009)

Automatic Question Generation for Educational Applications – The State of Art

Nguyen-Thinh Le[1], Tomoko Kojiri[2], and Niels Pinkwart[1]

[1] Humboldt Universität zu Berlin, Germany
{nguyen-thinh.le,niels.pinkwart}@hu-berlin.de
[2] Kansai University, Japan
kojiri@kansai-u.ac.jp

Abstract. Recently, researchers from multiple disciplines have been showing their common interest in automatic question generation for educational purposes. In this paper, we review the state of the art of approaches to developing educational applications of question generation. We conclude that although a great variety of techniques on automatic question generation exists, just a small amount of educational systems exploiting question generation has been developed and deployed in real classroom settings. We also propose research directions for deploying the question technology in computer-supported educational systems.

Keywords: automatic question generation, educational technology.

1   Introduction

Recently, the research area of automatic question generation for educational purposes has attracted attention of researchers from different disciplines. Question generation is defined by Rus et al. (2008) as follows: "Question generation is the task of automatically generating questions from various inputs such as raw text, database, or semantic representation". This definition indicates that the type of input for question generation can vary: it can be, for example, a sentence, a paragraph or a semantic map. According to Piwek and Boyer (2012), research on question generation has a long tradition and can be traced back to the application of logic to questions. One of the first works on questions was proposed by Cohen (1929) to represent the content of a question as an open formula with one or more unbound variables. While research on question generation has been being conducted for long time, deploying automatic question generation for educational purposes has raised interests in different research communities in recent years.

Studies have reported that deploying questions in teaching encourages students to self-explain, which has been shown to be highly beneficial for learning (Chi et al., 1994). With novice computer scientists, asking effective questions during the early phases of planning a solution can support the students' comprehension and decomposition of the problem at hand (Lane, 2005). Asking targeted, specific questions is

T.V. Do et al. (eds.), *Advanced Computational Methods for Knowledge Engineering*,
Advances in Intelligent Systems and Computing 282,
DOI: 10.1007/978-3-319-06569-4_24, © Springer International Publishing Switzerland 2014

useful for revealing knowledge gaps with novices, who are often unable to articulate their questions (Tenenberg & Murphy, 2005). In addition, in the view of improving meta-cognitive abilities, asking students to generate questions by themselves may enable students to construct meaningful knowledge and to employ various meta-cognitive strategies by themselves (Yu et al., 2005).

In this paper, we survey the state of the art of educational systems which deploy questions. The goal of this paper is to answer the following questions: 1) which methodologies can be applied to generate questions? 2) How can questions be deployed in educational settings?

2   Method

In order to answer the research questions introduced in the previous section, we will select peer-reviewed scientific reports on question generation systems for educational purposes. That is, not only complete systems but also work in progress will be taken into account in this paper.

We will group educational applications of question generation into the same class if they follow the same educational purpose and discuss their technical approaches. Since the aim of this review is to find out the current deployment of educational applications of questions, evaluation studies of existing works will be summarized.

3   Question Generation

3.1   Educational Applications of Question Generation

In this subsection, we review the current educational systems deploying question generation. The systems are classified into three classes according to their educational purposes: 1) knowledge/skills acquisition, 2) knowledge assessment, and 3) educational systems that use questions to provide tutorial dialogues.

First class: Knowledge/skills acquisition
The purpose of the first class of educational applications of question generation includes knowledge/skills acquisition. One of the first automatic question generation systems which have been developed to support learning was proposed by Wolfe (1976). The author proposed a system called AUTOQUEST to help novices learn English. Questions are generated from reading sources provided to students. Kunichika et al. (2001) applied a similar approach based on syntactic and semantic information extracted from an original text. Their educational purpose was to assess the grammar and reading comprehension of students. The extracted syntactic features include subject, predicate verb, object, voice, tense, and sub-clause. The semantic information contains three semantic categories (noun, verb and preposition) which are used to determine the interrogative pronoun for the generated question. For example, in the noun category, several noun entities can be recognized including person, time,

location, organization, country, city, and furniture. In the verb category, bodily actions, emotional verbs, thought verbs, and transfer verbs can be identified. The system is also able to extract semantic relations related to time, location, and other semantic categories, when an event occurs. Evaluations showed that 80% of the questions were considered as appropriate for novices learning English by experts and 93% of the questions were semantically correct.

Mostow and Chen (2009) developed an automated reading tutor which deploys automatic question generation to improve the comprehension capabilities of students while reading a text. The authors investigated how to generate self-questioning instruction automatically on the basis of statements about mental states (e.g., belief, intention, supposition, emotion) in narrative texts. The authors proposed to decompose the instruction process into four steps: describing a comprehension strategy, modeling its use, scaffolding its application, and prompting the child (who is the user of the reading tutor) to use it. The step of describing a comprehension strategy aims at explaining the user when to apply self-questioning. Then, the reading tutor poses a question about the sentence of a reading text in order to illustrate the use of self-questioning. During the step of scaffolding the application of self-questioning, the tutor system helps the child construct a question by choosing from four characters in the context of the given reading text from the on-screen menu (e.g., the town mouse, the country mouse, the man of the house, the cat), three question types (Why, What, How), and three question completers on a menu-driven basis. The reading tutor gives positive feedback in case the child constructed a correct question and invites the child to try again in case she/he created counterfactual questions. The step of prompting the child to use the self-questioning strategy encourages her/him to develop a question and to find an appropriate answer from the given text. The reading tutor has been evaluated with respect to the acceptability of menu choices (grammatical, appropriate, and semantically distinct), to the acceptability of generated questions, and to the accuracy of feedback. The authors reported that only 35.6% of generated questions could be rated as acceptable. 84.4% of the character choices and 80.9% of the question completer choices were classified as acceptable. However, the accuracy of detection of counterfactual questions was 90.0% which is high for generating plausible feedback. Applying a similar approach, Chen and colleagues (Chen et al., 2009) developed a reading tutor for informational texts for which another set of question templates need to be defined.

Also in the same class of educational applications, Liu and colleagues (Liu et al., 2012) introduced a system (G-Asks) for improving students' writing skills (e.g., citing sources to support arguments, presenting the evidence in a persuasive manner). The approach to generating questions deployed in this system is template-based. It takes individual sentences as input and generates questions for the following citation categories: opinion, result, aim of study, system, method, and application. The process of generating questions consists of three stages. First, citations in an essay written by the student are extracted, parsed and simplified. Then, in the second stage, the citation category is identified for each citation candidate. In the final stage, an appropriate question is generated using pre-defined question templates. For example, for the citation category "opinion", the following question templates are available: "Why

+subject_auxiliary_inversion()?", "What evidence is provided by +subject+ to prove the opinion?", "Do any other scholars agree or disagree with +subject+?". In order to instantiate these question templates, the "+subject_auxiliary_inversion()" operation places the auxiliary preceding a subject, and the "+subject+" operation replaces the place holder with a correct value. The evaluation has been conducted with 33 PhD students-writers and 24 supervisors. Each student has been asked to write a research proposal. Each proposal was read by a peer and a supervisor, who both were asked to give feedback in form of questions. In total, questions were produced from four sources for each student's proposal: questions generated from a supervisor, from a peer, from the G-Asks system, and from a set of five generic questions: 1) Did your literature review cover the most important relevant works in your research field? 2) Did you clearly identify the contributions of the literature reviewed? 3) Did you identify the research methods used in the literature reviewed? 4) Did you connect the literature with the research topic by identifying its relevance? 5) What were the author's credentials? Were the author's arguments supported by evidence? Each question producer generated a maximum of five questions. Students evaluated 20 questions at most for each student's proposal based on five quality measures: 1) grammatical correctness, 2) clearness, 3) appropriateness to the context, 4) helpfulness for reflecting what the author has written, 5) usefulness. Evaluation studies have reported that the system could generate questions as useful as human supervisors and significantly outperformed human peers and generic questions in most quality measures after filtering out questions with grammatical and semantic errors (Liu et al., 2012).

In the contrast to approaches to generating questions using text as an input, Jouault and Seta (2013) proposed to generate semantics-based questions by querying semantic information from Wikipedia database to facilitate learners' self-directed learning. Using this system, students in self-directed learning are asked to build a timeline of events of a history period with causal relationships between these events given an initial document (that can be considered a problem statement). The student develops a concept map containing a chronology by selecting concepts and relationships between concepts from the given initial Wikipedia document to deepen their understandings. While the student creates a concept map, the system also integrates the concept to its map and generates its own concept map by referring to semantic information of Wikipedia. The system's concept map is updated with every modification of the student's one. In addition, the system extracts semantic information from DBpedia (Bizer et al., 2009) and Freebase (Bollacker, 2008) which contains semantic representation of Wikipedia in order to select and add related concepts into the existing map. Thus, the system's concept map always contains more concepts than the student's map. Using these related concepts and their relationships the system generates questions for the student to lead to a deeper understanding without forcing to follow a fixed path of learning.

Second Class: Knowledge Assessment
The second class of educational applications of question generation aims at assessing knowledge of students. Heilman and Smith (2010) developed an approach to generating questions for assessing students' acquisition of factual knowledge from reading

materials. The authors developed general-purpose rules to transform declarative sentences into questions. The approach includes an algorithm to extract simplified statements from appositives, subordinate clauses, and other constructions in complex sentences of reading materials. Evaluation studies have been conducted to assess the quality and precision of automatic generated questions using Wikipedia and news articles. The authors evaluated the question generation approach with 15 English speaking university students who were asked to rate the system generated questions with respect to a list of deficiencies (ungrammatical, no sense, vagueness, obvious answer, missing answer, wrong "WH"-word, formatting errors, e.g., punctuation, and others). The participants of the evaluation study were asked to read a text of an article and to rate approximately 100 questions generated from the text. The authors reported that their system achieved 43.3% precision-at-10[1] and 6.8 acceptable questions could be generated from a source text of 250 words (Heilman & Smith, 2010). However, no evaluation studies with respect to the contribution of generated questions for learning can be found in literature.

For the purpose of assessing vocabulary of students, there are several attempts to automatically generate multiple-choice closed questions. In general, the process of generating questions for vocabulary assessment is determining which words to remove from the source sentence. The purpose of this step is to emphasize which vocabulary students should learn. If multiple-choice questions are supposed to be generated, wrong alternative answers (also referred to as distracters) for a specific multiple-choice question are required. Mitkov and colleagues (Mitkov et al., 2006) developed a computer-aided environment for generating multiple-choice test items. The authors deployed various natural language processing techniques (shallow parsing, automatic term extraction, sentence transformation, and computing of semantic distance). In addition, the authors exploited WordNet, which provides language resources for generating distracters for multiple-choice questions. The question generation process of this system consists of three steps. First, key terms, which are nouns or noun phrases with a frequency over a certain threshold, are extracted using a parser. The second step is responsible for generating questions. For this purpose, a clause filtering module was implemented to identify those clauses to be transformed into questions. The clauses are selected if they contain at least a key term, are finite, and are of a Subject-Verb-Object structure or a Subject-Verb structure. In addition, transformation rules have been developed to transform a source clause to a question item. The third step is deploying hypernyms and coordinates (which are concepts with the same hypernym) in WordNet to retrieve concepts semantically close to the correct answer. If WordNet provides too many related concepts, only the ones which occur most frequently in the textbook (which is used for generating multiple-choice questions) are selected. The authors demonstrated that the time required for generating questions including manual correction was less than for manually creating questions alone (Mitkov et al., 2006): For 1000 question items, the development cost would

---

[1] "We calculate the percentage of acceptable questions in the top N questions, or precision-at-N. We employ this metric because a typical user would likely consider only a limited number of questions." (Heilman & Smith, 2010).

require 30 hours of human work using the system, while 115 hours would be required without using the system. In addition, the quality of test items which have been generated and post-edited by humans was scored better than those produced manually without the automatic support of the system.

For the educational purpose of assessing vocabulary, Brown and colleagues (Brown et al, 2005) developed the system REAP which is intended to provide students with texts to read according to their individual reading levels. The system chooses text documents which include 95% of words that are known to the student while the remaining 5% of words are new to the student and need to be learned. After reading the text, the student's understanding is assessed. The student's responses are used to update the student model in order to provide appropriate texts in the next lesson. The authors suggested six types of questions: definition, synonym, antonym, hypernym, hyponym, and cloze questions. In order to generate questions of these types, the system REAP uses data from WordNet. When a word is input in WordNet, it may appear in a number of synonym sets (or synsets): nouns, verbs, adjectives, and adverbs and a synset can be linked to other synsets with various relations (including synonym, antonym, hypernym, hyponym, and other syntactic and semantic relations). While the definition, synonym, antonym, hypernym, and hyponym question types can be created directly using appropriate synsets' relations, the cloze questions are created using example sentences or phrases retrieved from the gloss for a specific word sense in WordNet. Once a question phrase with the target word is selected, the present word is replaced by a blank in the cloze question phrase. In order to validate the quality of system-generated questions, the authors asked three researchers to develop a set of question types that could be used to assess different levels of word knowledge. Experimental results have reported that with automatically generated questions, students achieved a measure of vocabulary skill that correlates well with performance on independently developed human-generated questions.

Third Class: Tutorial Dialogues

The third class of educational applications of question generation includes providing tutorial dialogues in a Socratic manner. Olney and colleagues (Olney et al., 2012) presented a method for generating questions for tutorial dialogue. This involves automatically extracting concept maps from textbooks in the domain of Biology. This approach does not deal with the input text on a sentence-by-sentence basis only. Rather, various global measures (based on frequency measures and comparison with an external ontology) are applied to extract an optimal concept map from the textbook. The template-based generation of questions from the concept maps allows for questions at different levels of specificity to enable various tutorial strategies, from asking more specific questions to the use of less specific questions to stimulate extended discussion. Five question categories have been deployed: hint, prompt, forced choice question, contextual verification question, and causal chain questions. Studies have been conducted to evaluate generated questions based on a rating scale between 1 (most) to 4 (least). All questions have been rated based on five criteria: 1) Is the question of the target type? 2) Is the question relevant to the source sentence? 3) Is the question syntactically fluent? 4) Is the question ambiguous? 5) Is the question pedagogic? Results have been reported that the prompt questions (M=1.55) were significantly less to be of the appropriate type than the hint questions (M=1.2), and less to be of the appropriate type than the forced choice questions (M=1.27). Regarding the fluency, hint questions (M=1.56) were significantly more fluent than prompts (M=2.92), forced choice questions (M=2.64), contextual verification questions (M=2.25), and causal chain questions (M=2.4). With respect to pedagogy, hint questions were significantly more pedagogic than prompts (M=3.09), forced choice questions (M=3.21), contextual verification questions (M=3.3), and causal chain questions (M=3.18). With regard to the relevance of generated questions, there was no significant difference between the five question categories and the relevance lies between 2.13 and 2.88. Ambiguity scores (between 2.85 and 3.13) across the five question categories shows a tendency that questions were slightly ambiguous. Note that no results were available with regard to learning effectiveness through using generated questions.

Also with the intention of supporting students using conversational dialogues, Person and Graesser (2002) developed an intelligent tutoring system that improves students' knowledge in the areas of computer literacy and Newtonian physics using an animated agent that is able to ask a series of deep reasoning questions[2] according to the question taxonomy proposed by Graesser & Person (Graesser & Person, 1994). In each of these subjects a set of topics has been identified. Each topic contains a focal question, a set of good answers, and a set of anticipated bad answers (misconceptions). The system initiates a session by asking a focal question about a topic and the student is expected to write an answer containing 5-10 sentences. Initially, the system used a set of predefined hints or prompts to elicit the correct and complete answer. Graesser and colleagues (Graesser et al., 2008) reported that with respect to learning effectiveness, the system had a positive impact on learning with effect sizes of 0.8 standard deviation units compared with other appropriate conditions in the areas of computer literacy (Grasser et al., 2004) and Newtonian physics (VanLehn, Graesser et al., 2007). Regarding the quality of tutoring dialogues, the authors reported that conversations between students and the agent were smooth enough that no participating students left the tutoring session with frustration, irritation, or disgust. However, with respect to students' perception, the system earned averaged ratings, with a slightly positive tendency.

Lane & VanLehn (2005) developed PROPL, a tutor which helps students build a natural-language style pseudo-code solution to a given problem. The system initiates four types of questions: 1) identifying a programming goal, 2) describing a schema for attaining this goal, 3) suggesting pseudo-code steps that achieve the goal, and 4) placing the steps within the pseudo-code. Through conversations, the system tries to

---

[2] Categories of deep reasoning questions:
   Causal antecedent: What state or even causally led to an event or state?
   Causal consequence: What are the consequences of an event or a state?
   Goal-orientation: What are the goals or motives behind an agent's action?
   Instrumental/procedural: What instrument or goal allows an agent to accomplish a goal?
   Enablement: What object or resource allows an agent to perform an action?
   Expectational: Why did some expected event not occur?

remediate student's errors and misconceptions. If the student's answer is not ideal (i.e., it cannot be understood or interpreted as correct by the system), sub-dialogues are initiated with the goal of soliciting a better answer. The sub-dialogues will, for example, try to refine vague answers, ask students to complete incomplete answers, or redirect to concepts of greater relevance. For this purpose, PROPL has a knowledge source which is a library of Knowledge Construction Dialogues (KDCs) representing directed lines of tutorial reasoning. They consist of a sequence of tutorial goals, each realized as a question, and sets of expected answers to those questions. The KCD author is responsible for creating both the content of questions and the forms of utterances in the expected answer lists. Each answer is either associated with another KCD that performs remediation or is classified as a correct response. KCDs therefore have a hierarchical structure and follow a recursive, finite-state based approach to dialogue management. PROPL has been evaluated with the programming languages Java and C and it has been reported that students who used this system were frequently better at creating algorithms for programming problems and demonstrated fewer errors in their implementation (Lane & VanLehn, 2005).

Table 1. Existing question generation systems for educational purposes

Educational purpose	System	Support type	Evaluation
Developing knowledge/ skills	Wolfe (1976)	learning English	-
	Kunichika et al. (2001)	grammar and reading comprehension	Quality of questions
	Mostow & Chen (2009)	Reading tutor	Quality of questions
	Liu et al. (2012)	Academic Writing Support	Quality of questions
	Jouault & Seta (2013)	Self-directed learning support	-
Knowledge assessment	Heilman & Smith (2010)	Assessing factual knowledge	Quality of questions
	Mitkov et al. (2006)	Assessing vocabulary	Time effectiveness for generating questions
	Brown et al. (2005)	Assessing vocabulary	Quality of questions
Socratic dialogues	Olney et al. (2012)	Providing feedback in form of questions	Quality of questions
	Graesser et al. (2008)	Tutor for Computer literacy and Newtonian physics	Quality of questions, learning effectiveness, and students' perception
	Lane & VanLehn (2005)	Tutor for programming	Learning effectiveness

In summary, existing educational applications of question generation can be classified into three classes based on their educational purposes (Table 1): 1) question generation for knowledge and skills acquisition, 2) question generation for knowledge assessment, and 3) question generation for development of tutorial dialogues. From this table, we can notice that most educational applications of question generation fall into the first class. In addition, most evaluation studies focused rather on the quality of question generation than on the learning effectiveness contributed by the question generation component. In the next section, we compare different approaches to generating questions.

3.2   Approaches to Automatic Question Generation

Rus et al. (2008) regarded question generation as a discourse task involving the following four steps: 1) when to ask the question, 2) what the question is about, i.e., content selection, 3) question type identification, and 4) question construction.

The first issue involves strategies to pose questions. The second and the third issues are usually solved by most question generation systems in a similar manner using different techniques from the field of natural language processing such as parsing, simplifying sentences (Knight and Marcu, 2000), anaphor resolution (Kalady et al., 2010), semantic role labeling (Mannem, et al., 2010), or named entity recognizing (Ratinov and Roth, 2009).

While most question generation systems share common techniques on the second and third step of the process of question generation, their main difference can be identified when handling the fourth issue, namely constructing questions in grammatically correct natural language expression. Many question generation systems applied transformation-based approaches to generate well-formulated questions (Kalady et al., 2010; Heilman and Smith, 2009; Ali et al., 2010; Pal et al., 2010; Varga and Le, 2010). In principle, transformation-based question generation systems work through several steps: 1) Delete the identified target concept, 2) place a determined question key word on the first position of the question, and 3) convert the verb into a grammatically correct form considering auxiliary and model verbs. For example, the question generation system of (Varga and Le, 2010) uses a set of transformation rules for question formation. For subject-verb-object clauses whose subject has been identified as a target concept, a "Which Verb Object" template is selected and matched against the clause. For key concepts that are in the object position of a subject-verb-object, the verb phrase is adjusted (i.e., auxiliary verb is used).

The second approach for question formation, which is also employed widely in several question generation systems, is template-based (Wyse and Piwek, 2009; Chen et al., 2009; Sneiders, 2002). The template-based approach relies on the idea that a question template can capture a class of questions, which are context specific. For example, Chen et al. (2009) developed the following templates: "what would happen if <X>?" for conditional text, "when would <X>?" and "what happens <temporal-expression>?" for temporal context, and "why <auxiliary-verb> <X>?" for linguistic modality, where the place-holder <X> is mapped to semantic roles annotated by a semantic role labeler. Note that these templates have been devised to generate questions from an informational text. For narrative texts, Mostow and Chen developed another set of question templates (2009). Wyse and Piwek (2009) developed a similar approach which consists of rules and templates. Rules are represented in form of regular expressions and are used to extract key concepts of a sentence. Pre-defined questions templates are used to generate questions for those concepts. No evaluation results have been documented for this system.

Table 2 shows the evaluation results of different existing question generation systems. From the table we can notice that the template-based systems (Chen et al., 2009; Mostow and Chen, 2009) achieved considerable results, whereas there seems to be room for improvement of the transformation-based systems (Kalady et al., 2010; Pal et al., 2010; Varga and Le, 2010).

Table 2. Evaluation results of existing question generation systems

System	Question type	Evaluation Results
Kalady et al. (2010)	Yes-No, Who, Whom, Which, Where, What, How	Recall=0.68; Precision=0.46
Ali et al. (2010)	Yes-No, Who, Which, Where, What, When, How, Why	Recall=0.32; Precision=0.49
Varga & Le (2010)	Who, Whose, Whom, Which, What, When, Where, Why, How many	Relevance[3]=2.45(2.85); Syntactic Correctness & Fluency=2.85(3.1)
Mannem et al. (2010)	Who, When, What, Where, why, How	Low acceptance. No statistic data available.
Pal et al. (2010)	Yes-No, Who, When, Which, What, Why, How many, How much	Satisfactory results. No statistic data available
Chen et al. (2009)	Question templates for informational text	79.9% plausible questions[4]
Mostow & Chen (2009)	Question templates for narrative text	71.3 % plausible questions

In addition to texts as input for question generation, structured database can also be used. Sneiders (2002) developed question templates whose answers can be queried from a structured database. For example, the template "When does <performer> perform in <place>?" has two entity slots, which represent the relationship (Performer-perform-place) in the conceptual model of the database. Thus, this question template can only be used for this specific entity relationship. For other kinds of entity relationships, new templates must be defined. Hence, this template-based question generation approach is mostly suitable for applications with a special purpose. However, to develop high-quality templates, a lot of human involvement is expected. Another type of structured database as input for question generation is using semantic representation. Jouault and Seta (2013) deployed semantic representation of Wikipedia for question

---

[3] The evaluation criteria Relevance and Syntactic correctness and fluency are rated by from 1 to 4, with 1 being the best score. Values outside and inside in the brackets indicate ratings of the 1st and 2nd human.

[4] The evaluation results are calculated as the average of the plausibility percentage of three different question types: 86.7% (condition), 65.9% (temporal information), 87% (modality).

generation. They use ontological engineering and linked open data (LOD) techniques (Heath & Bizer, 2011) in order to generate semantics-based adaptive questions and to recommend documents according to Wikipedia to help students create concept maps for the domain of history. One of the great advantages of adopting semantic information rather than natural language resources is that the system can give adequate advice based on the machine understandable domain models without worrying about ambiguity of natural language.

4   Directions for Question Generation in Educational Systems

In Section 3.1 we have reviewed educational applications of question generation. We have identified eleven systems for knowledge and skill acquisition purposes, and only two of them have been evaluated with respect to learning effectiveness. Furthermore, successful deployment of these educational systems in educational settings was not documented.

Although automatic question generation can be achieved using a variety of natural language processing techniques which have gained wide acceptance, there is a lack of strategies for deploying question generation into educational systems. A similar finding has also been identified by Mostow and Chen (2009) especially for the purpose of training reading comprehension: most existing work in this field rather randomly chooses sentences in a text to generate questions than posing questions in an educational strategic manner.

Research on question generation with focus on educational settings needs to develop further. In this paper, we propose three directions for deploying question generation in educational settings. First, in the area of Intelligent Tutoring Systems, several research questions can be investigated, e.g., if the intent of the questions is to facilitate learning, which question taxonomy should be deployed? Given a student model in an Intelligent Tutoring System, which question type is appropriate to pose the next questions to the student? The second research direction is deploying semantic information available on the Internet (e.g., Wikipedia, WordNet) to generate questions. The goal of generating semantics-based questions might include stimulating students' brainstorming, reminding students to additional information, and supporting students solving problems. The third research direction focuses on developing meta-cognitive skills of students. Using questions in teaching is known to be beneficial. Asking students to generate questions helps students recall knowledge and deepen learned content. In addition, it might also develop thinking skills. Deploying automatic question generation in educational systems may use model questions to help students improve the skill of creating questions and thus, meta-cognitive skills of students.

5   Conclusions

In this paper, we have reviewed numerous educational applications of question generation and technical approaches to generating questions. From the technical point of view, many technical approaches for generating questions are successful. Although question generation has a long history, the number of prototypes of question generation for educational purposes is still small.

For the research area in deploying question generation for educational purposes, we propose three research directions. First, question generation should be deployed in Intelligent Tutoring Systems in order to support students in problem solving. The second research direction is deploying semantic information available on the Internet (e.g., Wikipedia, WordNet) to generate semantics-based questions in self-directed or constructivist learning environments. The third research direction promotes applying automatic question generation in order to develop meta-cognitive skills of students, especially the skill of generating questions.

References

1. Ali, H., Chali, Y., Hasan, S.A.: Automation of question generation from sentences. In: Boyer, K.E., Piwek, P. (eds.) Proceedings of the 3rd Workshop on Question Generation, held at ITS 2010, pp. 58–67 (2010)
2. Bizer, C., Lehmann, J., Kobilarov, G., Auer, S., Becker, C., Cyganiak, R., Hellmann, S.: DBpedia - A crystallization point for the Web of Data. Web Semantics: Science, Services and Agents on the World Wide Web 7(3), 154–165 (2009)
3. Bollacker, K., Evans, C., Paritosh, P., Sturge, T., Taylor, J.: Freebase: a collaboratively created graph database for structuring human knowledge. In: Proceedings of the SIGMOD International Conference on Management of Data, pp. 1247–1250. ACM (2008)
4. Brown, J., Frishkoff, G., Eskenazi, M.: Automatic question generation for vocabulary assessment. In: Proceedings of Human Language Technology Conference and Conference on Empirical Methods in Natural Language Processing (2005)
5. Chen, W., Aist, G., Mostow, J.: Generating questions automatically from Informational text. In: Craig, S.D., Dicheva, D. (eds.) Proceedings of the 2nd Workshop on Question Generation, held at AIED 2009, pp. 17–24 (2009)
6. Chi, M.T.H., Lee, N., Chiu, M.H., LaVancher, C.: Eliciting Self-Explanations Improves Understanding. Cognitive Science 18(3), 439–477 (1994)
7. Cohen, F.S.: What is a Question? The Monist 39, 350–364 (1929)
8. Graesser, A.C., Person, N.K.: Question Asking during Tutoring. American Educational Research Journal 31(1), 104–137 (1994)
9. Graesser, A.C., Lu, S., Jackson, G.T., Mitchell, H.H., Ventura, M., Olney, A., Louwerse, M.M.: AutoTutor: a tutor with dialogue in natural language. Behavioral Research Methods, Instruments, and Computers 36(2), 180–192 (2004)
10. Graesser, A.C., Rus, V., D'Mello, S.K., Jackson, G.T.: AutoTutor: Learning through natural language dialogue that adapts to the cognitive and affective states of the learner. In: Robinson, D.H., Schraw, G. (eds.) Recent Innovations in Educational Technology that Facilitate Student Learning, pp. 95–125. Information Age Publishing (2008)
11. Heath, T., Bizer, C.: Linked Data: Evolving the Web into a Global Data Space. Morgan & Claypool Publishers (2011)
12. Heilman, M., Smith, N.A.: Question generation via over-generating transformations and ranking. Report CMU-LTI-09-013, Language Technologies Institute, School of Computer Science, Carnegie Mellon University (2009)

13. Jouault, C., Seta, K.: Building a Semantic Open Learning Space with Adaptive Question Generation Support. In: Proceedings of the 21st International Conference on Computers in Education (2013)
14. Kalady, S., Elikkottil, A., Das, R.: Natural language question generation using syntax and keywords. In: Boyer, K.E., Piwek, P. (eds.) Proceedings of the 3rd Workshop on Question Generation, held at ITS 2010, pp. 1–10 (2010)
15. Knight, K., Marcu, D.: Statistics-based summarization – step one: Sentence compression. In: Proceedings of the 17th National Conference of the American Association for AI (2000)
16. Kunichika, H., Katayama, T., Hirashima, T., Takeuchi, A.: Automated Question Generation Methods for Intelligent English Learning Systems and its Evaluation. In: Proceedings of the International Conference on Computers in Education, pp. 1117–1124 (2001)
17. Lane, H.C., Vanlehn, K.: Teaching the tacit knowledge of programming to novices with natural language tutoring. Journal Computer Science Education 15, 183–201 (2005)
18. Liu, M., Calvo, R.A., Rus, V.: G-Asks: An Intelligent Automatic Question Generation System for Academic Writing Support. Dialogue and Discourse 3(2), 101–124 (2012)
19. Mannem, P., Prasady, R., Joshi, A.: Question generation from paragraphs at UPenn: QGSTEC system description. In: Boyer, K.E., Piwek, P. (eds.) Proceedings of the 3rd Workshop on Question Generation, held at ITS 2010, pp. 84–91 (2010)
20. Mitkov, R., Ha, L.A., Karamanis, N.: A computer-aided environment for generating multiple-choice test items. Journal Natural Language Engineering 12(2), 177–194 (2006)
21. Mostow, J., Chen, W.: Generating instruction automatically for the reading strategy of self-questioning. In: Proceeding of the Conference on Artificial Intelligence in Education, pp. 465–472 (2009)
22. Olney, A.M., Graesser, A., Person, N.K.: Question Generation from Concept Maps. Dialogue and Discourse 3(2), 75–99 (2012)
23. Pal, S., Mondal, T., Pakray, P., Das, D., Bandyopadhyay, S.: QGSTEC system description - JUQGG: A rule-based approach. In: Boyer, K.E., Piwek, P. (eds.) Proceedings of the 3rd Workshop on Question Generation, held at ITS 2010, pp. 76–79 (2010)
24. Person, N.K., Graesser, A.C.: Human or Computer? AutoTutor in a Bystander Turing Test. In: Cerri, S.A., Gouardéres, G., Paraguaçu, F. (eds.) ITS 2002. LNCS, vol. 2363, pp. 821–830. Springer, Heidelberg (2002)
25. Piwek, P., Boyer, K.E.: Varieties of Question Generation: introduction to this special issue. Dialogue & Discourse 3(2), 1–9 (2012)
26. Ratinov, L., Roth, D.: Design Challenges and Misconceptions in Named Entity Recognition. In: Proceedings of the 13th Conference on Computational Natural Language Learning (2009)
27. Rus, V., Cai, Z., Graesser, A.: Question Generation: Example of A Multi-year Evaluation Campaign. In: Rus, V., Graesser, A. (eds.) Online Proceedings of 1st Question Generation Workshop, NSF, Arlington, VA (2008)
28. Sneiders, E.: Automated question answering using question templates that cover the conceptual model of the database. In: Proceedings of the 6th Int. Conference on Applications of Natural Language to IS, pp. 235–239 (2002)
29. Tenenberg, J., Murphy, L.: Knowing What I Know: An Investigation of Undergraduate Knowledge and Self-Knowledge of Data Structures. Journal Computer Science Education 15(4), 297–315 (2005)
30. Varga, A., Le, A.H.: A question generation system for the QGSTEC 2010 Task B. In: Boyer, K.E., Piwek, P. (eds.) Proceedings of the 3rd Workshop on Question Generation, held at ITS 2010, pp. 80–83 (2010)

31. Wyse, B., Piwek, P.: Generating questions from OpenLearn study units. In: Craig, S.D., Dicheva, D. (eds.) Proceedings of the 2nd Workshop on Question Generation, held at AIED 2009, pp. 66–73 (2009)
32. Yu, F.-Y., Liu, Y.-H., Chan, T.-W.: A Web-Based Learning System for Question-Posing and Peer Assessment. Innovations in Education and Teaching International 42(4), 337–348 (2005)
33. Vanlehn, K., Graesser, A.C., Jackson, G.T., Jordan, P., Olney, A., Rosé, C.P.: When are tutorial dialogues more effective than reading? Cognitive Science 31(1), 3–62 (2007)

Part VI

Nonlinear Systems and Applications

A New Approach Based on Interval Analysis and B-splines Properties for Solving Bivariate Nonlinear Equations Systems

Ahmed Zidna and Dominique Michel

Laboratory of Theoretical and Applied Computer Science
UFR MIM, University of Lorraine, Ile du Saulcy, 57045 Metz, France
{ahmed.zidna,domic62}@univ-lorraine.fr

Abstract. This paper is addressing the problem of solving non linear systems of equations. It presents a new algorithm based on use of B-spline functions and Interval-Newton's method. The algorithm generalizes the method designed by Grandine for solving univariate equations. First, recursive bisection is used to separate the roots, then a combination of bisection/Interval Newton's method is used to refine them. Bisection is ensuring robustness while Newton's iteration is guaranteeing fast convergence. The algorithm is making great benefit of geometric properties of B-spline functions to avoid unnecessary calculations in both root-separating and root-refining steps. Since B-spline functions can provide an accurate approximation for a wide range of functions (provided they are smooth enough), the algorithm can be made available for those functions by prior conversion/approximation to B-spline basis. It has successfully been used for solving various bivariate nonlinear equations systems.

1 Introduction

Solving nonlinear systems equations is critical in many research fields. Computing all solutions of a system of nonlinear polynomial equations within some finite domain is a fundamental problem in computer-aided design, manufacturing, engineering, and optimization. For example, in computer graphics, it is common to compute and display the intersection of two surfaces. Also the fundamental ray tracing algorithm needs to compute intersections between rays and primitive solids or surfaces [13]. This generally results in computing the roots of an equation or a system of equations, which in general case, is nonlinear. In global optimization, for equality constraints problem $min f(x)$ s.t. $g(x) = 0$, where g is nonlinear and x belongs to a finite domain. We may seek a feasible initial point to start a numerical algorithm. This requires to compute a solution of $\|g(x)\| = 0$. In [10,17,9], the authors investigated the application of binary subdivision and the variation diminishing property of polynomials in the Bernstein basis to eliciting the real roots and extrema of a polynomial within an interval. In [2,3], the authors extended this idea to general non uniform subdivision of B-splines. In [6], the authors use homotopy perturbation method for solving systems of nonlinear algebraic equations. In [11], a hybrid approach for solving system of nonlinear equations that combine a chaos optimization algorithm and quasi-Newton method has been proposed. The essence of the proposed method is to search an initial guess which should entry the convergent regions of quasi-Newton method. In the approach presented by the authors in [8], a system of nonlinear equations was converted to the minimization problem. Also they suggested to solve the minimization problem using a new particle swarm optimization algorithm. In [16], the authors use recursive subdivision to solve nonlinear systems of polynomial equations in Bernstein basis. They exploit the convex hull property of this basis to improve the bisection step. In the same context, in [1], the authors propose to use blending operators to rule out the no-root containing domains resulting from recursive subdivision. Again in [5], the authors use recursive subdivision to solve nonlinear systems of polynomial equations expressed in Bernstein basis. They use a heuristic sweep direction selection rule based on the greatest magnitude of partial derivatives in order to minimize the number of sub-boxes to be searched for solutions. In [7], the author proposes a method to find all zeroes of a univariate spline function using the interval Newton's method. Unlike the method proposed in [10], interval division is used to speed up bisection.

In this paper we focus on finding all zeroes of a function $f : x \rightarrow f(x)$, where x belongs to a n-dimensional box U, $f(x)$ belongs to $\mathbb{R}^n$ and f is C^1 over the inside of U. We make the assumptions that it is easy to compute bounds for f and all derivatives of f over any subset of U. Typically, these conditions are met by splines functions, which are commonly used as a low-computation cost extension of polynomials [4]. For this reason, in this paper, we use spline functions. The method described in this paper attempts to generalize the previous work done in [7]. It results in a hybrid algorithm that combines recursive bisection and interval Newton's method.

This paper is organized as follows : In section 2, we briefly recall the spline function model. In section 3, the main properties of arithmetic interval are presented. In section 4, the new algorithm is described. In section 5, some numerical examples are presented along with performance results. Finally, in the last section, we conclude and give some tips about possible improvements of the algorithm and possible applications.

2   B-spline Functions

Let m and k be two integer numbers such that $1 \leq k \leq m + 1$. Let $U = \{u_0, u_1, u_2, \ldots, u_{m+k}\}$ be a finite sequence of real numbers such that :

1. $u_i \leq u_{i+1}$ $\forall i = 0 \ldots m + k - 1$
2. $u_i < u_{i+k}$ $\forall i = 0 \ldots m$.

Under these hypotheses, the u_i are called knots and U a knot vector. Then, for any integer numbers r and i such that $1 \leq r \leq k$ and $0 \leq i \leq m + k - r$, a

B-spline function $N_i^r : \mathbb{R} \rightarrow \mathbb{R}$ is defined as follows:

$$N_i^1(u) = \begin{cases} 0, \text{ for } \quad u_i \leq u < u_{i+1} \\ 1, \text{ otherwise} \end{cases} \tag{1}$$

and for $2 \leq r \leq k$, we have

$$N_i^r(u) = \frac{u - u_i}{u_{i+r-1} - u_i} N_i^{r-1}(u) + \frac{u_{i+r} - u}{u_{i+r} - u_{i+1}} N_{i+1}^{r-1}(u). \tag{2}$$

The functions $u \mapsto N_i^k(u)$ for $i = 0 \ldots m$ satisfy the following properties :

1. Each B-spline N_i^k is finitely supported on $[u_i, u_{i+k}[$
2. Each B-spline is positive in the interior of its support.
3. The $\{N_i^k\}_{i=0}^m$ form a partition of unity, i.e, $\sum_{i=0}^m N_i^k(u) = 1$ when u draws $[u_{k-1}, u_{m+1}[$.

Given a second knot vector $V = \{v_0, v_1, v_2, \ldots, v_{n+l}\}$, a bivariate B-spline function $f : \mathbb{R}^2 \rightarrow \mathbb{R}^2$ can be defined by tensor product as follows :

$$f(u, v) = \sum_{i=0}^m \sum_{j=0}^n c_{ij} N_i^k(u) N_j^l(v) \tag{3}$$

where the coefficients c_{ij} are given in the two-dimensional space. $u \mapsto f(u, v)$, resp. $v \mapsto f(u, v)$, is a piecewise polynomial of degree $k - 1$ in u, resp $l - 1$ in v. This above definition can straightforwardly be generalized to higher dimensional cases. Let us recall that f can be evaluated or subdivided by using Cox-De Boor algorithm [2,4] and that the partial derivatives of f are still expressed as B-spline functions. From properties 2 and 3, it is obvious that, for any (u, v) in $[u_{k-1}, u_{m+1}[\times[v_{l-1}, v_{n+1}[, f(u, v)$ is a convex linear combination of the coefficients c_{ij}, thus the graph of f entirely lies in the convex hull of its coefficients c_{ij}. Consequently, the ranges of f and of its derivatives can be calculated at low computational cost [12]. This last property makes the B-spline functions very attractive whenever interval arithmetics is involved.

3   Interval Arithmetic

An interval number is represented by a lower and upper bound, $[a, b]$ and corresponds to a range of real values [14]. An ordinary real number can be represented by a degenerate interval $[a, a]$. In the Interval Newton's method, the basic operations $+, -, *, /$ are used on intervals instead of floating point numbers. Let us remind these 4 operations : let a, b, c, d be 4 real numbers (finite or infinite) such as $a \leq b$ and $c \leq d$.

1. $[a, b] + [c, d] = [a + b, c + d]$
2. $[a, b] - [c, d] = [a - d, b - c]$
3. $[a, b] * [c, d] = [min(a * c, a * d, b * c, b * d), max(a * c, a * d, b * c, b * d)]$

4. $\dfrac{[a,b]}{[c,d]} = \begin{cases}]-\infty, min(\dfrac{a}{c}, \dfrac{b}{d})] \cup [max(\dfrac{a}{d}, \dfrac{b}{c}), +\infty[, & \text{if } 0 \in]c,d[\\ [a,b] * [\dfrac{1}{d}, \dfrac{1}{c}], & \text{otherwise.} \end{cases}$

We define the absolute value of a real interval $[a,b]$ as the mignitude function,

that is : $|[a,b]| = \begin{cases} a & \text{if} & 0 \leq a \\ -b & \text{if} & b \leq 0 \\ 0 & \text{otherwise.} \end{cases}$

Let us point out that $*$ and $/$ dramatically increase the interval span. When used repeatedly, they yield useless results. Associativity is not true with interval operators $+$ and $*$, so different algorithms for inverting an $m \times m$ interval matrix M might not find the same matrix M^{-1}.

4   Root-Finding Algorithm for Bivariate System in B-spline Form

4.1   Univariate Interval Newton Method

In [7], the author proposed an algorithm to find all roots of a spline function f over an interval I in the one dimensional case. Starting from the interval I, the algorithm builds a binary tree of searching intervals, using the following iteration as a unique rule.

$$[a_1, b_1] = [a,b] \cap \left(m_{[a,b]} - \frac{f(m_{[a,b]})}{f'([a,b])} \right). \tag{4}$$

Where $[a,b]$ is the current searching interval, $[a_1, b_1]$ is the next searching interval, $m_{[a,b]}$ is the center of $[a,b]$ and $f'([a,b])$ is an interval to be sure to contain $f'(x)$ when x draws the interval $[a,b]$. If $f'([a,b])$ contains a zero then the division yields two intervals and consequently the resulting interval $[a_1, b_1]$ is the union of two disjoint intervals. This enables the algorithm to separate the roots. Eventually, the iteration rule either produces an empty interval, meaning no root or zeroes in on a tiny interval containing a single root. In this case Newton's Method quadratically converges to the root. The sharpness of the interval $f'([a,b]$ plays a crucial role in the convergence rate of the method.

4.2   Multivariate Interval Newton Method

The straightforward generalization of the equation (4) to the n-dimensional case is :

$$I_1 = I_0 \cap \left(m_{I_0} - [f'(I_0)]^{-1} f(m_{I_0}) \right) \tag{5}$$

where I_0 denotes the current searching hyper-interval, I_1 denotes the next searching hyper-interval, and $[f'(I_0)]^{-1}$ is $n \times n$ interval matrix resulting from calculating the inverse of the interval jacobian matrix of f over I_0. Meaning $[f'(I_0)]$ is such as if x lies in I_0, all the elements of the jacobian matrix $[f'(x)]$ belong to the matching interval-elements of the matrix $[f'(I_0)]$. Accordingly, for all x in the interval I_0, all the elements of the inverse of the jacobian matrix belong to the matching elements of the $n \times n$ interval matrix $[f'(I_0)]^{-1}$. The critical step of the equation (5) is the computation of the inverse of the jacobian matrix. Obviously, the algorithm to be used to compute the inverse of $[f'(I_0)]$ is the Gauss-Jordan algorithm, keeping in mind that every operation is done in interval arithmetic. The jacobian matrix might not be invertible. In the one dimensional version, this situation was used to split the searching interval. In multidimensional case, when the jacobian matrix is not invertible, the Gauss-Jordan algorithm produces a matrix of which the interval elements have infinite bounds which makes the results useless. Therefore in multidimensional case, we use recursive bisection to separate the roots. The Gauss-Jordan method is only used when the jacobian matrix is invertible.

Let $\{I_0^1, I_0^2, \ldots, I_0^{2^n}\}$, be a partition of I_0, we propose the following hybrid iteration instead of the previous equation (5) :

$$\bigsqcup_{j=1}^{2^n} I_1^j = \bigsqcup_{j=1}^{2^n} \left(I_0^j \cap \left(m_{I_0} - [f'(I_0)]^{-1} f(m_{I_0}) \right) \right). \tag{6}$$

The above formula is only used when the jacobian matrix is invertible, otherwise it is replaced by a simple bisection :

$$\bigsqcup_{j=1}^{2^n} I_1^j = \bigsqcup_{j=1}^{2^n} I_0^j. \tag{7}$$

In fact all the hyper-intervals I_0^j are not used. A filter based on geometric properties of B-spline is used to eliminate the irrelevant ones. Let us point out that using B-spline functions makes the computation of both the interval jacobian matrix and the function value easier. The interval Jacobian matrix can easily be computed since it is also a spline function which is contained in the convex hull of its B-spline coefficients.

The general algorithm for solving nonlinear equations systems in B-spline form with the interval Newton-Raphson method is described in Algorithm. 1. Where the method **Succ** performs the hybrid iteration formula is given by Algorithm. 2. For inverting the $n \times n$ interval jacobian matrix, we propose to use the Gauss-Jordan (see Algorithm. 3) which has arithmetic complexity of $\mathcal{O}(n^3)$. The numerical precision of the elements of the resulting matrix is improved by using total pivot. This strategy implies, at each step, to seek the input matrix element of greatest mignitude[14]. In the following we give some details about inverting the interval jacobian matrix.

4.3   Gauss-Jordan Algorithm for Inverting an Interval-Matrix

Let us briefly remind the definition of the inverse of a real interval matrix. Let M is a $m \times m$ real interval matrix (i.e. each element of M is a real finite interval), an inverse of M (denoted M^{-1}) is a $m \times m$ interval matrix such as : for any real

---

Algorithm 1. Recursive root finding algorithm

Function **Solve**(I_0,f);
Input: I_0 : an n-hyperinterval, $f : I_0 \to \mathbb{R}^n$
Output: A list of hyper-intervals, each of them contains one solution
begin
    if $I_0 = \emptyset$ *or* $(0, ..., 0) \notin f(I_0)$ **then**
      | **return** $\emptyset$
    end
    else if $| I_0 | \leq \varepsilon$ **then**
      | **return** I_0
    end
    else
      (* Let L be a list of hyper-intervals of dimension n *)
      $L \leftarrow Succ(I_0, f)$
      return $\cup_{l \in L} solve(l, f)$
    end
end

Algorithm 2. Newton-Raphson iteration algorithm

Function **Succ**(I_0,f);
Input: I_0 : an n-hyperinterval, $f : I_0 \to \mathbb{R}^n$
Output: A list of hyper-intervals
begin
    try{
    (* Let J be the jacobian matrix and $Inv(J)$ its inverse *);
    $J \leftarrow f'(I_0)$;
    (* Let $\mathcal{L}$ be the list of 2^n hyper-intervals resulting from the bisection of I_0 *);
    $\mathcal{L} \leftarrow dichotomie(I_0)$;
    (*Compute the inverse of J with the Gauss-Jordan method*);
    $Inv(J) \leftarrow inverseGaussJordan(J)$;
    $R \leftarrow mid_{I_0} - Inverse(J) \times f(mid_{I_0})$;
    return $\sqcup_{L \in \mathcal{L}}(L \cap R)$;
    }
    catch(Exception){
    (* Capture the exception produced when computing the inverse *);
    (* in case of division by an interval which contains 0 *);
    return $\mathcal{L}$;
    }
end

$m \times m$ matrix N of which all elements belong to the matching elements of M, then N^{-1} exists and all its elements belong to the matching elements of M^{-1}.

We use the classical Gauss-Jordan elimination technique with total pivot. We use exactly the same algorithm as the one working with real numbers. Actually, thanks to C++ features like template and operators overloading, we coded a unique algorithm and separately tested it against real matrices and interval matrices. Thanks to C++ function and operator overloading, the four operators $+, -, *, /$ and the absolute value function $x \rightarrow |x|$ working with real numbers are simply replaced by the four homologous operators and by the homologous absolute value function working with real intervals.

Let $M = (M_{ij})$ be a $m \times m$ real interval matrix (i.e. each element of M is a real finite interval) to be inverted. Let I be the $m \times m$ indentity matrix wih real interval elements (all the elements are set to $[0, 0]$ except for the diagonal elements which are set to $[1, 1]$). Let A and B be two other $m \times m$ real interval matrices. Let X be an array of m integer elements. The algorithm initiates with A set to M, B set to I and X set to $\{1, 2, ..., m\}$. It performs the same operations on both A and B. It ends up with A equalling I and B equalling M^{-1}. The purpose of X is to record all the column swapings (it records the positions of the unknowns). The algorithm works in three steps : first, it changes A into a upper-right triangular matrix, then it solves the resulting triangular system then it sorts out the lines of B according to X.

5   Numerical Results

The proposed algorithm has been implemented in $C++$, using finite precision floating point numbers and the images related to the results have been written in the Java programming language. On the next few examples, the algorithm has been tested against various functions in the univariate case and in the bivariate case. In each case, the average number of iterations is given.

Example 1 : f is a univariate non parametric cubic spline function defined over a uniform knot vector. The shape of f is defined by 11 coefficients (control points) as shown in Fig. 1. This cubic spline is constructed by an interpolation process in such a way it crosses five times the x-axis ($x = 0.1, 0.3, 0.5, 0.7, 0.9$) The average number of iterations is 7.

Exemple 2 : In this example we calculate the intersection points between a hyperbola and an ellipse which are respectively given by the following two equations :

$$\begin{cases} -2.077u^2 - 2.077v^2 + 5.692uv - 5.384u - 5.384v + 17.846 = 0 & \text{(Hyperbola)} \\ 23.692u^2 + 23.692v^2 - 41.23uv - 21.538u - 21.538v + 39.385 = 0 & \text{(Ellipse)} \end{cases}$$

The two curves intersect at four points : $(1.5, 2.5), (2.5, 1.5), (4.5, 5.5), (5.5, 4.5)$. The equations have been translated in B-spline form through quasi-interpolation techniques [15,12]. The average number of iterations is about 11.

Algorithm 3. Gauss-Jordan algorithm for inverting an interval-matrix

Function **InverseGaussJordanMatrix**();
Input: $M : m \times m$ matrix
Output: $R = M^{-1}$: the inverse $m \times m$ matrix of M
begin

$A \leftarrow M$; $B \leftarrow I$; $X \leftarrow \{0, 1, \ldots, m\}$
First step: Triangulation. A is changed into a upper right triangular matrix
for $r = 1$ *to* m **do**

 find the total pivot A_{kl} on the sub-matrix (A_{ij}) $r \leq i \leq m$; $r \leq j \leq m$
 if $|A_{kl}| = 0$ **then**
  | abort the algorithm; throw an exception : M is not invertible
 end
 swap lines A_r and A_k; swap lines B_r and B_k;
 swap columns A_r and A_l; swap elements $X[r]$ and $X[l]$;
 $A_r \leftarrow A_r/A_{rr}$; $B_r \leftarrow B_r/A_{rr}$; (*dividing by the pivot*)
 for $j = r + 1$ *to* m **do**
  | $A_j \leftarrow A_j - A_{jr} * A_r$; $B_j \leftarrow B_j - A_{jr} * B_r$
 end

end
Second step : Solving the triangular system
for $r = m$ *downto* 2 **do**

 for $j = r - 1$ *downto* 1 **do**
  | $A_j \leftarrow A_j - A_{jr} * A_r$; $B_j \leftarrow B_j - A_{jr} * B_r$
 end

end
Last step : Sorting out the lines of B
Let R be a $m \times m$ interval matrix
for $r = 1$ *to* m **do**
| $R_{x[r]} \leftarrow B_r$
end
return R (*the resulting matrix containing M^{-1}*)

end

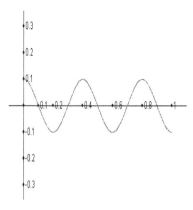

Fig. 1. Intersection of a non-parametric spline function and the x-axis

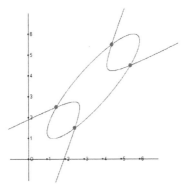

Fig. 2. Intersection of an ellipse and a hyperbola

Example 3 : In this example, the algorithm has been used to find the extrema of a functional $f(u, v)$. Here f is bivariate spline function of degree 2×2 defined by a grid of 5×5 control points (real numbers) on uniform knot vectors. The grid is $[(0, -1, 0, 1, 0)(-1, -2, 0, 2, 1)(0, 0, 0, 0, 0)(1, 2, 0, -2, -1)(0, 1, 0, -1, 0)]$. The system of equations to be solved is $\frac{\partial f}{\partial u} = 0$ and $\frac{\partial f}{\partial v} = 0$. The functional (see Fig. 3) has 4 extrema located at $(2.466, 2.466)$, $(0.534, 0.534)$, $(2.466, 0.534)$, $(0.534, 2.466)$, and a saddle point at $(1.5, 1.5)$. The average iteration number is about 6.

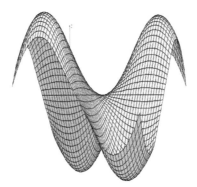

Fig. 3. Computation of extrema of a functional

Robustness and Convergence Rate Considerations : Bisection splits I_0 into a partition, therefore no root can be missed or found twice. Let us suppose that every root is a single root. Then beyond a certain recursion depth it must be contained in a searching hyper-interval I_0 over which $(f'(I_0))$ is invertible. Therefore, from that recursion depth until it terminates, the algorithm uses the first form of the iteration (equation (6)) to converge towards the solution. So as recursive subdivision, the algorithm might be linearly convergent at the beginning, but beyond a few iterations it becomes quadratically convergent as Newton method is.

6   Conclusion

In this paper we have proposed a new approach for solving a nonlinear system of equations in B-spline form over a finite hyper-interval. It is based on Interval Newton's Method and recursive subdivision. Newton's method brings fast convergence while bisection brings robustness. The proposed hybrid method strongly exploits the B-spline properties such as convex hull and diminishing variation features. This is useful in both the separating and the refining step of the new algorithm. The numerical results show the efficiency of the proposed method. Further work is to apply the algorithm to systems of any dimension, the main task being to represent a multi tensor product of B-spline functions in computer memory.

References

1. Barton, M.: Solving polynomial systems using no-root eliminations blending schemes. Computer Aided Design 43, 1870–1878 (2001)
2. Boehm, W.: Inserting new knots into B-spline curves. Computer Aided Design 12, 199–201 (1980)
3. Cohen, E., Lyche, T., Riesenfeld, R.F.: Discrete B-splines and subdivision techniques in computer-aided geometric design and computer graphics. Comput. Graphic Image Processing 14, 87–111 (1980)
4. De Boor, C.: A Practical Guide to Splines. Springer, New York (1978)
5. Garloff, J., Smith, A.P.: Solution of Systems of Polynomial Equations by Using Bernstein Expansion. In: Alefeld, G., Rump, S., Rohn, J., Yamamoto, T. (eds.) Symbolic Algebraic Methods and Verification Methods, pp. 87–97. Springer (2001)
6. Golbabai, A., Javidi, M.: A new family of iterative methods for solving system of nonlinear algebric equations. Applied Mathematics and Computation 190, 1717–1722 (2007)
7. Grandine, T.A.: Computing zeroes of spline functions. Computer Aided Geometric 6, 129–136 (1989)
8. Jaberipour, M., Khorram, E., Karimi, B.: Particle swarm algorithm for solving systems of nonlinear equations. Computers and Mathematics with Applications 62, 566–576 (2011)
9. Kiciak, P., Zidna, A.: Recursive de Casteljau bisection and rounding errors. Computer Aided Geometric Design 21, 683–695 (2004)
10. Lane, J.M., Riesenfeld, R.F.: Bounds on a polynomial. BIT: Nordisk Tidskrift for haformations-Behandling 21, 112–117 (1981)
11. Luo, Y.Z., Tang, G.Z., Zhou, L.N.: Hybrid approach for solving systems of nonlinear equations using chaos optimization and quasi-Newton method. Applied Soft Computing 8, 1068–1107 (2008)
12. Michel, D., Mraoui, H., Sbibih, D., Zidna, A.: Computing the range of values of real functions using B-spline form. Appl. Math. Comput. (2014), http://dx.doi.org/10.1016/j.amc.2014.01.114

13. Mitchell, D.P.: Robust ray intersection with interval arithmetic. In: Proceedings on Graphics Interface 1990, pp. 68–74 (1990)
14. Moore, R.E., Kearfott, R.B., Cloud, M.J.: Introduction to Interval Analysis, 223 pages. SIAM (2009)
15. Sablonnière, P.: Univariate spline quasi-interpolants and applications to numerical analysis. Rend. Sem. Mat. Univ. Pol. Torino 63(3) (2010)
16. Sherbrooke, E.C., Patrikalakis, N.M.: Computation of the solutions of nonlinear polynomial systems. Computer Aided Geometric Design 10, 379–405 (1993)
17. Zidna, A., Michel, D.: A two-steps algorithm for approximating real roots of a polynomial in Bernstein basis. Mathematics and Computers in Simulation 77, 313–323 (2008)

Application of Sigmoid Models for Growth Investigations of Forest Trees

Zoltán Pödör[1], Miklós Manninger[2], and László Jereb[3]

[1] Faculty of Forestry, University of West Hungary, Sopron, H-9400 Ady Endre út 5
[2] Forest Research Institute, National Agricultural Research and
Innovation Centre, Budapest, H-1027 Frankel Leó u. 1
[3] The Simonyi Karoly Faculty of Engineering, Wood Sciences and Applied Arts,
University of West Hungary, Sopron, H-9400 Bajcsy-Zsilinszky u. 9

Abstract. Radial growth of the trees and its relationships with other factors is one of the most important research areas of forestry for a long time. For measuring intra-annual growth there are several widely used methods: one of them is measuring the girth of trees regularly, preferably on weekly basis. However, the weekly measured growth data may have bias due to the actual precipitation, temperature or other environmental conditions. This bias can be reduced and using adequate growth functions the discrete growth data can be transformed into a continuous curve.

In our investigations the widely used logistic, Gompertz and Richards sigmoid growth models were compared on intra-annual girth data of beech trees. To choose the best model two statistical criteria, the Akaike weight and the modified coefficient of determination were examined. Based on these investigations and the view of applicability, Gompertz model was chosen for later applications. However, we came to the conclusion that all three models can be applied to the annual growth curve with sufficient accuracy.

The modified form of the Gompertz function gives three basic curve parameters that can be used in further investigations. These are the time lag, the maximum specific growth rate and the upper asymptotic value of the curve. Based on the fitted growth curves several other completely objective curve parameters can be defined. For example, the intersection of the tangent drawn in the inflection point and the upper asymptote, the distance between upper intersection point and the time lag, the time and value of the inflection point etc. and even different ratios of these parameters can be identified. The main advantages of these parameters are that they can be created uniformly and objectively from the fitted growth curves. This paper demonstrates the application opportunities of the curve parameters in a particular study of tree growth.

Keywords: sigmoid growth functions, curve fitting, intra-annual tree growth.

1 Introduction

The description of the intra-annual radial/circumferencial tree growth with mathematical functions is a really important task in forestry. Using fitted data the

T.V. Do et al. (eds.), *Advanced Computational Methods for Knowledge Engineering,*   353
Advances in Intelligent Systems and Computing 282,
DOI: 10.1007/978-3-319-06569-4_26, © Springer International Publishing Switzerland 2014

bias of the original data can be reduced and the discrete growth data can be transformed into a continuous curve. The function fitting allows the replacement of the missing data and it is possible to define other objective attributes directly from the basic curve parameters or indirectly from the fitted curve.

In forest research different growth curves and other derived data and parameters are used. Avocado tree growth cycles were examined by Thorp at al. over two growing seasons [1]. Logistic curves were fitted to enable statistical comparisons of growth cycles. They studied the realized increment of the avocado trees on different periods covered in the proportion of total growth. Bouriaud at al. fitted Gompertz function to the intra-annual growth data of Norway spruce [2]. These fitted growth curves were used to define the exact date of that growth size. A specially modified Weibull function was used by Wipfler at al. for describing the intra-annual tree growth processes of beech and Norway spruce [3]. They got the standardized annual growth profiles to obtain curve parameters for different statistical tests on the effects of ozone. To eliminate further non-climatic variations from the pentad increments a Gompertz function was applied by Seo at al. [4]. The fitted sigmoid trend represents the intra-annual biological growth dynamics of Scots pine.

Gruber at al. tested the hypothesis that intra-annual radial stem growth in *Pinus cembra* is influenced by different climate variables along the treeline ecotone in the Austrian Alps [5]. Additionally, time of maximum increment growth was determined throughout the treeline ecotone by applying Gompertz modelled growth functions. Radial stem growths were continuously followed by band dendrometers and were modelled using Gompertz functions to determine the time of maximum growth by Oberhuber and Gruber [6] and Oberhuber at al. [6,7]. Michelot at al. weekly monitored the stem radial increment using dendrometers, and Gompertz functions were used to fit the cumulative ring width of the beech, oak and pine trees [8]. To assess tracheid production dynamics, the number of cells over time was fitted using a Gompertz function by Cuny at al. [9]. Deslauirers at al. studied daily stem radial growth of balsam fir between 1998 and 2001 using automated point dendrometers to investigate meteorological influence [10]. Richards and Gompertz function were fitted to the cumulative stem radius increment. Gompertz function fitted well to the weekly measured cumulative cell production according to Zhai at al. [11].

The contributions of the paper can be summarized as follows. Section 2 provides a brief description of the functions used in the fitting procedures and of the the analysis approaches applied in this study. In Section 3 the comparison results are summarized and the applicability of the Gompertz function is shown. Then some parameters obtained from the Gompertz function are defined and several results illustrate how these parameters can be used for the growth investigations of forest trees. Finally, the paper is concluded and some further investigation directions are identified.

2   Data and Methods

2.1   Fitting of Functions

A vast number of natural and human-generated or artificial phenomena exist that initially indicate exponential characteristics of cell growth, for example the growth of a tumor or some ecological processes (plant growth progress, etc.). However, these curves cannot exceed a certain threshold level of saturation due to practical reasons. Thus, these constrained exponential growth processes cannot be defined by traditional exponential functions due to the unlimited increase in the number of parameters. For solving these fitting problems many solutions, functions have been generated during the last 150 years.

Based on [12] as well as [13] studies and practical considerations, the growth functions may be divided into three different groups (see Fig.1):

- functions without inflection points,
- functions with one inflection point,
- functions with two inflection points.

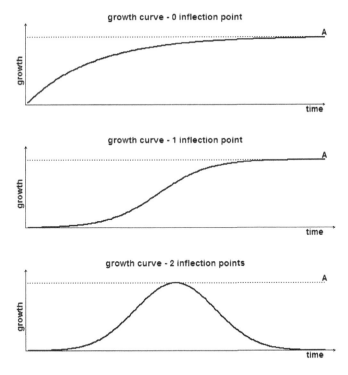

Fig. 1. Characteristics of growth curves

The intra-annual tree growth process can be well approximated with the growth functions of one inflection point. Such is called sigmoid function and characteristically shows an S shape. The main characteristic of the sigmoid functions is that they have a relatively short initial growth period. Then it is followed by an intensive exponential growth phase, where the growth rate accelerates to a maximal value. Eventually the growth curves contains a final phase in which the growth rate decreases and finally approaches zero. Mathematically, the value of the maximal growth rate can be defined as the slope of the curve at the inflection point.

Hereinafter – referring to the above mentioned reasons – we deal with those functions having only one inflection point. Based on forestry literature we selected three often used sigmoid functions: logistic (y_L), Gompertz (y_G) and Richards (y_R). We applied for each function the special converted forms that were were published by Zwietering at al. for the biological interpretation of the curve parameters [14].

The modified saturation functions have usually three typical parameters, where A is the upper asymptote of the curve, μ is the maximal growth rate and λ is the so-called time lag. The Richards function has the fourth parameter ν that determines the value of the function at the inflection point. According to [14] the modified functions are as follows:

$$y_L = \frac{A}{1 + e^{2+4\mu(\lambda-t)/A}}$$

$$y_G = Ae^{-e^{(1+\mu e(\lambda-t)/A)}}$$

$$y_R = A(1 + \nu e^{(1+\nu)}e^{\nu(1+\nu)(1+1/\nu)(\lambda-t)/A})^{-1/\nu}$$

During the investigation of growth functions, the data of the Forestry Research Institute were used. The sample plot (M01) marked for this study belongs to the ICP-Forest intensive monitoring system and is located in Mátra Mountains. The selected tree species is beech, because the growth curve of this species is relatively smooth. From all the measured trees the growth data of the best growing 6 dominant and co-dominant trees were used from 2002 to 2011. Girth bands were applied with weekly readings from March till November and biweekly readings in the dormancy period to determine the circumferential growth of the trees. Meteorological parameters were measured by automatic meteorological stations. Precipitation data were collected at the nearby open-air field, and the temperature data were measured inside the stand.

The fitting was carried out by using the R (R Development Core Team) software grofit package. The grofit package includes the above mentioned three sigmoid functions, uses non-linear least squares method (Gauss-Newton) for fitting. The initial values of the fitting method are determined automatically.

The models are ranked by the modified Akaike information criteria weight w. The value of w is calculated by the AICcmodavg package of the R and the bigger w indicates better model [15]. The goodness of the fitting is described by adjusted R-square ($\hat{R}^2$) values. This parameter takes into account the number of regression parameters similar to Akaike weight.

2.2   Data Analysis Method

The objective of this research was to establish relationships between the radial (circumferential) tree growth and the basic climatic parameters (monthly precipitation sum and average temperature). Moving windows and evolution techniques are often applied to show the varying effects of the climatic components [16,17,18,19] and specially developed periodic data are also used [20,21,22]. However, these methods are based on special experimental observations and are not intended to use in general conditions.

The use of the proposed data transformation procedure allows more intensive studies by the systematic expansion of the basic data set due to the applied special window technique CReMIT [23,24]. The transformation procedure combines the essence of moving intervals and evolution techniques, the systematic movement of the windows and the systematic creation of windows with different widths. In CReMIT the maximum time shifting, the maximum window width and the applied transformation function are defined by the user. The procedure ensures the combination of the time shifting and width values of the windows in a single process, significantly widening the sphere of the analysing possibilities. The procedure is integrated in an analysis frame that includes the preparation of the data, as well as the data transformation and the analysis procedural modules. The analysis basically searches the connections between the time series including the most frequent correlation and the regression analyses.

3   Results

3.1   Fitting of Sigmoid Functions

Data were analyzed including all 10 years (n_1), all 6 trees (n_2) with all the three growth functions (logistic=L, Gompertz=G and Richards=R) using the data measured between 1 March and 31 October. In Table 1 and 2 the rows of best function show how many times function G, L or R gave the best Akaike weight in the given year for all individual trees or for the given tree out of all years. In the other two rows for each sigmoid function the average w and average $\hat{R}^2$ values are presented. Table 3 compares all data obtained for the three functions for all 6 trees in all 10 years.

The results indicated that there were no significant differences between the three models based on adjusted ($\hat{R}^2$). Either model can be used in the fitting task. However, the Akaike Information Criterion (AIC) based examination showed, that the logistic function is less accurate than the other two. The Gompertz and Richards functions showed similar results compared to each other. The Richard function provided the best Akaike weight results. However, taking into account other practical aspects (number of parameters and the difficulties of fitting) and the reviewed literature we have opted for the Gompertz function to further analysis. The advantage of this selection is that only the three parameters of Gompertz function provide sufficient flexibility to our investigations, because all these parameters of the curve can be well interpreted and their possible biological meaning can be further investigated.

Table 1. Results of fitting by years

$n_1 = 6$		years									
		2002	2003	2004	2005	2006	2007	2008^1	2009	2010	2011
G	best function	0	3	0	2	1	3	3	3	4	4
	average w	0.06	0.53	0.26	0.34	0.19	0.44	0.37	0.45	0.49	0.56
	average $\hat{R}^2$	0.9958	0.9956	0.9978	0.9979	0.9918	0.9970	0.9986	0.9966	0.9978	0.9978
L	best function	0	0	0	0	3	0	0	0	0	0
	average w	0.00	0.00	0.00	0.01	0.36	0.03	0.00	0.04	0.04	0.00
	average $\hat{R}^2$	0.9880	0.9899	0.9967	0.9967	0.9929	0.9953	0.9953	0.9942	0.9959	0.9960
R	best function	6	3	6	4	2	3	4	3	2	2
	average w	0.94	0.47	0.74	0.65	0.45	0.53	0.63	0.51	0.47	0.44
	average $\hat{R}^2$	0.9968	0.9958	0.9983	0.9983	0.9933	0.9973	0.9987	0.9968	0.9982	0.9980

Table 2. Results of fitting by trees

$n_2 = 10$		trees					
		2	4	19	21	22	26^1
G	best function	6	2	4	6	2	3
	average w	0.41	0.26	0.48	0.50	0.16	0.41
	average $\hat{R}^2$	0.9973	0.9946	0.9970	0.9972	0.9974	0.9967
L	best function	1	1	0	0	1	0
	average w	0.07	0.07	0.00	0.01	0.11	0.03
	average $\hat{R}^2$	0.9950	0.9926	0.9919	0.9935	0.9975	0.9940
R	best function	3	7	6	4	7	8
	average w	0.52	0.67	0.52	0.50	0.73	0.56
	average $\hat{R}^2$	0.9977	0.9951	0.9973	0.9974	0.9985	0.9969

Table 3. Summarized results of fitting for all cases

altogether $n = 60^1$	**G**	**L**	**R**
best function	23	3	35
average $\hat{R}^2$	0.9967	0.9941	0.9971

3.2 Application of the Curve Parameters

Defining the Curve Parameters

Based on the fitted Gompertz curve we can define further parameters objectively, thus the curve parameters can be divided into two groups: (i) basic and (ii) derived parameters.

---

[1] In 2008 the Gompertz and Richards functions have the same weight for tree 26.

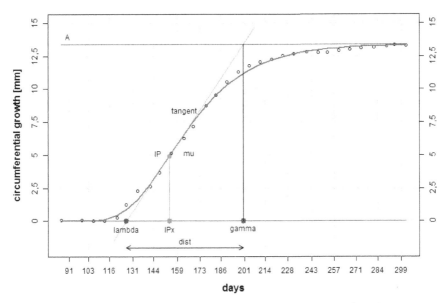

Fig. 2. Basic and derived parameters of the Gompertz function

The basic curve parameters are the time lag (λ), the maximal growth rate of the curve (μ) and the upper asymptote (the cumulative growth value) of the curve (A). These parameters are generated directly from the equation: μ is the slope of the curve at the inflection point, while λ is determined by the intersection of the tangent drawn at inflection point and the lower asymptote $y = 0$.

Using these characteristics some other parameters can be derived, too (see Fig. 2). These are the intersection of the tangent drawn at the inflection point and the upper asymptote (γ), the distance (in day) between γ and λ, $dist = \gamma - \lambda$. Based on the fitted Gompertz curves it is easy to identify the location and the value of the inflection point:

$$IPx = \frac{A}{\mu e} ln(\frac{\mu e}{A} \lambda + 1), \qquad IPy = \frac{A}{e}.$$

Relationships with the Meteorological Parameters

We investigated the relationships between the monthly meteorological variables and the following basic and derived curve parameters:

- the upper asymptote of the curve (A),
- the time lag (λ),
- the maximal growth rate of the curve (μ),
- the location of the inflection point (IPx),
- the intersection of the tangent drawn at inflection point and the upper asymptote (γ),
- the distance (in day) between (γ) and (λ), $dist = \gamma - \lambda$.

Table 4. Summary of the relationships of A for all windows based on calendar

Upper asymptote, A		M01-Mátra	
		Temperature	Precipitation
windows of the previous year	sign of significant relationships	positive	negative (with exceptions)
	number of significant relationships	few (25 of 270, mostly dependent on trees)	several (72 of 270)
	number of general relationships	very few (1 of 45)	several (12 of 45)
	highlighted periods	p8-p9	p4-p8, p4-p8, p6-p8, p6-p9
windows of the overlapping period	sign of significant relationships	positive	positive
	number of significant relationships	few (20 of 432, only for one tree)	several (88 of 432)
	number of general relationships	none	few (14 of 72)
	highlighted periods	none	p10-c3, p10-c4, p10-c5, p10-c6
windows of the previous year	sign of significant relationships	negative (with exceptions)	positive (with exceptions)
	number of significant relationships	several (60 of 216)	several (59 of 216)
	number of general relationships	several (11 of 36)	several (11 of 36)
	highlighted periods	c5-c6, c5-c7, c5-c8, c6-c6	c1-c4, c1-c5

The raw meteorological data were transformed to monthly sum of precipitation and monthly average of temperature values by the TR function of CReMIT taking into consideration the later upscaling possibilities of the results. The maximum time shifting and window width were set to 17 months equally covering two vegetation periods (from April to the next August). Simple linear correlation analysis was applied and the significance of correlation was evaluated by Student's t-test at level of significance $\alpha = 0.05$ with the number of degrees of freedom $n - 2$, where n is the length of our data series.

The output consists of 12 spreadsheets with 153 rows (the number of the different windows) and the columns of the selected trees as dependent variables in each with the significant regression coefficients. For the demonstration of the relationships the upper asymptote (A) of the curve from the basic and IPx from derived parameters were chosen and the results are summarized in Tables 4 and 5, where p the previous, c the current year and the number next to the letter indicates the month. In the case of IPx the last month of the current year periods is June, because of the expected time of the inflection point. Trees are evaluated individually. If in a given window most of the trees showed significant

Table 5. Summary of the relationships of IPx for all windows based on calendar

IPx		M01-Mátra	
		Temperature	Precipitation
windows of the previous year	sign of significant relationships		varying
	number of significant relationships	none	several (32 of 270)
	number of general relationships	none	few (2 of 45)
	highlighted periods	none	p6-p7 (negative)
windows of the overlapping period	sign of significant relationships	negative	positive
	number of significant relationships	few (30 of 324)	several (121 of 324)
	number of general relationships	very few (1 of 54)	several (13 of 54)
	highlighted periods	p12-c6	p9-c6, p10-c4, p10-c5, p10-c6, p11-c4, p11-c6, p12-c4, p12-c5, p12-c6
windows of the previous year	sign of significant relationships	negative (with exceptions)	positive (with exceptions)
	number of significant relationships	many (65 of 126)	many (79 of 126)
	number of general relationships	many (12 of 21)	many (11 of 21)
	highlighted periods	c3-c5, c3-c6, c4-c6, c5-c5 c5-c6, c6-c6	c1-c4, c1-c5, c1-c6, c2-c4, c2-c5, c2-c6, c3-c4, c3-c5, c3-c6, c4-c6

relationship, the relationship is considered general and out of them the best ones are highlighted.

The precipitation of the previous year usually has a negative impact on the annual tree growth. There are several windows covering partially or fully the previous vegetation period that show significant relationships.

On the other hand, the precipitation of the overlapping period has a positive impact and the precipitation of the wide windows (e. g. $p10 - c4$) containing almost the entire dormancy period increases the yearly tree growth. These relationships emphasize the importance of the dormancy period. The precipitation of the current year periods have a positive impact on the annual tree growth. In contrast to the expectations there is no any significant window in the vegetation period.

The impact of the temperature of the previous year and the overlapping time periods are slightly positive on the annual tree growth, however, with the exception of period $p8 - p9$, there is no any window to be highlighted. The correlations between the current year temperature and the annual tree growth usually show negative relationships, mainly for the windows at the very beginning of the growing period.

To illustrate the applicability of the derived parameters the location of the infection point (IPx) was selected, which is the time of the maximum growth rate as well.

There are some significant relationships between the previous year precipitations and IPx with varying signs. However there is only one window ($p6 - p7$) when this negative relationship is generally valid. The precipitation in the overlapping periods and at the beginning of the current year has similar positive impact on IPx as in the case of the annual tree growth. The higher the precipitation in these periods (mainly in the dormancy periods) the later the IPx is.

There are no and very few significant relationships between the previous year temperature and IPx and between the overlapping period and IPx, respectively. In the latter case only one window ($p12 - c6$) can be identified when the increasing temperature increases the IPx value as well. Contrary to the impact of the previous year many relationships were found in the current year and most of them show negative signs. According to these experiences the higher temperature of the months before IPx (from March to June) delays the location of the inflection point in time and therefore the maximal growth rate occurs later.

4   Conclusions

During this research three widely used sigmoid growth models (logistic, Gompertz, Richards) were compared on intra-annual growth data of beech trees. Based on the Akaike-weight (w), the modified coefficient of determination ($\hat{R}^2$) and some practical reasons, the Gompertz function was selected for modeling the intra-annual tree growth. The Gompertz function modified by Zwietering et al. (1990) has three parameters: the upper asymptote of the curve A, the maximal growth rate of the curve μ and the so-called time lag λ. Using the fitted growth curve other parameters can be defined exactly and objectively from mathematical point of view. Additionally, three other parameters (measured in day) were defined: the location of the inflection point IPx, the intersection of the tangent drawn at inflection point and the upper asymptote γ and the distance between γ and λ, ($dist = \gamma - \lambda$). Based on the basic and derived parameters several analyses related to meteorological variables (precipitation and temperature) were performed. The effect of the different time periods (windows) ranging 1 to 17 months was examined by CReMIT.

Fitting of the sigmoid functions and using their parameters proved to be a promising tool in the investigations to describe the relationships between tree growth and meteorological variables. However, mostly from biological (forestry) point of view, further investigations are necessary (i) with the parameters already available or (ii) other curve parameters that will be defined in the future.

Acknowledgments. This work was supported by the TAMOP-4.2.2.C-11/1/ KONV-2012-0015 (Earth-system) project sponsored by the EU and European Social Foundation.

References

1. Thorp, T.G., Anderson, P., Camilerri, M.: Avocado tree growth cycles - a quantitative model. In: Proceedings of The World Avocado Congress III, pp. 76–79 (1995)
2. Bouriaud, O., Leban, J.-M., Bert, D., Deleuze, C.: Intra-annual variations in climate influence growth and wood density of Norway spruce. Tree Physiology 25, 651–660 (2005)
3. Wipfler, P., Seifert, T., Biber, P., Pretzsch, H.: Intra-annual growth response of adult Norway spruce (Picea abies [L.] KARST.) and European beech (Fagus sylvatica L.) to an experimentally enhanced, free-air ozone regime. Eur. J. Forest Res. 128, 135–144 (2009)
4. Seo, J.-W., Eckstein, D., Jalkanen, R., Schmitt, U.: Climatic control of intra- and inter-annual wood-formation dynamics of Scots pine in northern Finland. Environmental and Experimental Botany 72(3), 422–431 (2011)
5. Gruber, A., Zimmermann, J., Wieser, G., Oberhuber, W.: Effects of climate variables on intra-annual stem radial increment in Pinus cembra (L.) along the alpine treeline ecotone. Ann. For. Sci. 66, 11 (2009)
6. Oberhuber, W., Gruber, A.: Climatic influences on intra-annual stem radial increment of Pinus sylvestris (L.) exposed to drought. Trees (Berl West) 24(5), 887–898 (2010)
7. Oberhuber, W., Gruber, A., Kofler, W., Swidrak, I.: Radial stem growth in response to microclimate and soil moisture in a drought-prone mixed coniferous forest at an inner Alpine site. Eur. J. Forest Res. (published online: January 3, 2014)
8. Michelot, A., Bréda, N., Damesin, C., Dufrêne, E.: Differing growth responses to climatic variations and soil water deficits of Fagus sylvatica, Quercus petraea and Pinus sylvestris in a temperate forest. Forest Ecology and Management 265, 161–171 (2012)
9. Cuny, E.H., Rathgeber, B.H.C., Lebourgeois, F., Fortin, M., Fournier, M.: Life strategies in intra-annual dynamics of wood formation: example of three conifer species in a temperate forest in north-east France. Tree Physiology 32, 612–625 (2012)
10. Deslauirers, A., Morin, H., Urbinati, C., Carrer, M.: Daily weather response of balsam fir (Abies balsamea (L.) Mill.) stem radius increment from dendrometer analysis in the boreal forests of Quebec (Canada). Trees 17, 477–484 (2002)
11. Zhai, L., Bergeron, Y., Huang, J.-G., Berninger, F.: Variation in intra-annual wood formation, and foliage and shoot development of three major Canadian boreal tree species. Am. J. Bot. 99, 827–837 (2012)
12. Zeide, B.: Analysis of Growth Equations. Forest Science 39(3), 594–616 (1993)
13. Sit, V., Poulin-Costello, M.: Catalog of curves for curve fitting. BC Ministry of Forests Biometrics Information Handbook No. 4, p. 110 (1994)
14. Zwietering, M.H., Jongerburger, I., Rombouts, F.M., Van't Riet, K.: Modeling of the Bacterial Growth Curve. Applied and Environmental Microbiology 56(6), 1875–1881 (1990)

15. Mazerolle, M.J.: AICcmodavg, Model selection and multimodel inference based on (Q)AIC(c) (2012), http://CRAN.R-project.org/package=AICcmodavg
16. Büntgen, U., Frank, C.D., Schmidhalter, M., Neuwirth, B., Seifert, M., Esper, J.: Growth/climate response shift in a long subalpine spruce chronology. Trees 20, 99–110 (2006)
17. Zhang, Y., Wilmking, M., Gou, X.: Changing relationships between tree growth and climate in Northwest China. Plant Ecol. 201, 39–50 (2008)
18. Zhang, Y., Wilmking, M., Gou, X.: Dynamic relationships between Picea crassifolia growth and climate at upper treeline in the Qilian Mts., Northeast Tibetan Plateau, China. Dendrochronologia 29, 185–199 (2011)
19. Macias, M., Andreu, L., Bosch, O., Camaero, J.J., Gutierrez, E.: Increasing aridity is enhancing silver fir (abies alba mill.) Water stress in its south-western distribution limit. Climatic Change 79, 289–313 (2006)
20. Carrer, M., Urbinati, C.: Assessing climate-growth relationships: a comparative study between linear and non-linear methods. Dendrochronologia 19(1), 57–65 (2001)
21. Briffa, K.R., Osborn, J.T., Schweingruber, H.F., Jones, D.P., Shiyatov, G.S., Vaganov, A.E.: Tree-ring width and density data around the Northern Hemisphere: Part 1, local and regional climate signals. The Holocene 12(6), 737–757 (2002)
22. Novák, J., Slodicák, M., Kacálek, D., Dusek, D.: The effect of different stand density on diameter growth response in Scots pine stands in relation to climate situations. Journal of Forest Science 56(10), 461–473 (2010)
23. Edelényi, M., Pödör, Z., Manninger, M., Jereb, L.: Developing of method for transforming time series data and demonstration on forestry data. Acta Agraria Kaposváriensis 15(3), 39–49 (2011) (in Hungarian)
24. Edelényi, M., Pödör, Z., Jereb, L.: Use of transformed data sets in examination of relationship between growth of trees and weather parameters. Agrárinformatika / Agricultural Informatics 2(1), 39–48 (2011) (in Hungarian)

Part VII
Computational Methods for Mobile and Wireless Networks

GPS Tracklog Compression by Heading-Based Filtering

Marcell Fehér and Bertalan Forstner

Department of Automation and Applied Informatics,
Budapest University of Technology and Economics,
Budapest, Hungary
{marcell.feher,bertalan.forstner}@aut.bme.hu

Abstract. In the era of ubiquitous computing, a great amount of location data is recorded by a wide variety of devices like GPS trackers or smartphones. The available datasets are proven to be very good ground for data mining and information extraction in general. Since the raw data usually carries a huge rate of redundancy, the first step of the process is to reduce dataset size. Our research focuses on mining valuable information from GPS tracklogs on smartphones, which environment is well known to be extremely sensitive to resource-demanding operations, therefore it is crucial to reduce input data size as much as possible. In this paper we introduce a heading-based filtering method, which is able to drastically decrease the number of GPS points needed to represent a trajectory. We evaluate and test it by simulations on a real-world dataset, *Geolife Trajectories*. We show that using this filter, data is reduced to an average of 39 percent of the original size per trajectory, which causes a 250% speedup in the total runtime of Douglas-Peucker line generalization algorithm.

Keywords: algorithm, Douglas-Peucker, improvement, GPS, heading, data mining.

1   Introduction and Related Work

Nowadays an extremely large volume of trajectory data is recorded and available for further processing. Low cost of GPS receivers made it possible for smartphone manufacturers to incorporate a positioning module in virtually every device they produce. With the extreme penetration trend of smartphones and other GPS-enabled personal equipment, the generated trajectory data is not only huge in amount, but also very *noisy* and mostly *oversampled*, as demonstrated in Fig. 1. Accuracy of positioning modules used in consumer devices is ranging approximately from five meters (GPS, best case) to 300-500 meters (cellular positioning, worst case), which causes a great deal of error in the recorded data. Oversampling is also a challenge from the perspective of data mining, since algorithms used to process tracklogs usually dispose of $O(n * logn)$, $O(n^2)$ or even higher complexity. These two qualities are always present in real-life trajectory data and

T.V. Do et al. (eds.), *Advanced Computational Methods for Knowledge Engineering*,
Advances in Intelligent Systems and Computing 282,
DOI: 10.1007/978-3-319-06569-4_27, © Springer International Publishing Switzerland 2014

should be addressed before extracting valuable information with heavy-weight algorithms. This phase of data mining is usually referred as data *preparation*, *cleaning* and *reduction*.

Our research focuses on processing trajectories in mobile environment, recorded using GPS and/or cellular technology with a smartphone. Our main goal is to extract movement patterns of individuals based on constantly recorded movement during their daily lives, and use these so-called *routines* for a variety of services, like car-pooling recommendation, precisely targeted advertisement or finding the optimal meeting point with friends. The most important properties of the mobile environment are the lack of resources (relatively low computational power, memory and storage space) and the fact that they operate on a battery. Therefore we are concerned about algorithms with the following attributes to reduce raw trajectory data:

- Least possible complexity to prevent battery drain and long execution
- Leave out as many points as possible while preserving enough information of trajectories so that the subject's real movement can be projected onto the street network level (e.g. it should be possible to replay the movement in turn-by-turn navigation style)

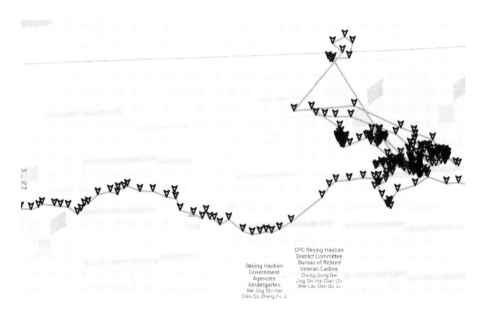

Fig. 1. Snippet of a GPS tracklog from *Geolife Trajectories* dataset. It is clearly visible that the recorded real-life data is both noisy and oversampled.

Dealing with noise present in raw data is usually achieved by using a filtering technique. The most common ones in this field are *mean* and *median* filters, *Kalman filtering* [5] and *particle filtering* [6]. Although data cleaning is out of scope of this paper, in mobile environment we suggest using median filter because of its massive simplicity over Kalman and particle filtering (requiring less computational power) and introducing less delay than mean filter.

Reducing the size of data while preserving valuable information is the main topic of our contribution in this paper. There are many existing methods for polyline simplification, which can be used for reducing the size of GPS tracklogs. The algorithms are classified into two main groups based on time of applicability: *batch* and *online* compression techniques. The first group is suitable when the whole trajectory is available (i.e. after the data collection is finished), while online methods are able to compress data during the location tracking process.

The most simple and only linear complexity reduction method is called *Uniform Sampling* [8]. The main idea of this technique is that the original trajectory is also a sample of the true path, therefore keeping only an uniform sample of it is still an approximation of the real trajectory. Therefore it keeps only every i-th location (e.g. 10th, 15th, 30th) of the original trajectory and deletes others. It is a computationally efficient method, but it doesn't necessarily capture valuable parts of the trajectory.

A more sophisticated and very well-known batch method is the *Douglas-Peucker* algorithm, which was invented independently by numerous researchers: [1], [2], [3], [7], [9], [10]. It operates by approximating a sequence of input points by a single line which goes from the first to the last point. In case the farthest point of the original set is more distant from the approximated segment than a predefined tolerance threshold, the sequence is splitted to two groups by the most erroneous point, and the method recursively calls itself with the subproblems. Otherwise the sequence is substituted by its endpoints. Fig. 2 demonstrates the algorithm. Regarding its time complexity, in best case it requires $\Omega(n)$ steps, the worst case cost is $\Theta(nm)$ while the expected time is $\Theta(n*logn)$, where n denotes the original and m stands for the size of approximated points.

Our previous work includes a modification of this algorithm. In [4], we proposed a method for efficiently specifying the threshold input parameter of Douglas-Peucker in meters, a much more convenient way, while keeping the running time low. We are going to use this version of the line generalization method when evaluating our algorithm by simulations.

In this paper we propose a filtering algorithm, which can be used both in batch and online modes to drastically reduce the size of a trajectory before applying a more complex line generalization algorithm. Our method exploits the fact that a near-linear movement can be represented only by the start and endpoint, therefore intermediate points can be eliminated. Despite its simplicity, this concept proved an average of 61% compression rate of trajectories on real GPS data, while keeping running time as low as $O(n)$. Applying this filter before our modified parameter list Douglas-Peucker line generalization, an $O(n*logn)$ algorithm, we measured a 250-300% speedup of the whole process.

For simulation purposes we used *Geolife Trajectories* [11,12,13], a well known real-life GPS dataset collected by Microsoft Research Asia. It consists of approximately 18.000 trajectories, recorded over four years by 178 subjects.

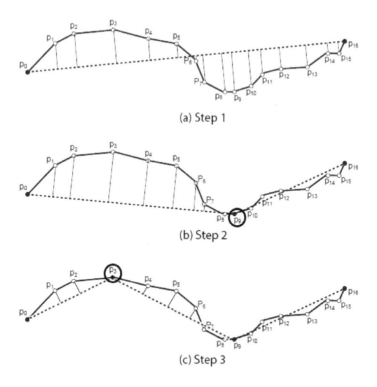

(a) Step 1

(b) Step 2

(c) Step 3

Fig. 2. The first two steps of *Douglas-Peucker* algorithm. It produces the approximated trajectory recursively, adding the most distant point in each step. In Step 1, the
algorithm start at the endpoints, p_0 and p_{15}. In step 2, p_9 is selected and added to the
set, then p_3 is added in Step 3. Image credit: Wang-Chien Lee and John Krumm

2   Contribution

We are proposing a linear time filtering technique, which can be used to drastically reduce the size of an average person's daily trajectories before further
processing. The main idea exploits two facts. First, most human movement GPS
tracklogs are oversampled in a sense that they consist of considerably more points
than needed for a very good approximation of the original route. The reason for
this is the simple fact that the vast majority of people live in cities and move
along the street network, which is mostly intersecting straight streets. Secondly,
a line can be represented only by the start and end points, while intermediate
measurements can be discarded. This assumption is only valid if we are not interested in the speed changes along the straight segment, since this information
is lost when deleting intermediate points.

According to these two assumptions, our algorithm works as follows (see Algorithm 1).

Algorithm 1. *headingFilter(originalTrajectory, angleTreshold)*

---

1. initialize *reducedPath* empty array
2. add first point of *originalTrajectory* to *reducedPath*
3. *startIndex* ← 0
4. *midIndex* ← 1
5. **for** (size of *originalTrajectory* - 2) **do**
6.   *startPoint* ← *originalData[startIndex]*
7.   *midPoint* ← *originalData[midIndex]*
8.   *lastPoint* ← *originalData[midIndex + 1]*
9.   **if** (*startPoint* equals *midPoint* OR *midPoint* equals *lastPoint*) **then**
10.   *midIndex* ← *midIndex* + 1
11.   continue
12.   **end if**
13.   *angle* ← calculate difference of slopes between line segments *startPoint-midPoint* and *midPoint-endPoint*
14.   **if** (*angle* >= *angleTreshold*) **then**
15.   add *midPoint* to *reducedPath*
16.   *startIndex* ← *midIndex*
17.   **end if**
18.   *midIndex* ← *midIndex* + 1
19. **end for**
20. add last point of *originalTrajectory* to *reducedPath*
21. **return** *reducedPath*

The function takes two parameters, the original trajectory and opening angle threshold. In our simulation case, the input trajectory is an array that consists of GPS latitude-longitude pairs in decimal format (as of the raw data of *Geolife Trajectories*). Second input parameter is the maximum decimal degree value of the angle of two consecutive line segments (as illustrated in Fig.3).

The method starts by initializing an empty array which will hold the points of the reduced trajectory and act as the return value of the method. Since we are filtering on angles, which only occur between inner line segments of the trajectory, the first and last point of the original track will always be included in the reduced set. This is done in lines 2 and 20.

In the main **for** loop (between lines 5 and 19) the algorithm operates on two consecutive line segments defined by *startPoint*, *midPoint* and *endPoint*. The conditional block at line 9 was necessary because we often found equal consecutive points in the simulation dataset. In this case, the algorithm simply steps forward the *midPoint* and subsequently *endPoint*. After line 9, where we calculate slope difference of the current two segments, this value is compared to the algorithm input parameter *angleThreshold*. This angle exceeds the threshold when the heading change at *midPoint* is more than what we want to allow, therefore *midPoint* must be included in the output trajectory and the next iteration of the main **for** loop must continue with this point as *startPoint*. Whether or not the angle limit was exceeded, we increase *midIndex* to step to the next segment pair at line 18.

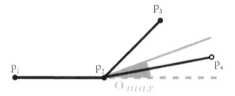

Fig. 3. Illustration of angle threshold, denoted by α_{max} on the figure. The algorithm discards p_4, since it lays inside the *safe zone*, but includes p_3 in the reduced trajectory, because the slope difference between segments $p_1 - p_2$ and $p_2 - p_3$ exceeds the given threshold.

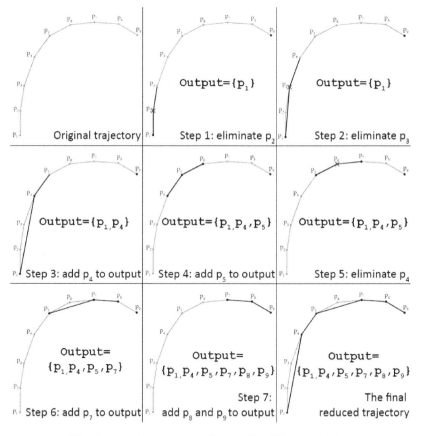

Fig. 4. Sample process of the *headingFilter* technique

We consider a great advantage of our method that it can be applied both in batch (*e.g. the whole trajectory is available at the time of reduction and the algorithm is allowed to peek the future*) and online (*running during data collection, no information about the future*) reduction modes. Since each step requires

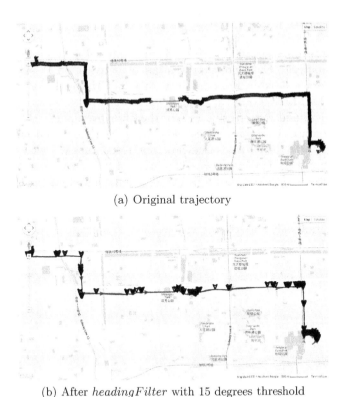

(a) Original trajectory

(b) After *headingFilter* with 15 degrees threshold

Fig. 5. Reducing a real-life GPS trajectory with *headingFilter*

information only about the present and the past, there is no restriction of using *headingFilter* in real time when recording a trajectory.

A common challenge of line generalization algorithms is dealing with slow turning of the trajectory, since the deviation between consecutive segments in the original trajectory is smaller than the given threshold (either euclidean distance, difference of slope or any arbitrary distance). In this case certain algorithms tend to cut off all points of the turn, resulting a high error rate in the reduced trajectory. Our *headingFilter* algorithm deals with this scenario by fixing the start point of the first edge until the midpoint is added to the result set. This behavior ensures that the maximal angle difference in reduced trajectory does not exceed the user given threshold. An illustration of this case is shown in Fig. 4.

Running the algorithm on a real-life trajectory is illustrated in Fig. 5. The original route consists of 1158 points, the angle threshold was set to 15 degrees. After our algorithm eliminated the unnecessary points, size of the reduced trajectory was 542 (46.8% of the original).

3   Simulation Results and Discussion

To evaluate the proposed filter, we used randomly selected subsets of *Geolife Trajectories* dataset. The algorithm was implemented in PHP, simulation was ran on an average desktop workstation with a 3.6GHz Intel Core i5 processor and 8 GB of RAM.

First we measured compression rate of the filter by running it on all trajectories of the dataset with angle threshold setting 10, 15 and 20 decimal degrees. Over approximately 58.00 measurements, our results showed that the average size of the compressed trajectory was 39 percent of the original one. A histogram of the measurement cycle is shown in Fig. 6. (Please note that the high value at 100% is caused by very short trajectories of the dataset (less than 20 points), which cannot be reduced either by *headingFilter*, or Douglas-Peucker. Taking into consideration only trajectories with at least 20 points, the average compression rate is 37%.)

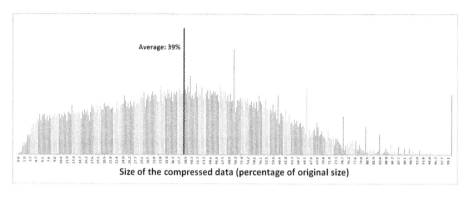

Fig. 6. Compression rate histogram of *headingFilter* method over 58.000 measurements with different angle threshold settings

In the next measurement setting, we used *headingFilter* before our modified Douglas-Peucker algorithm, which takes threshold parameter in meters (introduced in section 1.). First we generated approximately 120.000 test cases with random trajectories from *Geolife*, using angle threshold 10, 15 and 20 degrees, and maximum distance parameter 30, 40, 50 and 60 meters for the modified Douglas-Peucker. We compared the total runtime and peak memory usage (measured by *memory_get_peak_usage* PHP function) of the DP algorithm alone with applying our proposed filter before running the $O(n * log n)$ line generalization algorithm. Our results showed a 2.5 times speedup when using *headingFilter* (illustrated by Fig. 7), although there was no notable change in the maximum memory footprint of the algorithms (see Fig. 8).

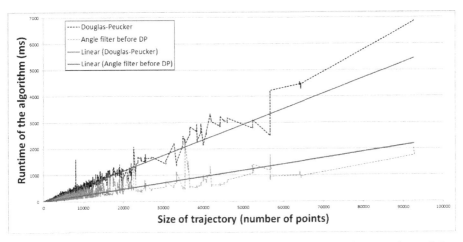

Fig. 7. Comparing runtime of Douglas-Peucker algorithm alone and applying *headingFilter* before it. The results show an average of 2.5 times speedup when using our proposed filter.

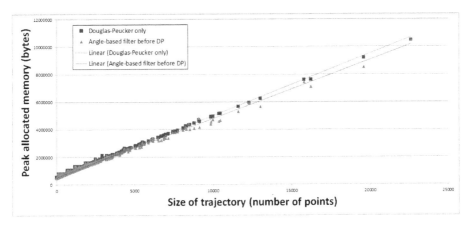

Fig. 8. Comparing peak memory usage of Douglas-Peucker algorithm alone and applying *headingFilter* before it. There is no notable difference between the two setups.

It can be stated that applying our proposed filter improved the performance of Douglas-Peucker line generalization in every configuration we tested. However, using this filter is recommended only when the following assumptions are true.

- Input trajectory data is oversampled and dense
- Information of speed changes does not have to be preserved

4   Conclusions

In this paper we introduced a linear time filtering method, which can be used to drastically reduce the size of oversampled trajectories. We presented the pseudo-code of the algorithm as well as demonstrated its effectiveness with several simulations. Our results shown an average of over 60 percent size reduction on a well known GPS tracklog dataset.

Acknowledgment. This work was partially supported by the European Union and the European Social Fund through project FuturICT.hu (grant no.: TAMOP-4.2.2.C-11/1/KONV-2012-0013) organized by VIKING Zrt. Balatonfüred.

This work is connected to the scientific program of the "Development of quality-oriented and harmonized R+D+I strategy and functional model at BME" project. This project is supported by the New Széchényi Plan (Project ID: TÁMOP-4.2.1/B-09/1/KMR-2010-0002).

References

1. Ballard, D.H.: Strip trees: A hierarchical representation for curves. Communications of the ACM 24(5), 310–321 (1981)
2. Douglas, D.H., Peucker, T.K.: Algorithms for the reduction of the number of points required to represent a digitized line or its caricature. Cartographica: The International Journal for Geographic Information and Geovisualization 10(2), 112–122 (1973)
3. Duda, R.O., Hart, P.E., et al.: Pattern classification and scene analysis, vol. 3. Wiley, New York (1973)
4. Feher, M., Forstner, B.: Self-adjusting method for efficient gps tracklog compression. In: 2013 IEEE 4th International Conference on Cognitive Infocommunications (CogInfoCom), pp. 753–758. IEEE (2013)
5. Gelb, A.: Applied optimal estimation. The MIT Press (1974)
6. Hightower, J., Borriello, G.: Particle filters for location estimation in ubiquitous computing: A case study. In: Mynatt, E.D., Siio, I. (eds.) UbiComp 2004. LNCS, vol. 3205, pp. 88–106. Springer, Heidelberg (2004)
7. Pavlidis, T.: Structural pattern recognition. Springer Series in Electrophysics, vol. 1. Springer, Berlin (1977)
8. Potamias, M., Patroumpas, K., Sellis, T.: Sampling trajectory streams with spatiotemporal criteria. In: 18th International Conference on Scientific and Statistical Database Management, pp. 275–284. IEEE (2006)
9. Ramer, U.: An iterative procedure for the polygonal approximation of plane curves. Computer Graphics and Image Processing 1(3), 244–256 (1972)
10. Turner, K.: Computer perception of curved objects using a television camera. Ph.D. thesis, University of Edinburgh (1974)
11. Zheng, Y., Chen, Y., Li, Q., Xie, X., Ma, W.Y.: Understanding transportation modes based on gps data for web applications. ACM Transactions on the Web (TWEB) 4(1), 1 (2010)
12. Zheng, Y., Xie, X., Ma, W.Y.: Geolife: A collaborative social networking service among user, location and trajectory. IEEE Data Eng. Bull. 33(2), 32–39 (2010)
13. Zheng, Y., Zhang, L., Xie, X., Ma, W.Y.: Mining interesting locations and travel sequences from gps trajectories. In: Proceedings of the 18th International Conference on World Wide Web, pp. 791–800. ACM (2009)

Optimal Path Planning for Information Based Localization

Francis Celeste and Frédéric Dambreville*

[1] Direction Générale de l'Armement
7-9 rue des Mathurins, 92221 Bagneux
`francis.celeste@intradef.gouv.fr`
[2] Lab-STICC UMR CNRS 6285, ENSTA Bretagne
2 rue Franois Verny, 29806 Brest Cedex 9, France
`submit@fredericdambreville.com`

Abstract. This paper addresses the problem of optimizing the navigation of an intelligent mobile in a real world environment, described by a map. The map is composed of features representing natural landmarks in the environment. The vehicle is equipped with a sensor which implies range and bearing measurements from observed landmarks. These measurements are correlated with the map to estimate the mobile localization through a *filtering algorithm*. The optimal trajectory can be designed by adjusting a measure of performance for the filtering algorithm used for the localization task. As the state of the mobile and the measurements provided by the sensors are random data, criterion based on the estimation of the *Posterior Cramer-Rao Bound* (PCRB) is a well-suited measure. A natural way for optimal path planning is to use this measure of performance within a (constrained) Markovian Decision Process framework and to use the *Dynamic Programming* method for optimizing the trajectory. However, due to the functional characteristics of the PCRB, Dynamic Programming method is generally irrelevant. We investigate two different approaches in order to provide a solution to this problem. The first one exploits the Dynamic Programming algorithm for generating feasible trajectories, and then uses Extreme Values Theory (EV) in order to extrapolate the optimum. The second one is a rare evnt simulation approach, the *Cross-Entropy* (CE) method introduced by Rubinstein & al. [9]. As a result of our implementation, the CE optimization is assessed by the estimated optimum derived from the EV.

Keywords: planning, cross entropy, estimation, PCRB, Extreme Value Theory.

1 Introduction

In this paper we are concerned with the task of finding a plan for a vehicle moving around in its environment. The goal is to reach a given target position

---

* Corresponding author.

T.V. Do et al. (eds.), *Advanced Computational Methods for Knowledge Engineering*,   377
Advances in Intelligent Systems and Computing 282,
DOI: 10.1007/978-3-319-06569-4_28, © Springer International Publishing Switzerland 2014

from an initial position. In many application, it is crucial to be able to estimate accurately the state of the mobile during its motion. In that case, it is necessary to take into consideration the execution stage in the planning activities. One way to achieve that objective is to propose trajectories where a level of performance of the localization algorithm should be guaranteed. This kind of problem has been well studied in the robotics community and in most approaches the first step is to define a model by discretizing the environment, and to define a graph whose nodes correspond to particular area and edges are actions to move from one place to another. Some contributions address the problem within a Markov Decision Process (MDP) modeling where a graph representation is also used for the mobile state space and provide theoretical framework to solve it in some cases.

In the present paper, we also consider the MDP framework as introduced in [10]. The reward parameter which is based on the Posterior Cramer-Rao bound metric. However, we remind that the nature of the objective function for the path planning makes it impossible to optimize by means of MDP. To solve the problem, we propose two steps. First we take advantage of the Dynamic Programming algorithm to approximate the probability distribution of the cost function over the admissible paths space. Then we perform an analysis of the tail of this distribution via Extreme Values Theory (EV). At this point we obtain an approximation of the optimum and measure how far are the best simulated path from the optimal path. Following the first step, we implement an optimization approach based on the Cross-Entropy method originally used for *rare-events* simulation.

In the second section we introduce the problem formalisms. Section 3 deals with the Posterior Cramèr Rao Bound and its properties. Section 4 introduce a Markov Decision Process framework for generating feasible trajectories and extreme values theory is use in order to extrapolate the optimum. Last section is dedicated to the application of the Cross-Entropy algorithm, and the two approaches are confronted.

2   Problem Statement and Notation

We tackle the problem of finding an optimal path for a mobile vehicle. The vehicle has a reference map $\mathcal{M}$ describing its environment. It is also equipped with sensors to collect local measurements from landmarks of $\mathcal{M}$ while moving. These measurements are "correlated" with the map to provide estimate of its state during the evolution. The optimality concern is to find one path which should guarantee a good level of performance of the localization algorithm used during the path execution. In this paper, the Bayesian estimation approach is taken into account as the noise models adopted in the system model are considered randomized.

Mobile Dynamic System Model. In the following, we make some assumptions on the mobile state evolution so that it can be described by a discrete-time dynamic model:

$$\mathbf{p}_{k+1} = f(\mathbf{p}_k, u_k) + \mathbf{e}_k \text{ and } \mathbf{p}_0 \sim \pi_0(\overline{\mathbf{p}}_0) . \tag{1}$$

where $f(.)$ is (in general) a nonlinear function of the state $\mathbf{p}_k$ and the input vector u_k at time t_k and $\{\mathbf{e}_k\}$ is a sequence of mutually independent random vectors. The error model of the initial state $\mathbf{p}_0$ is Gaussian defined by mean vector $\overline{\mathbf{p}}_0$ and a covariance matrix P_0. In our application, the state $\mathbf{p}_k$ and the input (or control) u_k are vectors in $\mathbf{R}^3$. $\mathbf{p}_k$ contains the position and global orientation of the mobile and u_k is composed of the velocity components and the angular variation at time k of the mobile.

$$\mathbf{p}_k \overset{\Delta}{=} [r_k^x, r_k^y, \theta_k]^T \text{ and } u_k \overset{\Delta}{=} [v_k^x \quad v_k^y \quad \gamma_k] .$$

More precisely, the state evolution process is:

$$
\begin{aligned}
r_{k+1}^x &= r_k^x + v_k^x \, \cos(\theta_k + \gamma_k) \, \delta_k \, + \, e_x^k \\
r_{k+1}^y &= r_k^y + v_k^y \, \sin(\theta_k + \gamma_k) \, \delta_k \, + \, e_y^k \\
\theta_{k+1} &= \theta_k + \gamma_k \, + \, e_\theta^k
\end{aligned}
\tag{2}
$$

where $\mathbf{e_k} \overset{\Delta}{=} \left[e_x^k, e_y^k, e_\theta^k\right]^T$ follows $\mathcal{N}(0, Q_k)$ with Q_k a 3×3 covariance matrix.

Map Model and Measurements. The vehicle is equipped with sensors which allow it to interact with its environment. While moving, the vehicle makes range and bearing measurements to features in the environment provided in the given known map $\mathcal{M}$. The map is composed of N_f landmarks $(m_i)_{1 \leq i \leq N_f}$ with respective position $(x_i, y_i) \in \mathbb{R}^2$. At time t_k, due to the sensor capabilities, only a few landmarks can be observed. So, the observation process depends on the number of visible features. Moreover, we do not consider data association and non detection problem for the moment. That is to say, each observation is made from one landmark represented in the map with $P_D = 1$. If we denote $I_v(p_k)$ the indexes of visible landmarks at time t_k, the measurement vector received at time t_k is the stacked vector $Z_k = [z^k(i_1) \, z^k(i_2) \, \cdots \, z^k(i_{n_k})]^T$ with n_k the number of visible landmarks at time k and $i_j \in I_v(p_k)$ for $j = 1 : n_k$:

$$
z^k(i_j) = h(\mathbf{p}_k, m_{i_j}) + \mathbf{w_k} \overset{\Delta}{=}
\begin{cases}
z_r^k(i_j) = \sqrt{(r_k^x - x_{i_j})^2 + (r_k^y - y_{i_j})^2} + w_r^k(i_j) , \\
z_\beta^k(i_j) = \arctan(\frac{y_{i_j} - r_k^y}{x_{i_j} - r_k^x}) - \theta_k + w_\beta^k(i_j) ,
\end{cases}
\tag{3}
$$

where $\{w_r^k(i_j)\}$ and $\{w_\beta^k(i_j)\}$ are considered as white Gaussian noises mutually independent with variances σ_{r,i_j}^2 and σ_{β,i_j}^2. The sensor visible area is supposed to be a sector of a corona (figure 1(a)) where the robot direction is its axis of symmetry.

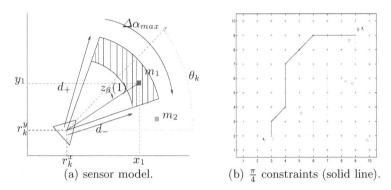

(a) sensor model.   (b) $\frac{\pi}{4}$ constraints (solid line).

Fig. 1. Observation model and Constrained move

Path Model for the Planning Task. To formalize the problem as a discrete feasible planning task, both the mobile continuous orientation space $]-\pi,\pi]$ and position space $X \subset \mathbf{R}^2$ are discretized as in [10]. So a grid is defined over $\mathcal{M}$. Let us denote by N_x and N_y the number of points griding the map along each axis of $\mathcal{R}_0$. So the total number of points griding the map is $N_s = N_x N_y$. For each point $\tilde{p}(i,j) = (x(i,j), y(i,j))$ on that grid, we define a unique auxiliary state $s = g(i,j) \in \mathcal{S} \triangleq \{1, \cdots, N_s\}$, typically $g(i,j) = i + (j-1)N_x$. The mobile orientation is also supposed to be restricted to eight main directions which corresponds to eight actions or decisions $a \in \mathcal{A} \triangleq \{1, \cdots, N_a\}$ (with $N_a = 8$). Therefore from each point of the grid, there are at most 8 reachable neighbors. The actions are clockwise numbered from 1 (”go up and right”) to 8 (”go up”). The discretization approach allows us to search paths which are sequences of displacements with constant headings and constant velocity:

$$u_k = u(a_k) = \begin{bmatrix} v_k^x & v_k^y & \gamma_k \end{bmatrix} = \left[v \, \mathrm{sgn}(\cos \gamma_k), v \, \mathrm{sgn}(\sin \gamma_k), (2-a)\frac{\pi}{4} \right], \quad (4)$$

where $\mathrm{sgn}(\omega) = \omega/|\omega|$ if $\omega \neq 0$, $= 0$ else. Then given trajectory defined by $\tilde{p}_{1:k}$ is equivalent to the sequence $s_{1:k} = \{s_1, \cdots, s_k\}$ and $a_{1:k-1} = \{a_1, \cdots, a_{k-1}\}$. In that way, a graph structure can be used to describe our problem, where the vertices and edges are respectively the states s and the actions a. Our problem is thus a *sequence of decisions* [7].

We also make the assumptions that the starting p_i and goal p_f (or more generally the goal set X_f) positions are on the grid and corresponds to states s_i and s_f (if not the nearest positions on the grid are used). Moreover state $s = 0$ is added to S, in order to take into account actions which cause a move outside the map or toward states inside possible obstacles. In practice, some constraints on the control input u_k must be considered. For example, the heading variation may be bounded which reduces the number of attainable state at the current time. So, at each time k, the chosen decision depends from the previous one taken at time $k-1$. This can be done by defining a specific $\delta(a_k, a_{k-1})$ *authorized transition matrix* which indicates actions that can be chosen at time k according to the choice made at time $k - 1$. For example, if we assume that the heading control is bounded by $\frac{\pi}{4}$ that is $|\gamma_k| \le \frac{\pi}{4}$, such a matrix can be expressed as below :

$$\delta(a_k, a_{k-1}) = \begin{cases} 1 \text{ if } a_k - a_{k-1} \in \{-1, 0, 1\} + 8\,\mathbb{Z}, \\ 0 \text{ else.} \end{cases} \tag{5}$$

Figure 1(b) shows two trajectories. The headings constraint is satisfied for the first but not for the second.

3 Estimation Process and Posterior Cramer Rao Bound

The main objective of our problem is to find trajectories which give good estimation performance of the mobile position $\tilde{p}_k$. The estimation can be achieved by using probabilistic filtering algorithms like extended Kalman or particle filtering. The Posterior Cramèr Rao Bound (PCRB) is a convenient way for evaluating the optimal performance of filtering methods. The PCRB is given by the inverse of the Fisher Information Matrix (FIM) [5,1], and it provides an accurate estimate and lower bound of the minimum mean square errors of the estimate.

Tichavsky PCRB Recursion. For filtering algorithm applied to system modeled by equations 1 and 3, the FIM $F_k(\mathbf{p}_{0:k}, \mathbf{u}_{0:k-1}, Z_{1:k})$ along the the whole trajectory $\mathbf{p}_{0:k}$ up to time k with the control sequence $\mathbf{u}_{0:k-1}$ and $Z_{1:k}$ is given by :

$$F_k(\mathbf{p}_{0:k}, \mathbf{u}_{0:k-1}, Z_{1:k}) \triangleq E\left[-\Delta^{\mathbf{p}_{0:k}}_{\mathbf{p}_{0:k}} \log\ (p(\mathbf{p}_{0:k}, \mathbf{u}_{0:k-1}, Z_{1:k}))\right], \tag{6}$$

where $p(\mathbf{p}_{0:k}, \mathbf{u}_{0:k-1}, Z_{1:k})$ is the joint probability of states and measures in regards to the control trajectory. The FIM is a $((k + 1)r \times ((k + 1)r$ matrix where r is the state dimension. It sets a lower limit for the covariance of the estimate $\hat{p}_{0:k}(\mathbf{u}_{0:k-1}, Z_{1:k})$ of the whole state history up to time k. It can be shown [12] that :

$$E\{(\hat{p}_k(\mathbf{u}_{0:k-1}, Z_{1:k}) - \mathbf{p}_k)\ (\hat{p}_k(\mathbf{u}_{0:k-1}, Z_{1:k}) - \mathbf{p}_k)^T\} \succeq J_k^{-1} \tag{7}$$

where J_k^{-1} is the $(r \times r)$ bottom-right block of $F_k(\mathbf{p}_{0:k}, \mathbf{u}_{0:k-1}, Z_{1:k})$.

A remarkable contribution of Tichasky et al. [12] was to introduce a Ricatti like recursion for computing J_k. In our problem, this recursion takes the form:

$$J_{k+1} = D_k^{22} - D_k^{21}(J_k + D_k^{11})^{-1}D_k^{12}, \tag{8}$$

where:

$$D_k^{11} = E_{\mathbf{p}_k}\left\{[\nabla_{\mathbf{p}_k} f^T(\mathbf{p}_k)]\,Q_k^{-1}\,[\nabla_{\mathbf{p}_k} f^T(\mathbf{p}_k)]^T\right\}, \tag{9}$$

$$D_k^{12} = [D_k^{21}]^T = -E_{\mathbf{p}_k}\left\{[\nabla_{\mathbf{p}_k} f^T(\mathbf{p}_k)]\right\}Q_k^{-1},$$

$$D_k^{22} = Q_k^{-1} + J_{k+1}(Z)\,,\,J_0 = E\left\{\Delta^{p_0}_{p_0} \log\ (\pi_0)\right\}.$$

The *observations contribution* $J_{k+1}(Z)$ is given by the following expression :

$$J_k(Z) = E_{\mathbf{p}_k}\{ \sum_{j \in I_v(\mathbf{p}_k)} H(\mathbf{p}_k, j)^T R_k^{-1} H(\mathbf{p}_k, j) \}, \tag{10}$$

where $H(\mathbf{p}_k, j) \triangleq \nabla_{p_k} h(\mathbf{p}_k, m_j)$ and $R_k = diag[\sigma_{r,i_j}^2, \sigma_{\beta,i_j}^2]$. Obviously, there is no explicit expression for D_k^{11}, D_k^{12} and $J_k(Z)$ because the system dynamic models are nonlinear. In practice, it is necessary to resort to Monte Carlo simulation to approximate D_k^{11}, D_k^{12} and $J_k(Z)$. This is done by sampling several real trajectories, on the basis of the planned theoretical trajectories $(s_{1:k}, a_{1:k-1})$, by setting $u_{1:k} = u(a_{1:k-1})$ as in equation (4). Figure 2(a) illustrates the generation of random realizations for the computation of PCRB. For the two paths the PCRB history are evaluated. The lower bound of the covariance position estimate along both paths can then be deduced by using a projection on the position subspace, denoted $\mathcal{B}$ (figure 2(b)).

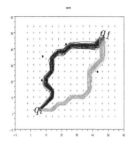
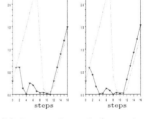

(a) Random paths realizations.

(b) Lower bound (x and y) of error estimate for the position.

Fig. 2. Monte Carlo computation of the PCRB

Criterion for Path Planning. We will optimize a tradeoff on the trajectory localization error:

$$\mathbf{u}_{1:K-1}^{\diamond} \in arg \max_{\mathbf{u}_{1:K-1}} \phi_2(J_{1:K}(\mathbf{u}_{1:K-1})) ,$$

$$\text{where: } \phi_2(J_{1:K}(\mathbf{u}_{1:K-1})) = - \sum_{k=1}^{K} w_k \, \det(\mathcal{B}^T J_k^{-1}(\mathbf{u}_{1:K-1})\mathcal{B}) . \tag{11}$$

A particular case is to optimize the final localization:

$$\mathbf{u}_{1:K-1}^{*} \in arg \max_{\mathbf{u}_{1:K-1}} \phi_1(J_{1:K}(\mathbf{u}_{1:K-1})) ,$$

$$\text{where: } \phi_1(J_{1:K}(\mathbf{u}_{1:K-1})) = - \det(\mathcal{B}^T J_K^{-1}(\mathbf{u}_{1:K-1}\mathcal{B}) . \tag{12}$$

It is recalled that $\mathcal{B}$ is the projection onto the position subspace.

4   A Stochastic Approach Based on the MDP Framework

Generating Valid Trajectories. The structure of the mobile dynamic is obviously related to the Markov Decision Process (MDP) framework. In the MDP context when the transition function and the *local* reward are completely known, the optimal policy can be computed using Dynamic Programming (DP) technique or similar algorithms such as Value Iteration [11]. These approach are based on the principle of optimality due to Bellman [2]. In our problem, the plan must take into account operational constraints on control input. As a consequence, the original MDP algorithm must be extended like in [10], so as to take into account constraints like heading control bounds. However, the implementation of the DP with cost functional of the PCRB is quite restricted as shown in [6]. The cost functional of the PCRB must satisfy the "Matrix Dynamic Programming Property". For most general functional such as the "determinant used" in our work, this property is not verified [6]. Dynamic Programming approach cannot be used as a direct solution to our optimization.

On the other hand, it is perfectly possible to use the DP for generating random valid trajectories, and evaluate them by means of the PCRB:

1. Generate a random cost series $(c_{s's}(a))$ by means of the uniform distribution $\mathcal{U}_{[0,1]}$ on $[0,1]$,
2. Compute by dynamic programming the path τ, which minimize the cumulative sampled cost,
3. Compute the PCRB along τ and derive $\phi_i(\tau)$.

This sampling process may be used in order to find rough suboptimal solution to the mobile trajectory in regard to the criterion $\phi_i(\tau)$. We will see that the application of Extreme Value (EV) theory makes possible an extrapolation of the law of the performance around the optimal trajectories. By the way, this provides an interesting way to assess the performance of an optimization.

Extrapolate the Optimum Using Extreme Value Analysis. Let $X_{1:N}$ be a sample of a random variable X drawn from an unknown continuous cumulative Density Function (CDF) F having finite and unknown right endpoint $\omega(F)$. The EVT tries to model the behavior mainly outside the range of these available observations in the (right) tail of F [4]. Let u be a certain threshold and F_u the distribution of exceedances, i.e. of the values of x above u:

$$F_u(y) \triangleq Pr(X - u \leq y | X > u), \qquad 0 \leq y \leq \omega(F) - u \qquad (13)$$
$$= \frac{F(u+y) - F(u)}{1 - F(u)} = Pr(u < x \leq u + y) .$$

We are mainly interested in the estimation of F_u. A great part of the main results of the EVT is based on the following fundamental theorem:

Theorem 1. *[8] For a large class of underlying distributions F the exceeding distribution F_u, for u enough large, is well approximated by*

$$F_u(y) \approx G_{\xi,\sigma}(y) , \quad for \ 0 \leq y \leq \omega(F) - u ,$$

where $G_{\xi,\sigma}$ is the so-called Generalized Pareto Distribution (GPD) defined as:

$$G_{\xi,\sigma}(y) = \begin{cases} 1 - (1 + \frac{\xi}{\sigma}y)^{-\frac{1}{\xi}} & \text{if } \xi \neq 0 , \\ 1 - e^{y/\sigma} & \text{if } \xi = 0 . \end{cases} \tag{14}$$

At this point, this theorem indicates that, under some particular conditions on F, the distribution of exceeding observations can be approximated by a parametric model with two parameters. Inference can be made from this model to make an analysis of the behaviour of F around its maximum endpoints. Hence the $p - th$ quantile x_p, more often called the Value at Risk, can be evaluated. It corresponds to the value of x which is exceeded probability p, that is $1 - p = F(x_p)$. The analytical expression of x_p can be derived from equations 13 and 14:

$$x_p = u + \frac{\sigma}{\xi} \left[\left(\frac{N}{N_u} p \right)^{-\xi} - 1 \right] , \tag{15}$$

where N_u is the number of observations above the threshold u or exceedances. The GPD parameters $\zeta \triangleq (\xi, \sigma)$ can be estimated from the Likelihood, which is defined as :

$$L(\xi, \sigma) = \sum_{i_l \in L} \log \frac{\partial G_{\xi,\sigma}}{\partial y} (y_{i_l}) ,$$

where $(x_{i_l})_{i_l \in L}$ with $card(L) = N_u$ are the observations above the threshold and $(y_{i_l})_{i_l \in L}$ the associated exceedances. For example, a $1 - \alpha$ confidence interval for the estimate $\hat{\xi}$ is given by values of ξ satisfying:

$$CI(\hat{\xi}) = \left\{ \xi | - 2 \left[\max_{\sigma} L(\xi, \sigma) - L(\hat{\xi}, \hat{\sigma}) \right] \leq \chi^2_{1,\alpha} \right\} ,$$

where $\chi^2_{1,\alpha}$ is the $1-\alpha$ quantile of the χ^2 distribution with one degree of freedom. From equation (15), it is also possible to deduce a $1-\alpha$ confidence interval $CI(\hat{x}_p)$ for $\hat{x}_p$.

Experiment. To illustrate the above method, we consider two different experiments. The map is defined on the set $[-4, 54] \times [-4, 54]$ and the grid resolution is $d_x = d_y = 4$, so the dimension of the discretized state space $\mathcal{S}$ is $N_s = 225$. In those scenarios, no obstacles are present. There are 4 landmarks in the a priori map and the model noises are made identical:

	m_0	m_1	m_2	m_3
x	7.2	7.8	22.8	43.1
y	20.4	35.2	42.8	25.8

$$P_0 = Q_k = \begin{pmatrix} 1 & 0 & 0 \\ 0 & 1 & 0 \\ 0 & 0 & 0.5 \end{pmatrix}, \quad R_k = \begin{pmatrix} 1 & 0 \\ 0 & 0.5 \end{pmatrix} ,$$

Landmarks.   Noises.

The variances on orientation and bearings measurements are expressed in degree here for clarity but have to be converted in radian. The constant velocity at each step is $4\,m.s^{-1}$ and the mobile can only apply $\{-\frac{\pi}{4}, 0, \frac{\pi}{4}\}$ between two consecutive times so the $\delta(.,.)$ matrix is exactly the same as defined in 2. For the observation model, a landmark m_j is visible at time k provided that:

$$r_- \leq z_r^k(j) \leq r_+ \qquad \text{and} \qquad |z_\beta^k(j)| \leq \theta_{max} \ .$$

where $r_- = 0.01$ m., $r_+ = 12$ m. ($3 \times$ grid resolution) and $\theta_{max} = 90$ deg. We consider trajectories with at most $T_{max} = 30$ elementary moves and the initial and final positions are respectively $\mathbf{p}_i = (6; 2; 45)$ and $\tilde{p}_f = (46; 46)$. For the PCRB estimate computation from Monte Carlo simulation, $N_{mc} = 800$ noisy realizations of the trajectories are used. For the extreme values analysis, we generate 40000 paths and evaluate the cost function ϕ_2 for each one. We considered 2% of the sample to make inference and estimate the parameters $(\hat{\xi}, \hat{\sigma})$ of the GPD distribution, that is to say the 800 best plans are considered. The criterion is normalized:

$$c_2^j = \phi_2^j - m[\phi_2]/\sigma[\phi_2] \ .$$

Figure 3(a) shows the best paths among the 20000 runs. The associated cost is $c_2^{max} \approx 0.8815$. In figure 3(b) we plotted the histogram of the 800 extremes values of c_2. The threshold corresponding to 2% is $u = 0.8479$. the estimation of the quantile x_p and its related confidence interval can be calculated. For $p = 2.10^{-5}$, we found $\hat{x}_p = 0.8804$ and the 95% confidence interval given by the profile likelihood and delta methods are respectively $[0.879; 0.8829]$ and $[0.8758; 0.8849]$. The values of $1 - F_u(x)$ are representing on figure (3(c)) with $\hat{x}_p$ and the both confidence intervals.

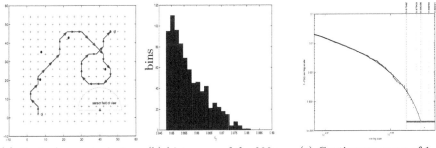

(a) Best drawn trajectory among the 40000.

(b) histogram of the 800 extremes values for c_2.

(c) Continuous curve of $1 - G_{\hat{\xi},\hat{\sigma}}$ fitting the 800 extreme values (dots). $\hat{x}_p$ (red) with its 95% intervals of confidence based on Profile Likelihood (green).

Fig. 3. Best trajectory and extremes values

5   The Cross Entropy Approach

The previous approach does not exploit the possibility to tune the sampling process so as to sample more interesting trajectories. Rare event simulation algorithms, like the Cross Entropy algorithm used here [9,3], implement this idea. In the context of our application, the sampling of a trajectory is derived from a local sampler:

$$(s, a) \mapsto P^t(a|s) \text{ is the local sampler at step } t .$$

Of course, the generation of a trajectory has to take into account the various constraints of the mobile dynamic. A trajectory $\tau = (s_i, a_0, s_1, \cdots, s_{K-1}, a_{K-1}, s_K)$ is generated as follows:

1. Set $k = 0$ and generate a_0 by means of $P(a|s_i)$,
2. Compute s_{k+1} from s_k and a_k,
3. If s_{k+1} is a forbidden state or $k \geq T_{max}$, then **stop** and set $K = k$,
4. Generate a_{k+1} by means of $P(a|s_{k+1}, \delta(a, a_k) = 1)$ and repeat from 2.

In general, $s_K \neq s_f$, so that this trajectory generation process does not satisfies the arrival point constraint. In order to handle this constraint, a penalty function is added to the evaluation criterion:

$$\phi_f(\tau, s_K, s_f) = -\mu_f \ f(dist(s_K, s_f)) . \tag{16}$$

Being given a selective rate ρ and a smoothing parameter α, the CE simulation for our example is as follows:

1. Initialize $P^0(\,|\,)$ to a flat law and $sett = 0$
2. Generate trajectories $\tau_{1:N}^t$ by means of $P^t(\,|\,)$ and compute the evaluations $(\phi_i + \phi_f)(\tau_{1:N}^t)$,
3. Compute γ_t as the $(1 - \rho)$-quantile of $(\phi_i + \phi_f)(\tau_{1:N}^t)$,
4. Define $P^{t+1}(\,|\,) = \alpha P^t(\,|\,) + (1 - \alpha)P^{t+1|t}(\,|\,)$ where $P^{t+1|t}(\,|\,)$ is learned from the samples of the quantile:

$$P^{t+1|t}(a|s) \propto \sum_{n=1:N} I[(\phi_i + \phi_f)(\tau_n^t) > \gamma_t] \times I[\tau_n^t \text{ contains sequence } sa] , \tag{17}$$

5. Set $t \leftarrow t + 1$ and repeat from step 2 until convergence.

Experiments. We consider here the same scenario as in section 4 and apply the approach based on the cross entropy algorithm. The PCRB is approximated by means of $N_{mc} = 800$ samples. At each step of the recursion, the quantile cost $\gamma(t)$ and the cost $c_2^{max}(t)$ of the best drawn path is saved. To compare with the results obtained with the MDP framework, we normalized the cost with the same mean $m[c_2]$ and standard deviation $\sigma[c_2]$ computed by means of EV method.

For the CE algorithm 4000 trajectories are sampled at each iteration of the algorithm. The selection rate is $\rho = 0.1$, and the 400 best paths contributes

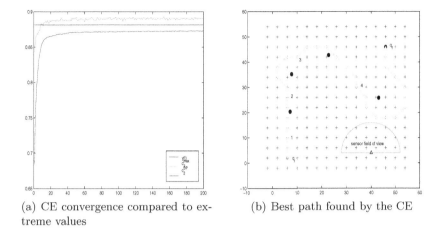

(a) CE convergence compared to extreme values

(b) Best path found by the CE

Fig. 4. CE convergence compared to MDP&EV

to the computation of $P^{t+1|t}(\,|\,)$. The smoothing parameter is $\alpha = 0.4$. Figure 4(a) shows the evolution of γ and $c_2^{max}(t)$ for the performance function ϕ_2. The best cost obtained with the MDP&EV approach is also plotted. The convergence is fast. Indeed the $\gamma(t)$ and $c_2^{max}(t)$ become stable around 20 steps of the CE. Moreover, the CE outperforms the MDP&EV approach. The best trajectory at the last iteration of the CE is represented in figure (4(b)). Its associated cost is 0.8903 whereas the high estimated quantile for $p = 2^{-5}$ is 0.8804, and is better the approximated reachable cost obtained from the MDP&EV. It is also noticed that the final sampler $P(\cdot|\cdot)$ is multi-modal (it is not a Dirac) for 4 states among 31, say $s_{1:4}$, and offers equivalent choices:

$$P(1|s_1) = 1 - P(8|s_1) = 0.37 \qquad P(7|s_2) = 1 - P(8|s_2) = 0.22 , \qquad (18)$$
$$P(1|s_3) = 1 - P(2|s_3) = 0.71 \qquad P(3|s_4) = 1 - P(4|s_4) = 0.48 . \qquad (19)$$

6   Conclusion and Perspectives

In this paper, we present an algorithmic solution to a path planning task which aimed to guarantee the performance of its pose estimation. We are concerned with localization given an a priori metric map of the environment composed of landmarks. Cost criterion based on the Posterior Cramèr Rao Bounds is derived from the system dynamic and the measurement models, so as to evaluate the performance of localization on the trajectory. The optimization problem cannot be solved by means of Dynamic Programming (DP) approaches. However, we use DP for simulating feasible trajectories with the perspective to extrapolate the planning optimum by implementing an Extreme Value (EV) estimation. In the last section of the paper, we propose an optimization of the planning based on the Cross Entropy (CE) algorithm. The CE produced solutions which are comparable to the EV estimation of the optimum.

This work is an example of convergence assessment based on the EV estimation. Of course, our implementation of the EV is cannot be used in practice as a criterion for ending the CE process, since it is based at this time on a large population simulation. However, the samples generated by the importance sampling of the cross entropy may be used as an entry for the extreme value estimation. Thus, we intend to combine CE and EV so as to have a reasonable criterion in the CE process for evaluating the convergence.

Acknowledgement. We express our appreciation to the late Jean-Pierre Le Cadre, whose contribution to this work was of essential significance.

References

1. Bergman, N.: Recursive bayesian estimation: navigation and tracking applications, Ph.D Thesis (1999)
2. Bertsekas, D.: Dynamic Programming. Wiley, New York (1968)
3. de Boer, P., Kroese, D., Mannor, Rubinstein, R.: A tutorial on the cross-entropy method (2003), http://www.cs.utwente.nl/~ptdeboer/ce/
4. Coles, S.: Extreme value theory and applications (June 15, 1999)
5. Van Trees, H.L.: Detection, Estimation and Modulation Theory. Wiley, New York (1968)
6. Le Cadre, J.P., Trémois, O.: The matrix programming property and its applications. SIAM J. Matrix Anal. Appl. 18(4), 818–826 (1997)
7. LaValle, S.: Planning Algorithms. Cambridge University Press (2006)
8. Pickands, J.: Statistical Inference Using Extreme Order Statistics. Annals of Statistics 3, 119–131 (1975)
9. Rubinstein, R.Y., Kroese, D.: The Cross-Entropy Method: A uniform approach for Combinatorial Optimization, Monte-Carlo Simulation and Machine Learning. Information Science and Statistics (2004)
10. Paris, S., Le Cadre, J.P.: Trajectory planning for terrain-aided navigation. In: Fusion 2002, vol. 7-11, pp. 1007–1014 (2002)
11. Sutton, R., Barto, A.: Reinforcement Learning. An Introduction. A Bradford Book (2000)
12. Tichavsky, P., Muravchik, C., Nehorai, A.: Posterior cramér-rao bounds for discrete-time nonlinear filtering. IEEE Transactions on Signal Processing 46(5), 1386–1396 (1998)

Formal Security Verification of Transport Protocols for Wireless Sensor Networks

Vinh-Thong Ta[1], Amit Dvir[4], and Levente Buttyán[2,3]

[1] INRIA, CITI/INSA-Lyon, F-69621, Villeurbanne, France
[2] Laboratory of Cryptography and System Security (CrySyS), BME, Hungary
[3] MTA-BME Information Systems Research Group,
Magyar tudósok körútja 2, 1117 Budapest, Hungary
[4] Computer Science School, The College of Management - Academic Studies, Israel
`vinh-thong.ta@inria.fr, azdvir@gmail.com, buttyan@crysys.hu`

Abstract. In this paper, we address the problem of formal security verification of transport protocols for wireless sensor networks (WSN) that perform cryptographic operations. Analyzing this class of protocols is a difficult task because they typically consist of complex behavioral characteristics, such as launching timers, performing probabilistic behavior, and cryptographic operations. Some of the recently published WSN transport protocols are DTSN, which does not include cryptographic security mechanism, and two of its secured versions, SDTP and STWSN[1]. In our previous work, we formally analyzed the security of Distributed Transport for Sensor Networks (DTSN) and Distributed Transport Protocol for Wireless Sensor Networks (SDTP), and showed that they are vulnerable against packet modification attacks. In another work we proposed a new Secure Transport Protocol for WSNs (STWSN), with the goal of eliminating the vulnerability of DTSN and SDTP, however, its security properties have only been informally argued. In this paper, we apply formal method to analyze the security of STWSN.

1 Introduction

Wireless Sensor Networks [2] consist of a large number of resource constrained sensor nodes and a few more powerful base stations. The sensors collect various types of data from the environment and send those data to the base stations using multi-hop wireless communications. Some typical applications that require the use of a transport protocol for ensuring reliable delivery and congestion control are: reliable control and management of sensor networks; remotely programming/retasking sensor nodes over-the-air. It is widely accepted that transport protocols used in wired networks (e.g., the well-known TCP) are not applicable in WSNs, because they perform poorly in a wireless environment and they are

---

[1] STWSN and SDTP$^+$ are the same. SDTP$^+$ is the protocol name used in [1], however, because it is not only a minor extension of SDTP, but is based on a completely different security solutions, it is better to change the name to STWSN.

T.V. Do et al. (eds.), *Advanced Computational Methods for Knowledge Engineering,* 389
Advances in Intelligent Systems and Computing 282,
DOI: 10.1007/978-3-319-06569-4_29, © Springer International Publishing Switzerland 2014

not optimized for energy consumption. Therefore, a number of transport protocols specifically designed for WSNs have been proposed in the literature (see, e.g., [3] for a survey). The main design goal of those transport protocols is to achieve reliability and energy efficiency. However, despite the fact that WSNs are often envisioned to operate in hostile environments, existing transport protocols for WSNs do not address security issues at all and, as a consequence, they ensure reliability and energy efficiency only in a benign environment where no intentional attack takes place [4].

Attacks against WSN transport protocols can be attacks against reliability and energy depleting attacks. In the first case, the attackers can cause undetected or permanent packet lost, while in the second case the attackers make the sensor nodes deplete their battery by performing futile computations. Some of the recently published WSN transport protocols are DTSN [5], and two of its security extensions, SDTP [6] and STWSN [1]. Many tricky attack scenarios have been found against WSN transport protocols, and even secure WSN transport protocols that were believed to be secure, have turned out to be vulnerable. The main reason is that the designers reason about the security of their protocol based only on manual analysis, which is error-prone. DTSN and SDTP have been analyzed based on formal method in [7], and were shown to be vulnerable against packet modification attacks.

STWSN is designed with the goal of eliminating the vulnerability of DTSN and SDTP, however, its security properties have been argued only informally. Our goal is to apply formal method to analyze WSN transport protocols, which provide a more reliable and systematic way of security proof. In this paper, we apply formal method to analyze the security of WSN transport protocol. In particular, using the process algebra language, $crypt_{time}^{prob}$, proposed in [7], we showed that the STWSN protocol is secure against packet modification attacks.

2   Related Works

We provide a brief overview of the DTSN and SDTP protocols, which are two recent and representative WSN transport protocols, and they are closely related to the STWSN protocol.

DTSN [5] is a reliable transport protocol developed for sensor networks where intermediate nodes between the source and the destination of a data flow cache data packets in a probabilistic manner such that they can retransmit them upon request. The main advantages of DTSN compared to the end-to-end retransmission mechanism is that it allows intermediate nodes to cache and retransmit data packets, hence, the average number of hops a retransmitted data packet must travel is smaller than the length of the route between the source and the destination. Intermediate nodes do not store all packets but only store packets with some probability p, which makes it more efficient. DTSN uses positive acknowledgements ($ACKs$), and negative acknowledgements ($NACKs$) to control caching and retransmissions. An ACK refers to a data packet sequence number n, and it should be interpreted such that all data packets with sequence number smaller than or equal to n were received by the destination. A $NACK$ refers to a base sequence number n and it also contains a bitmap, in which each bit represents a different sequence number starting from the base sequence number n. A $NACK$ should be interpreted such that all data packets with sequence number smaller than or equal to n were received by the destination and the data packets corresponding to the set bits in the bitmap are missing.

Reasoning about the security of DTSN: Upon receiving an ACK packet, intermediate nodes delete from their cache the stored messages whose sequence number is less than or equal to the sequence number in the ACK packet, because the intermediate nodes believe that acknowledged packets have been delivered successfully. Therefore, an attacker may cause permanent loss of some data packets by forging or altering ACK packets. This may put the reliability service provided by the protocol in danger. Moreover, an attacker can trigger unnecessary retransmission of the corresponding data packets by either setting bits in the bit map of the $NACK$ packets or forging/altering $NACK$ packets.

SDTP [6] is a security extension of DTSN aiming at patching the security holes in DTSN. SDTP protects data packets with MACs (Message Authentication Code) computed over the whole packet. Each MAC is computed using a per-packet key correspond to each packet. The intermediate and source nodes cache packets along with their MACs, and whenever the destination wants to acknowledge the first n packets it sends an ACK message with the key for the MAC corresponding to this packet. Similarly, when the destination request the retransmission of packets it sends a $NACK$ message with the keys corresponding the given packets. The rationale behind the security of SDTP is that only the source and destination knows the correct keys to be revealed.

Reasoning about the security of SDTP: The main security weakness of the SDTP protocol is that the intermediate nodes store the received data packets without any verification. Intermediate nodes do not verify the origin and the authenticity of the data packets or the ACK and the $NACK$ messages, namely, they cannot be sure whether the data packets that they stored were sent by the source node, and the control messages were really sent by the destination. Indeed, the security solution of SDTP only enables intermediate nodes to verify the matching or correspondence of the stored packets and the revealed $ACK/NACK$ keys. Hence, SDTP can be vulnerable in case of more than one attacker node (compromised node) who can cooperate.

In [7] we formally proved that an attacker may cause permanent loss of some data packets both in DTSN and STDP, by forging or altering the sequence number in ACK message. We also showed that an attacker can trigger futile retransmission of the corresponding data packets by either setting bits in the bit map of the $NACK$ packets or forging/altering $NACK$ packets. The first case violates the reliability while the second case violates the energy efficiency requirement. In [1] we proposed a new secured WSN transport protocol, called STWSN, and argued its security. However, the security of STWSN has not been analyzed based on formal method so far. In this paper, we perform formal security analysis of STWSN and prove that it is secure against the successful attack scenarios in case of DTSN and SDTP.

3 STWSN – A Secure Distributed Transport Protocol for WSNs

We proposed STWSN in [1], in order to patch the security weaknesses can be found in DTSN and SDTP. STWSN aims at authenticating and protecting the integrity of control packets, and is based on an efficient application of digital signature and authentication values, which are new compared to SDTP. The security mechanism of STWSN is based on the application of Merkle-tree [8] and hash chain [9], which have been used for designing different security protocols such as Castor [10], a scalable secure routing protocols for ad-hoc networks, and Ariadne [11]. Our contribution is applying Merkle-tree and hash chain in a new context. The general idea of STWSN is the following: two types of "per-packet" *authentication values* are used, ACK and $NACK$ authentication values. The ACK authentication value is used to verify the ACK packet by any intermediate node and the source, whilst the $NACK$ authentication value is used to verify the $NACK$ packet by any intermediate node and the source. The ACK authentication value is an element of a hash chain [9], whilst the $NACK$ authentication value is a leaf and its corresponding sibling nodes along the path from the leaf to the root in a Merkle-tree [8]. Each data packet is extended with one Message Authentication Code (MAC) value (the MAC function is HMAC), instead of two MACs as in SDTP. STWSN adopt the notion and notations of the pre-shared secret S, ACK, and $NACK$ master secrets K_{ACK}, K_{NACK}, which are defined and computed in exactly the same way as in SDTP [6]. However, in STWSN the generation and management of the per-packet keys $K_{ACK}^{(n)}$, $K_{NACK}^{(n)}$ is based on the application of hash-chain and Merkle-trees, which is different from SDTP.

3.1 The ACK Authentication Values

The ACK authentication values are defined to verify the authenticity and the origin of ACK messages. The number of data packets that the source wants to send in a given session, denoted by m, is assumed to be available. At the beginning of each session, the source generates the ACK master secret K_{ACK} and calculates a hash chain of size $(m+1)$ by hashing K_{ACK} $(m+1)$ times, which is illustrated in Figure 1. Each element of the calculated hash-chain represents a *per packet ACK authentication value* as follows: $K_{ACK}^{(m)}, K_{ACK}^{(m-1)} ..., K_{ACK}^{(1)}, K_{ACK}^{(0)}$, where $K_{ACK}^{(i)} = h(K_{ACK}^{(i+1)})$ and h is a one-way hash function. The value $K_{ACK}^{(0)}$ is the root of the hash-chain, and $K_{ACK}^{(i)}$ represents the ACK authentication value corresponding to the packet with sequence number i. When the destination wants to acknowledge the successful delivery of the i-th data packet, it reveals the corresponding $K_{ACK}^{(i)}$ in the ACK packet.

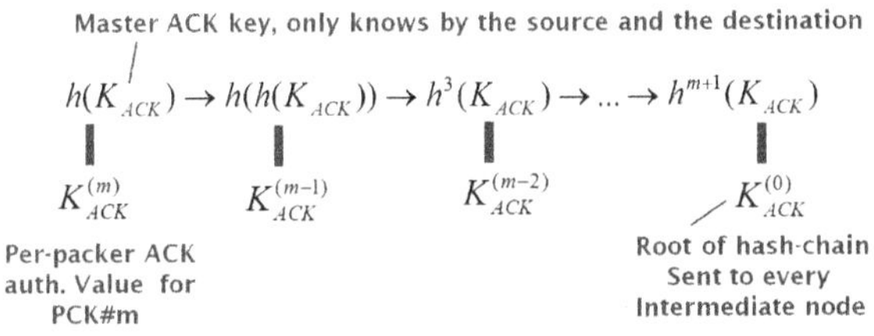

Master ACK key, only knows by the source and the destination

$$h(K_{ACK}) \rightarrow h(h(K_{ACK})) \rightarrow h^3(K_{ACK}) \rightarrow \dots \rightarrow h^{m+1}(K_{ACK})$$

$K_{ACK}^{(m)}$ $K_{ACK}^{(m-1)}$ $K_{ACK}^{(m-2)}$ $K_{ACK}^{(0)}$

Per-packer ACK
auth. Value for
PCK#m

Root of hash-chain
Sent to every
Intermediate node

Fig. 1. *The element $K_{ACK}^{(i)}$, $i \in \{1, \dots, m\}$, of the hash-chain is used for authenticating the packet with the sequence number i. The root of the hash-chain, $K_{ACK}^{(0)}$, which we get after hashing $(m+1)$ times the ACK master key K_{ACK}. This root is sent to every intermediate node in the open session packet, which is digitally signed by the source.*

3.2   The *NACK* Authentication Values

For authenticating the *NACK* packets, STWSN applies a Merkle-tree (also known as hash-tree), which is illustrated in Figure 2. When a session has started, the source computes, the *NACK* per-packet keys $K_{NACK}^{(n)}$ for each packet to be sent in a given session. Afterwards, these *NACK* per-packet keys are hashed and assigned to the leaves of the Merkle-tree: $K_{NACK}^{'(n)} = h(K_{NACK}^{(n)})$. The internal nodes of the Merkle-tree are computed as the hash of the (ordered) concatenation of its children. The root of the Merkle-tree, $H(h_p, S_p)$, is sent by the source to intermediate nodes in the same open session packet that includes the root of the hash-chain. For each $K_{NACK}^{(j)}$, $j \in \{j1, \dots, jm\}$, the so called sibling values $S_1^j, \dots, S_t^j$, for some t, are defined such that the root of the Merkle-tree can be computed from them. For instance, the sibling values of $K_{NACK}^{(j1)}$ are $K_{NACK}^{(j2)}$, $S_1, \dots, S_p$. From these values $H(h_p, S_p)$ can be computed.

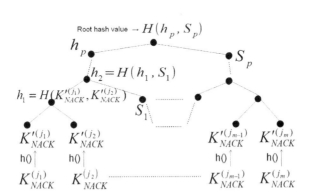

Fig. 2. *The structure of Merkle-tree used in STWSN. Each internal node is computed as the hash of the ordered concatenation of its children. The root of the tree, $H(h_p, S_p)$, is sent out by the source.*

3.3   The Operation of STWSN

In this section a short flow of the STWSN protocol is given, for more information please refer to [1]. When a session is opened, first, the source computes the ACK and $NACK$ master keys K_{ACK} and K_{NACK}, respectively. Then, the source calculates the hash-chain and the Merkle-tree for the session. Afterward, the source sends an open session message with the following parameters: the roots of the hash chain $(K_{ACK}^{(0)})$ and of the Merkle-tree $(H(h_p, S_p))$, the length of the hash chain $(m + 1)$, the session $SessionID$, the source and destination IDs. Before sending the open session packet, the source digitally signs it to prevent the attackers from sending fake open session packets. When the destination node receives an open-session packet sent by the source, it verifies the signature computed on the packet. Upon success, the destination starts to generate the same values as the source and sends an ACK packet to the source. Upon receipt of an open session packet and the corresponding ACK packet, an intermediate node verifies signature computed on the packet, and in case of success, it stores the root values of the hash chain and the Merkle-tree, the session ID, $SessionID$, and forwards the packet towards the destination. Otherwise, an intermediate node changes its probability to store packets in the current session to zero.

After receiving an ACK message corresponding to the session open packet, from the destination, the source starts to send data packets. Each data packet is extended with the MAC, computed over the whole packet (except for the EAR and RTX flags), using the shared secret between the source and the destination. Upon receipt of a data packet, an intermediate node stores with probability p the data packet and forwards the data packet towards the destination. Upon receiving a data packet with sequence number i, first, the destination checks MAC using the secret shared between the source and the destination. Upon success, the destination delivers the packet to the upper layer. Otherwise, the packet is ignored and dropped. Upon the receipt of a packet with a set EAR flag, the destination sends an ACK or a $NACK$ packet depending on the existence of gaps in the received data packet stream. The ACK packet that refers to sequence number i is composed of the pair $(i, K_{ACK}^{(i)})$. Similarly, the $NACK$ packet with base sequence number i is extended with the ACK authentication value $(K_{ACK}^{(i)})$, and if the destination wants to request for re-transmission of some packet j, then it also includes the corresponding $NACK$ authentication values $K_{NACK}^{(j)}, S_1^j, \ldots, S_q^j$ in the $NACK$ packet.

When an intermediate node receives an ACK packet, $(i, K_{ACK}^{(i)})$, it verifies the authenticity and the origin of the ACK message by hashing $K_{ACK}^{(i)}$ i times, and comparing the result with the stored root value of the hash chain. If the two values are equal, then all the stored packets with the sequence number less than or equal to i are deleted. Afterward, the intermediate node passes on the ACK packet towards the source. Otherwise, the ACK packet is ignored and dropped. In case of $NACK$ packet that refers to the sequence numbers i, the same is perform for the ACK part. From the $NACK$ part the root of the Merkle-tree is re-generated, and in case of equality, the stored data packets are re-transmitted and the $NACK$ is modify. Afterward, forwards the $NACK$ with the modified list towards the source. When the source node receives an ACK packet or $NACK$ packets it perform the same steps as the intermediate (with the fact the $p = 1$).

In DTSN and SDTP, the destination sends an ACK or a $NACK$ packet upon receipt of an EAR. In order to mitigate the effect of EAR replay or EAR forging attacks where the EAR flag is set/unset by an attacker(s), STWSN uses two new mechanisms: status timer and limiting the number of responses to EARs. The status timer is set at the destination and its duration could be a function of the source EAR timer. To counter that attackers always set the EAR bits, the destination limits the number of responses on receiving a set EAR flag. In the period of the destination's EAR timer the destination will not send more than X control packets.

4   Formal Security Analysis of STWSN Using $crypt_{time}^{prob}$

In this subsection, we perform a formal security analysis of STWSN based on a mathematically sound formal language, the $crypt_{time}^{prob}$ calculus [7], which has been used to analyze the security of DTSN and SDTP. We start with a brief description of $crypt_{time}^{prob}$, and provide the proof technique based on it, finally, we turn to prove the security of our proposed STWSN protocol.

4.1   The $crypt_{time}^{prob}$ Calculus

Due to lack of space we only provide a very brief overview of the $crypt_{time}^{prob}$ calculus, interested readers are referred to [7] for further details. $crypt_{time}^{prob}$ is a probabilistic timed calculus, for modeling and analyzing cryptographic protocols that involve clock timers and probabilistic behavioral characteristics. $crypt_{time}^{prob}$ has been successfully used for analyzing the security of DTSN and SDTP [7].

The basic concept of $crypt_{time}^{prob}$ is inspired by the previous works [12], [13], [14] proposing analysis methods separately for cryptographic, timed, and probabilistic protocols, respectively. Specifically, it is based on the concept of probabilistic timed automata, hence, the correctness of $crypt_{time}^{prob}$ comes from the correctness of the automata because the semantics of $crypt_{time}^{prob}$ is equivalent to the semantics of the probabilistic timed automata.

The formal syntax of $crypt_{time}^{prob}$ is composed of two main building blocks, namely, the *terms* which model protocol messages and their components, and *probabilistic timed processes* which describe the internal operation of communication partners according to their specification. Terms can be, for example, random nonces, secret keys, encryptions, hashes, MACs and digital signatures computed over certain messages, and they can represent entire messages as well. The set of terms also includes communication channels (denoted by c_i, for some index i) defined between participants, such that messages can be sent and received through these channels. We distinguish *public* and *private* channels, where the attackers can eavesdrop on public channels, while they cannot in the private case. The set of probabilistic timed processes defines the internal behavior of the participants, namely, each process defines an action can be performed by a given participant. For instance, *input* and *output processes* define the message receiving and sending actions, respectively. Processes also define the message verification steps that a communication partner should perform on the received messages (e.g., signature verification, comparisons, MAC verification). Finally, processes can also be used to define the whole protocol, composing of several communication partners running parallel. This is similar to the terminology of sub-procedures and main-procedure in programming languages (e.g., C, Java). We denote probabilistic timed processes by $procA^i$ with $i \in \{1, 2, \ldots, k\}$ for some finite k, and A can be the name of a communication partner or a protocol (e.g., *procSrc* represents the process that describes the behavior of the source node). Non-channel terms are given the name of a message or its components (e.g., ID is a term that models a message ID).

The operational semantics of $crypt_{time}^{prob}$ is built-up from a probabilistic timed labeled transition system (PTTS) defined specifically for this calculus. The PTTS contains the rules of form $s_1 \overset{\alpha,\, d}{\to} s_2$, where s_1 describes the current state of a given process, while s_2 represents the state we reach after some action α has been performed, which consumes d time units. For instance, α can be the message sending action, while s_1 and s_2 are the states before and after the message has been sent, respectively. There are three types of actions that can be performed by the communication partners: (i) silent (internal computation) action; (ii) message input or (iii) message output on a channel. Silent action (e.g., message verification steps) are not visible for an external observer (environment), while message input and outputs on a public channel are visible. Message input and output on a private channel can be seen as silent actions. The PTTS of $crypt_{time}^{prob}$ contains the rules that define all the possible actions of the communication partners, according to the protocol description.

In order to prove or refute the security of protocols and systems, $crypt_{time}^{prob}$ is equipped with the *weak probabilistic timed (weak prob-timed) labeled bisimilarity*. The bisimilarity definition is used to prove or refute the behavioral equivalence between two variants of a protocol or system. The definition of weak prob-timed bisimilarity is given as follows:

Definition 1. (*Weak prob-timed labeled bisimulation for* $crypt_{time}^{prob}$)
We say that two states $s_1 = (procA^1, v_1)$ and $s_2 = (procA^2, v_2)$ are weak prob-timed labeled bisimilar, denoted by $(s_1 \; \mathfrak{R}_t^p \; s_2)$

1. *if an observer who can eavesdrop on the network communication cannot distinguish the message output or input in states s_1 and s_2 (which we called as statically equivalence [7]);*
2. *if from s_1 we can reach the state s_1' after a silent (internal) action after d_1 time units, then s_2 can simulate this action via the corresponding silent action trace after d_2 time units, leading to some s_2', and $s_1' \; \mathfrak{R}_t^p \; s_2'$ holds again.*
3. *if from s_1 we can reach the state s_1' after a non-silent labeled transition (i.e., message input or ouput) after d_1 time units, then s_2 can simulate this action via the corresponding labeled transition trace after d_2 time units, leading to some s_2', and $s_1' \; \mathfrak{R}_t^p \; s_2'$ holds again,*

and vice versa. We say that two variants of a protocol Prot1 and Prot2 are weak prob-timed labeled bisimilar if their initial states (s_1^{init}, s_2^{init}) are weak prob-timed labeled bisimilar.

In this definition, s_j is a protocol (system) state, which is composed of the pair ($procA^j$, v_j), where $procA^j$ is a $crypt_{time}^{prob}$ process, representing the current behavior of the participants in the protocol, and v_j is the current timing value(s) of the clock(s). Hence, based on this definition whenever we refer to a (prob. timed) protocol A we mean the state s^{init}, $s^{init} = (procA^{init}, v^{init})$.

Intuitively, Definition 1 says that two versions of a given protocol are "behavioral" equivalent if any action (formally, any labeled transition) that can be performed by one protocol version can be simulated by the corresponding action(s) (formally, a corresponding trace of labeled transition) in the another version, and vice versa. More precisely, if s_1 and s_2 are in weak prob-timed labeled bisimulation, then the behavior of the two protocol versions $procA^1$ and $procA^2$ are equivalent for an observer (environment) who eavesdrops on the communication between every pair of partners.

4.2   Security Proof Technique Based on $crypt_{time}^{prob}$

We apply the proof technique that is based on Definition 1. Namely, we define an ideal version of the protocol run, in which we specify the ideal (i.e., secure) operation of the real protocol. This ideal operation, for example, can require that honest nodes always know what is the correct message they should receive/send, and always follow the protocol correctly, regardless of the attackers' activity. Then, we examine whether the real and the ideal versions, running in parallel with the same attacker(s), are weak prob-timed bisimilar.

Definition 2. Let the $crypt_{time}^{prob}$ processes procProt and procProt^{ideal} specify the real and ideal versions of some protocol Prot, respectively. We say that Prot is secure (up to the ideal specification) if (procProt, v^{init}) and (procProt^{ideal}, v_{ideal}^{init}) are weak prob-timed bisimilar:

$$(procProt, v^{init}) \approx_{pt} (procProt^{ideal}, v_{ideal}^{init}),$$

where v^{init} and v_{ideal}^{init} are the initial values of the clocks (typically in reset status). The strictness of the security requirement, which we expect a protocol to fulfill, depends on how "ideally secure" we specify the ideal version. Intuitively, Definition 2 says that Prot is secure if the attackers cannot distinguish the operation of the two instances based on the message outputs (and inputs). In the rest of this section, we refer to the source, intermediate and destination nodes as S, I and D, respectively.

Let us consider a simplified network topology for the STWSN protocol. We assume the network topology $S-I-D$, where " $-$ " represents a bi-directional link. Moreover, public communication channels are defined between each node pair for message exchanges, c_{si} between S and I, c_{id} between I and D. We also assume the presence of an attacker or attackers who can eavesdrop on public communication channels, and can use the eavesdropped information in their attacks (e.g., modifying the ACK/NACK packets and forward them).

The main difference between the ideal and the real systems is that in the ideal system, honest nodes are always informed about what kind of packets or messages they should receive from the honest sender node. This can be achieved by defining private (hidden) channels between honest parties, on which the communication cannot be observed by the attacker(s). Figure 3 shows the difference in more details. In the ideal case, three private channels are defined which are not available to the attacker(s). Whenever S sends a packet pck on public channel c_{si}, it also informs I about what should I receive, by sending at the same time pck directly via private channel c_{privSI} to I, hence, when I receives a packet via c_{si} it compares the message with pck. The same happens when I sends a packet to D, and vice versa, from D to I and I to S. Whenever a honest node receives an unexpected data (i.e., not the same as the data received on private channel), it interrupts its normal operation. The channels c_{privSD} and c_{privID} can be used by the destination to inform S and I about the messages to be retransmitted.

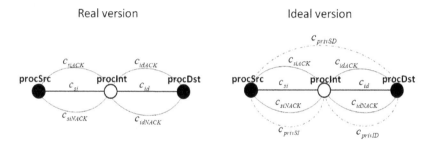

Fig. 3. $procSrc$, $procInt$ and $procDst$ denote the $crypt_{time}^{prob}$ processes that specify the operation of the source, the intermediate and the destination nodes, respectively. In the ideal version, the private (hidden) channels c_{privSD}, c_{privID} and c_{privSI} are defined between $procSrc$ and $procDst$, $procInt$ and $procDst$, $procSrc$ and $procInt$, respectively.

With this definition of the ideal system we ensure that the source and intermediate nodes are not susceptible to the modification or forging of data packets and $ACK/NACK$ messages since they make the correct decision either on retransmitting or deleting the stored packets. Namely, this means that the honest nodes only handle the messages received on public channels when they are equal to the expected messages received on private channels.

The Attacker Model. $\mathcal{M}_A$: We assume that the attacker(s) can intercept the information output by the honest nodes on public channels, and modify them according to its (their) knowledge and computation ability. The attacker's knowledge consists of the intercepted output messages during the protocol run and the information it can create. The attacker(s) can modify the elements of the $ACK/NACK$ messages, such as the acknowledgement number and the bits for retransmissions, as well as the EAR and RTX bits, and sequence numbers in data packets. The attacker can also create entire data or control packets based on the data it possesses. Further, an attacker can send packets to its neighborhood. We also consider several attackers who can share information with each other.

To describe the activity of the attacker(s), we apply the concept based on the so-called *environment*, used in the applied π-calculus [12] to model the presence of the attacker(s). Every message that is output on a public channel is available for the environment, that is, the environment can be seen as a group of attackers who can share information with each other, for instance, via a side channel. This definition of attackers allow us to apply the weak prob-timed bisimilarity in our security proofs.

4.3 Security Analysis of STWSN Based on the $crypt_{time}^{prob}$ Calculus

As already mentioned in Section 3, STWSN uses different cryptographic primitives and operations such as one-way hash, message authentication code (MAC), and digital signature. In $crypt_{time}^{prob}$, secret and public keys can be defined by (atomic) names, and cryptographic operations can be defined as functions (for modelling crypto-primitive generation) and equations (for cryptographic verification):

Keys: sk_{src}, pk_{src}, K_{ack}; K_{nack}; K_{sd};

Functions: $sign(t,\ sk_{src})$; $H(t)$;
Equation: $checksign(sign(t,\ sk_{src}),\ pk_{src}) = ok$;

Functions: $K(n,\ ACK)$; $K(n,\ NACK)$;
    $mac(t,\ K(n,\ ACK))$; $mac(t,\ K(n,\ NACK))$;
Equations:
  $checkmac(mac(t,\ K(n,\ ACK)),\ K(n,\ ACK)) = ok$;
  $checkmac(mac(t,\ K(n,\ NACK)),\ K(n,\ NACK)) = ok$.

where sk_{src} and pk_{src} represent the secret and public key of the source node. K_{ack}, K_{nack} and K_{sd} represent the $ACK/NACK$ master keys, and the shared key of the source and the destination for a given session, which are freshly generated at the beginning of each session. The functions $sign(t,\ sk_{src})$ and $H(t)$ define the digital signature computed on the message t using the secret key sk_{src}, and the one-way hash computed on t, respectively. The equation $checksign(sign(t, sk_{src}),\ pk_{src}) = ok$ defines the signature verification, using the corresponding public key pk_{src}. We do not define an equation for the hash function $H(t)$ in order to ensure its one-way property. Namely, $H(t)$ does not have a corresponding inverse function which returns t, and $H(t_1) = H(t_2)$ holds only when t_1 and t_2 are the same.

For simplicity, we assume the network topology $S-I-D$, where " $-$ " represents a bi-directional link, while S, I, D denote the source, an intermediate node, and the destination node, respectively (like in Figure 3). However, we emphasize that the security proofs in this simplified topology can also be applied (with some extension) and remain valid in the topologies that contain more intermediate nodes. We define symmetric public channels between the upper layer and the source, c_{sup}; the upper layer (i.e. the application that uses the protocol) and the destination, c_{dup}; the source and the intermediate node, c_{si}; the intermediate node and the destination c_{id}. The public channels c_{siACK}, c_{siNACK}, c_{idACK} and c_{idNACK} are defined for sending and receiving ACK and $NACK$ messages between the source and the intermediate, and between the destination and the intermediate nodes, respectively. Public channels c_{siOPEN} and c_{idOPEN} are defined for delivering the open-session messages between S and I, as well as I and D. To apply Definition 1 in our proofs, additional public channels $c_{badOPEN}$, c_{badPCK}, c_{error} and $c_{sessionEND}$ are defined for sending and receiving $bad\ packet$, $error$ and $session\text{-}end$ signals. Honest nodes send out these signals when error or session-end is detected at these nodes. Finally, we add a public channel c_{emptyC} for signalling that the cache has been emptied at a given node.

According to the proof technique, we define an ideal and a real version of the STWSN protocol. In the ideal version of the STWSN protocol the definition of the processes $procSrc$, $procInt$, and $procDst$ in $crypt_{time}^{prob}$ are extended (compared to the real version) with some additional equality checks between the messages received on the corresponding private and public channels. Specifically, processes $procDst$ and $procInt$ output the special constant $BadOpen$ on channel $c_{badOPEN}$ when they receive an unexpected open-session packet (i.e., when the message which the honest nodes receive on the public channels c_{siOPEN} and c_{idOPEN} is not equal to the corresponding packet received on the private channels c_{privSI} and c_{privID}, respectively). Process $procDst$ outputs the constant $BadData$ on c_{badPCK} when it receives an unexpected data packet. After receiving an unexpected $ACK/NACK$, the processes $procSrc$ and $procInt$ output the constant $BadControl$ on c_{badPCK}. Finally, after emptying the buffer $procSrc$ and $procInt$ output the constants $EmptyCacheS$ and $EmptyCacheI$, respectively.

We examined the security of STWSN regarding the following scenarios: In the first scenario SC-1, the attacker(s) modifies the ACK/NACK messages, while in the SC-2 the attacker(s) modifies the EAR and RTX bits in data packets. Let us name the $crypt_{time}^{prob}$ processes that define the real and ideal versions of STWSN by $procSTWSN$ and $procSTWSN^{ideal}$, respectively.

- **Scenario SC-1**: STWSN is not vulnerable to the attack scenario SC-1 because process $procSTWSN^{ideal}$ can simulate (according to Definition 1) every labeled transition produced by $procSTWSN$: Recall that in STWSN S verifies the $ACK/NACK$ packets by comparing the stored roots of the hash-chain and the Merkle-tree with the re-computed roots. For simplicity, we assume that the source is storing the first three packets for a given session, and the source wants to send four data packets in this session. This means that the root of the hash-chain is $H(H(H(H(H(K_{ack})))))$. Let assume that the source has received an ACK packet, where $ACK = (m, h_m)$, and according to the protocol, to make the source accept this ACK the m-time hashing on h_m must be $H(H(H(H(H(K_{ack})))))$. To empty the buffer, m must be at least 3. In case $m = 4$, h_m must be $H(K_{ack})$. This hash value cannot be computed by the attacker(s) because the source and the destination never reveal K_{ack}, and function $H(t)$ is defined to be one-way, so from $H(H(K_{ack}))$ the attacker cannot compute $H(K_{ack})$. Hence, the attacker(s) must receive or intercept $H(K_{ack})$ from a honest node, which means that $H(K_{ack})$ has already been revealed by the destination. m cannot be greater than four, otherwise, the attacker(s) must have K_{ack}, that is, K_{ack} must have been revealed by the source or destination, which will never happen according to the protocol. When $m = 3$, h_m must be $H(H(K_{ack}))$, which cannot be computed by the attacker(s) since according to the protocol, K_{ack} and $H(K_{ack})$ have not been revealed yet, and we did not define any equation for the hash function $H(t)$, hence, from $H(h_i)$ the value of h_i cannot be derived. To summarize, either the attacker sends a correct ACK or the ACK with incorrect authentication value, the ideal and the real systems can simulate each other. In the first case, the constant $EmptyCacheS$, while in the second case $BadControl$ is output in both systems. The reasoning is similar in case of $NACK$ message.

In the following, we examine whether the attackers can make the intermediate node incorrectly empty its buffer. Let us assume that I has already accepted the open session packet and has stored the hash-chain root, denoted by h_{root}, in it. The signature in the open session packet must be computed with the secret key of the source, sk_{src}. This is because after receiving a packet on channel c_{siOPEN}, the verification $[checksign(x_{sig}, pk_{src}) = ok]$ is performed within process $procInt$, and only the signature computed with sk_{src} can be verified with pk_{src}. However, this means that h_{root} must be generated by the source, that is, $h_{root} = H(H(H(H(H(K_{ack})))))$. Again, assume that in the current state I stores the first three data packets. From this point, the reasoning is similar to the source's case, namely, either the constant $EmptyCacheI$ or $BadControl$ is output in both the ideal and the real systems.

– **Scenario SC-2**: The formal proof regarding the second scenario is based on a similar concept as in case of analyzing fault tolerance systems. Namely, for reasoning about the scenario SC-2, we modify the ideal version of STWSN as follows: we limit the number of the $ACK/NACK$ that the destination sends when it receives a data packet in which the EAR flag is set to 1 by the attacker. Based on the concept of private channels where the destination is informed about the message sent by a honest node, in the ideal system, the destination is able to distinguish between the EAR flag set by the attacker and the flag set by a honest node. In the ideal system, we modify the specification of the destination such that within a session, the destination will only handle according to the protocol, the first MAX_{badear} packets in which the EAR flag is set to 1 by the attacker, and sends back the corresponding $ACK/NACK$ for them. Formally, for the first MAX_{badear} packets with an incorrectly set EAR flag, the destination does not output the constant $BadData$ on channel c_{badPCK}, but only from the following incorrect packet. The constant MAX_{badear} is an application specific security treshold.

In the STWSN protocol, to alleviate the impact of the *EAR* setting attack, the destination limits the number of responses for the packets with the *EAR* flag set by either an attacker or a honest node [1]. Within a finite period of time, called *destination EAR timer* (denoted by *dest_EAR_timer*), the destination node will not send more than D control packets in total, for some given security treshold D. Within a session, the destination launches the timer *dest_EAR_timer* when it has received the first packet containing a set *EAR* flag. Until the session ends, *dest_EAR_timer* is continually reset upon timeout. Let tmr_{dst} be the upperbound of the number of launching/resetting *dest_EAR_timer* within a session. The values of *dest_EAR_timer* and D are set such that $D \times tmr_{dst} \leq MAX_{badear}$.

The main difference between the ideal and the real STWSN specifications, $procSTWSN$ and $procSTWSN^{ideal}$, is that in the ideal case the destination does not need to launch the timer *dest_EAR_timer*, because it is aware of the packets with an incorrectly set *EAR* flag. Instead, the destination only limits the total number of responses for the incorrect packets to MAX_{badear}. To prove the security of STWSN regarding *SC-2*, we prove that the STWSN protocol is secure regarding the scenario *SC-2*, by showing that $procSTWSN^{ideal}$ weak prob-timed simulates the real system $procSTWSN$ (instead of resisting on bisimilarity). Intuitively, this means that the set of probabilistic timed transitions of $procSTWSN$ is a subset of the set of probabilistic timed transitions of $procSTWSN^{ideal}$. The reverse direction that $procSTWSN$ can simulate $procSTWSN^{ideal}$, is not required for this scenario.

We also examined the other possible attack attempts (e.g., when the attacker(s) modifies the data part in data packets, and the open-session packets) with the same method, based on the definition of weak prob-timed bisimilarity. Namely, we define the corresponding real and ideal versions, and we showed that they can simulate each other regarding the outputs of the constants *BadData* and *BadOpen*, respectively. Hence, based on the Definition 1, we showed that the two versions are weak prob-timed bisimilar. Due to lack of space we do not include them here, interested readers can find them in the report [15].

5   Conclusion and Future Works

We addressed the problem of formal security verification of WSN transport protocols that launch timers, as well as performing probabilistic and cryptographic operations. We argued that formal analysis of WSN transport protocols is important because informal reasoning is error-prone, and due to the complexity of the protocols, vulnerabilities can be overlooked. An example would be the case of SDTP, which was believed to be secure, but later, it turned out to be vulnerable. In this paper, using the algebra language $crypt^{prob}_{time}$, we formally proved that the STWSN protocol is secure against the packet modification attacks.

In this paper, we also demonstrated the expressive power of the $crypt^{prob}_{time}$ calculus, and showed that it is well-suited for analyzing protocols that may include timers, probabilistic behavior, and cryptographic operations. One interesting future direction could be examining the usability of $crypt_{time}^{prob}$ for other class of protocols (e.g., wired transport protocols). In addition, designing an automated verification method based on $crypt_{time}^{prob}$ also raises interesting questions.

References

1. Dvir, A., Buttyán, L., Ta, V.-T.: SDTP+: Securing a distributed transport protocol for wsns using merkle trees and hash chains. In: IEEE International Confenrence on Communications (ICC), Budapest, Hungary, pp. 1–6 (June 2013)
2. Yicka, J., Mukherjeea, B., Ghosal, D.: Wireless sensor network survey. Computer Networks 52(12), 2292–2330 (2008)
3. Wang, C., Sohraby, K., Li, B., Daneshmand, M., Hu, Y.: A survey of transport protocols for wireless sensor networks. Network 20(3), 34–40 (2006)
4. Buttyán, L., Csik, L.: Security analysis of reliable transport layer protocols for wireless sensor networks. In: IEEE Workshop on Sensor Networks and Systems for Pervasive Computing, Mannheim, Germany, pp. 1–6 (March 2010)
5. Marchi, B., Grilo, A., Nunes, M.: DTSN - distributed transport for sensor networks. In: IEEE Symposium on Computers and Communications, Aveiro, Portugal, pp. 165–172 (July 2007)
6. Buttyán, L., Grilo, A.M.: A Secure Distributed Transport Protocol for Wireless Sensor Networks. In: IEEE International Conference on Communications, Kyoto, Japan, pp. 1–6 (June 2011)
7. Ta, V.-T., Dvir, A.: On formal and automatic security verification of wsn transport protocols. ISRN Sensor Networks (December 2013) (accepted)
8. Merkle, R.C.: Protocols for Public Key Cryptosystems. In: Symposium on Security and Privacy, California, USA, pp. 122–134 (April 1980)
9. Coppersmith, D., Jakobsson, M.: Almost optimal hash sequence traversal. In: Blaze, M. (ed.) FC 2002. LNCS, vol. 2357, pp. 102–119. Springer, Heidelberg (2003)
10. Galuba, W., Papadimitratos, P., Poturalski, M., Aberer, K., Despotovic, Z., Kellerer, W.: Castor: Scalable Secure Routing for Ad-Hoc Networks. In: Infocom, Rio de Janeiro, Brazil, pp. 1–9 (2010)
11. Hu, Y.-C., Perrig, A., Johnson, D.B.: Ariadne: a secure on-demand routing protocol for ad hoc networks. Wireless Networks Journal 11(1-2), 21–38 (2005)
12. Fournet, C., Abadi, M.: Mobile values, new names, and secure communication. In: ACM Symposium on Principles of Programming, pp. 104–115 (2001)
13. Goubault-Larrecq, J., Palamidessi, C., Troina, A.: A probabilistic applied pi–calculus. In: Shao, Z. (ed.) APLAS 2007. LNCS, vol. 4807, pp. 175–190. Springer, Heidelberg (2007)
14. D'Argenio, P.R., Brinksma, E.: A calculus for timed automata. In: Jonsson, B., Parrow, J. (eds.) FTRTFT 1996. LNCS, vol. 1135, pp. 110–129. Springer, Heidelberg (1996)
15. Ta, V.-T., Dvir, A.: On formal and automatic security verification of wsn transport protocols. Cryptology ePrint Archive, Report 2013/014 (2013)

How to Apply Large Deviation Theory to Routing in WSNs

János Levendovszky and Hoc Nguyen Thai

Faculty of Electrical Engineering and Informatics
Budapest University of Technology and Economics
H-1117, Budapest, Magyar Tudosok krt 2, Hungary
levendov@hit.bme.hu, thaihocme@gmail.com

Abstract. This paper deals with optimizing energy efficient communication subject to reliability constraints in the case of Wireless Sensor Networks (WSNs). The reliability is measured by the number of packets needed to be sent from a node to the base station via multi-hop communication in order to receive a fixed amount of data. To calculate reliability and efficiency as a function of the transmission energies proves to be of exponential complexity. By using the statistical bounds of large deviation theory, one can evaluate the necessary number of transmitted packets in real time, which can later be used for optimal, energy aware routing in WSNs. The paper will provide the estimates on efficiency and test their performance by extensive numerical simulations.

Keywords: Network reliability and availability estimation, Moore–Penrose pseudo inverse, WSNs, Routing, Large deviation.

1   Introduction

Due to the recent advances in electronics and wireless communication, the development of low-cost, low-energy, multifunctional sensors have received increasing attention [4]. Although there are many applications of Wireless Sensor Networks – based health monitoring to civil engineering structures [5], WSNs still have some limitations, most notably (i) less reliable data transmission than wired sensing systems; (ii) a relatively short communication range, and (iii) operational constraints due to limited power. The paper addresses reliable packet transmission in WSNs when packets are to be received by the BS with a given reliability, in terms of keeping this probability is bigger than a given threshold. As result, the number of data packets transmitted from a sensor node to the BS need to be calculated carefully with special focus on the probability that the minimum residual energy $g(y)$ exceeds a given threshold at each sensor node $P\left(g(y) > LT\right)$ after the BS has received L packets successfully. Here T is a predefined residual energy at each sensor node, in order to increase the lifetime of the network and guarantee that the network works with the highest reliability. Fast methods for evaluating this probability can become the bases for novel energy aware routing algorithms. The new results have been tested by extensive simulations which evaluated the probability by Markov inequality and Chernoff inequality or some others large deviation techniques. But our concern is to derive the optimal transmission

T.V. Do et al. (eds.), *Advanced Computational Methods for Knowledge Engineering*,
Advances in Intelligent Systems and Computing 282,
DOI: 10.1007/978-3-319-06569-4_30, © Springer International Publishing Switzerland 2014

energies for each scenario needed to achieve a given reliability and yielding the longest possible lifespan of the network.

The results are given in the following structure:

(i)   in Section 2, the communication model is outlined;
(ii)   in Section 3, the quality of route R with regard to reliable packet transfer and energy awareness is discussed;
(iii)   in Section 4, a novel method is derived to approximate the weight vector;
(iv)   in Section 5, using large deviation techniques to estimate the remaining energy in each sensor node;
(v)   in Section 6, the performances of the new algorithms to calculate the probability are analyzed numerically.

2   The Model

WSNs contain sensor nodes communicating with each other by radio communication. Information is sent in the forms of packets and the task of the routing protocol is to ensure that all messages transfer to the BS successfully.

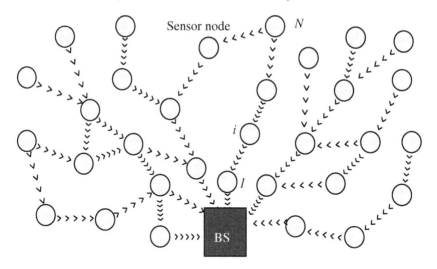

Fig. 1. Communication connectivity between sensor nodes and the BS in WSNs

From a 2D graph we often change to a 1D model, i.e. after the routing protocol has found the path to the BS, the nodes participating in the packet transfer can be regarded as one dimensional chain labeled by $i = 1,...,N$ and depicted by Figure 2.

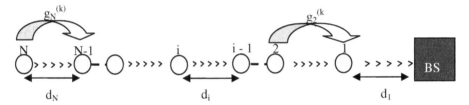

Fig. 2. One dimensional chain topology of WSNs packet forwarding

We assume network has the following properties [3]:

- There is only a single, stationary BS on a fixed location;
- The BS is energy abundant as it can be recharged or connected to an energy supply network, the BS is able to reach node of the WSN (even the furthest ones) by radio communication;
- The nodes are also assumed to be stationary.
- There are several nodes which do not have enough power to reach the BS directly, hence multi hops packet transmission is in use, where packet forwarding is determined by an addressing mechanism.
- The direction of communication is Node – to – BS (the data acquired by the sensor nodes must be collected by the BS).
- Let us assume that there is a Wireless Sensor Network (WSN) perceived as a 2D graph $G(V, E, e_{(u,v)}, (u,v) \in E)$ with V represents the set of wireless sensor nodes and E set of edges in the network [1]; e (u, v) is associated with link (u, v) denotes the energy by which node u send a packet to node v. The probability of successful packet transfer from node u to node v is determined by the Rayleigh fading model [2], given as:

$$p_u := \exp\left(-\frac{Qn_z^2}{P_0 d_u^{-n}}\right) \tag{1}$$

Where $Q(m^{-n})$ is the sensitivity threshold, $n_z^2(W)$ denotes the noise power, $d_u(m)$ is distance from node u to node v, with the value of n depends on fading model, $P_0(W)$ is a transmit power. The probability of successful packet transfer from node N^{th} to BS is determined by the Rayleigh fading model:

$$p_r := \prod_{i=1}^{N} \exp\left(-\frac{Qn_z^2}{P_0 d_i^{-n}}\right) \tag{2}$$

Let us now assume that one wants to send L packets from a given source node S to the BS according to a query process over a route R containing N hops, which has been chosen prior to the packet transfers. In order to make sure that a required L number of packets is transferred to the BS successfully, the node transfers M packets ($M \geq L$). The main question is to evaluate the number M for which the probability the remaining minimum residual energy of the nodes over the path still exceeding a given threshold will remain beyond a given limit.

Let binary vector $y = (y_1,...,y_N, y_{N+1},..., y_{2N},..., y_{(M-1)N+1},..., y_{MN}) \in \{0,1\}^{MN}$ denote the state of transferring M packets over the N hops route according to the definition:

$$y_{li} = \begin{cases} 1 & \text{if the network transferrs } l^{th} \text{ packet successfully} \\ 0 & \text{otherwise} \end{cases}$$

The size of the state space is $|S| = 2^{MN}$ with $Z = (Z_1,...,Z_N)$ where $Z_i = (y_{i1},..., y_{iN}) \in \{0,1\}^N$. Some transmission scenarios and the corresponding binary state vector are depicted by Table 1 and Table 2, respectively.

Table 1. Transferring data in the network while the number of transmissions $k < M$

Packets	Node N	Node N-1	...	Node 2	Node 1	Base Station received
1	1	1	...	1	1	1st successful
2	1	1	...	0	0	Unsuccessful
3	1	1	...	1	1	2nd successful
...	...	...	...	...	...	...
k	1	1	...	1	1	L^{th} successful
k+1	0	0	...	0	0	Stop receiving
...	...	...	...	...	...	Stop receiving
M	0	0	...	0	0	Stop receiving

In Table 1 we can see that the BS will stop transferring data whenever it has received L successful packets.

In order to measure the efficiency of packet transfers we introduce a function $g(y)$ with $y \in \{0,1\}^{MN}$ over the state space defined as follows:

$$g(y) = \begin{cases} Le_{residual} & \text{if } L \text{ successful packet transfer has occured} \\ e_{residual} & \text{if less than } L \text{ successful packet transfer has occured} \end{cases}$$

Our main concern is to evaluate $P\big(g(y) \geq LT\big) = \displaystyle\sum_{y:g(y)\geq LT} p(y)$. Since the state space is extremely large, the objective is to evaluate g(y) without visiting each $y \in \{0,1\}^{MN}$. To avoid large scale calculations, we use Monte Carlo Simulation techniques and Large Deviation Theory to obtain fast estimations of $P\big(g(\mathbf{y}) \geq LT\big)$.

3 Evaluating the Quality of Route R with Regard to Reliable Packet Transfer and Energy Awareness by Using LDT

In this section we investigate the possible calculation of the QoS parameter of the given route R.

3.1   Exhaustive Search

It is easy to see that [10, 11, and 12]

$$P\left(g\left(y\right)\geq LT\right)=\sum_{y:g(y)\geq LT}p\left(y\right)=\sum_{y:g(y)\geq LT}\prod_{i=1}^{MN}p_{i}^{y_{i}}\left(1-p_{i}\right)^{1-y_{i}} \tag{3}$$

This entails exponential complexity, as one cannot visit all $y \in \{0,1\}^{MN}$ binary state vectors to evaluate the probability above. Thus, we need efficient methods to solve this problem by using large deviation theories.

3.2   Evaluating $P\left(g\left(y\right)\geq LT\right)$ by Using LDT

In order to apply the standard inequalities of LDT (i.e. Markov and Chernoff bounds), first we approximate we approximate $g\left(y\right)$ with a linear combination of the components of the state vector, in the form of $g\left(\overline{y}\right)\approx\sum_{i=1}^{MN}w_{i}y_{i}$. In order to obtain the best approximation we sample the value of $g\left(y\right)$ by Monte Carlo simulation by obtaining a training set $\tau^{(k)}=\left\{\left(\overline{y}^{(i)},g(\overline{y}^{(i)})\right),i=1,...,MN\right\}$. Having MN free parameters the least square optimization problem $w_{opt}:\min_{w}\sum_{k=1}^{MN}\left(g\left(y^{(k)}\right)-\sum_{i=1}^{MN}w_{i}y_{i}^{(k)}\right)^{2}$ can be solved by making the squared sum zero and setting $\mathbf{w}_{opt}$ as solving the following set of linear equations $\sum_{j=1}^{MN}w_{j}y_{j}^{(i)}=g(\overline{y}^{(i)}),i=1,...,MN$.   Having   the   approximation $g\left(\overline{y}\right)\approx\sum_{i=1}^{MN}w_{iopt}y_{i}$ at hand, one can apply LDT to approximate $P\left(g\left(y\right)\geq LT\right)$, as shown by the following sections.

3.3   Using Markov Inequality to Evaluate the Energy in Each Sensor Nodes

$$P\left(g\left(y\right)\geq L*T\right)\approx P\left(\sum_{i=1}^{MN}w_{i}y_{i}\geq LT\right)\leq\frac{E\left(\sum_{i=1}^{MN}w_{i}y_{i}\right)}{L*T}$$

$$\begin{cases} P(y_i = 1) = P_{ri} \\ P(y_i = 0) = 1 - P_{ri} \end{cases} \qquad E(y_i) = 1P_{ri} + 0(1 - P_{ri}) = P_{ri}$$

$$P(g(y) \ge L*T) \le \frac{\sum_{i=1}^{MN} w_i E(y_i)}{LT} = \frac{\sum_{i=1}^{MN} w_i P_{ri}}{LT} \qquad (4)$$

3.4   Using Chernoff Inequality to Evaluate the Energy in Each Sensor Nodes

Let $X_1 \ldots X_n$ be n independent random variables, and $Y = \sum_{i=1}^{N} x_i$. Then, for any $C > 0$,

we have $\qquad P(Y \ge C) \le e^{\mu(s^*) - sC} \qquad (5)$

Where $\mu(s^*) = \ln\left[E(e^{sY})\right]$ and $s^* = \inf_s (\mu(s) - sC)$.

By the use of the Chernoff inequality, one can obtain

$$P(g(y) \ge LT) \le e^{\sum_{i=1}^{MN} w_i \mu_i(s^*) - s^* LT} \qquad (6)$$

Where: $\mu_i(s) = \ln\left((1 - P_i)e^{s0} + P_i e^{s1}\right) = \ln\left(1 - P_i + P_i e^s\right)$ and $s^* = \inf_s \sum_{i=1}^{MN} \left(w_i \, \mu_i(s) - sLT\right)$

4   The Computational Procedure to Evaluate the Energy Efficiency of Route R in WSNs

In this section, we describe the communication algorithm with 6 steps, as follows:

GIVEN. We set the initial variables: N, L, M, g_0, Q, σ^2, C,

STEP 1. Select an initial λ for generating distance vectors and calculate the probability of successful packets by Rayleigh fading model.

STEP 2. Generate vector y by the distribution of $\prod_{i=1}^{MN} p_i^{y_i} (1 - p_i)^{1 - y_i}$

STEP 3. Check the results of y vector, and counting the number of successful packets. One successful packet has all high bits.

STEP 4. Calculating the energy in each sensor nodes after network finish transferring data. The minimum energy in each sensor nodes will depend on the number of successful packets and the distances from that sensor node to the BS.

STEP 5. Evaluating the weight vector by linear approximation.

STEP 6. Using large deviation to calculate the probability: $P(g(y) > LT)$

5   Numerical Results

In this section the performance of the application large deviation in routing of WSNs described above are investigated by extensive simulations.

5.1   Network Description and the Propagation Model

The simulations were carried out in three different scenarios. All of them including 5 sensor nodes and placing them with three different scenarios following [3]:

- Random 1: The distances between the sensor nodes are generated subject to exponential distribution.
- Random 2: The distances between the sensor nodes are generated subject to 1D Poisson distribution.
- Equidistant case: The distances between the sensor nodes are chosen to be uniform.

     We set the initial parameters:

- The propagation model is determined by the Rayleigh fading, yielding

$$g_i = \frac{d_i^\alpha Q n_z^2}{-\ln(P_i)} + g_0 \; ;$$

- Conditioning energy needed by the electronics $g_0 = 50\mu W$;
- Threshold: $Q = 10$;
- Average noise energy: $\sigma^2 = 0.1$;
- Propagation parameter: $\alpha = 2$;
- Initial energy: $g_j(0) = 50mW; j = 1 \cdots N$;
- Number of packets which Base station received successfully $L = 500$;
- The maximum of transferred packets for one message $M = 1000$.

5.2   Performance Analysis

In this section, a detailed performance analysis of the energy in each sensor nodes is given. The aim is to calculate the level of energy and evaluate the probability of the minimum residual in each sensor nodes is bigger than a predefined energy. The following figure shows how the level energy in each sensor nodes changed in network with $N = 5$, the initial battery power was $C = 50$ (mW) and the distance based on 3 scenarios above.

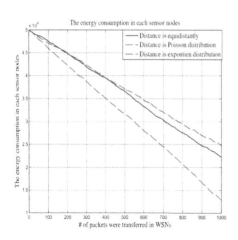
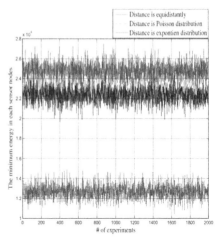

Fig. 3. The energy consumption in each sensor node

Fig. 4. The minimum residual energy in each sensor node

From Figure 3 and Figure 4, one can see that, the level of energy in each sensor node depends on the distances from the base station to sensor node and the number of successful packets per message. The sensor node $N = 5^{th}$ is furthest from the BS, has bigger energy consumption than any other sensor node.

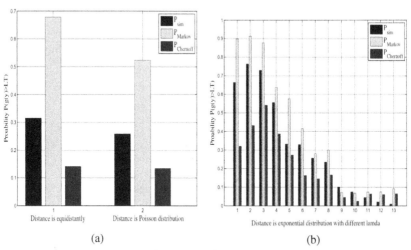

(a)   (b)

Fig. 5. The probability $P\left(g\left(y\right) > LT\right)$ with three different distance vectors

It can be seen that the exponential distribution has the most decreasing trend while the Poisson distribution and equidistance have the smaller energy consumption. It means that the minimum residual energy in each sensor node in exponential distribution will be the smallest level. So the network will work stably with high reliability parameters while the minimum residual energy in each sensor node still bigger than a predefined threshold in the exponential distribution.

It is clear that the probability of minimum residual energy in each sensor node is bigger than a predefined energy will depend on the structure of sensor node in WSNs. With the distance vectors are equidistantly and Poisson distribution on Figure 5 we can see that Chernoff inequality is the tightest bound which we can use for estimate the initial battery for each sensor node. The simulation results shows that in the case of equidistant arrangement ; for transferring $L = 500$ packets successfully, the network will have to transfer $M_1 = \dfrac{L}{P(g(y) > LT)} = \dfrac{500}{0.315} = 1588$ packets based on the result

of model, and $M_2 = 738$ packets based on Markov inequality and $M_3 = 3547$ packets based on Chernoff inequality. If the distance from source nodes to the BS is Poisson distribution, these values respectively will be $M_1 = 1938$ packets $M_2 = 957$ packets and $M_3 = 3732$ packets. With the results of simulations we can use them for estimating the initial battery in each sensor node. In the following table will describe the probability of minimum residual energy in each sensor node when we change the initial λ in the equation of distance.

6   Conclusions

In this paper, we managed to provide fast estimates by LDT for the probability of the residual energy exceeding a given threshold in each sensor node. By using the new results, one can evaluate the number of packets needed to be transferred by the network to the BS if a given number of successful packet transfers are required. The result can directly be used in any applications, where energy consumption and lifespan are of concern. One primary target field is related to biomedical applications, where the energy consumption of sensorial implants must be minimized (in order to avoid the hazard of galvanic recharging via the human body) In this case, reliable but low-energy packet transfers of measurements from the body to the monitoring BS proves to be   crucial [8].

References

1. Zhou, H., Wu, H., Xia, S., Jin, M., Ding, N.: A Distributed Triangulation Algorithm for Wireless Sensor Networks on 2D and 3D Surface - National Science Foundation under Award Number CNS-1018306
2. Haenggi, M.: IEEE- On Routing in Random Rayleigh Fading Networks. IEEE Transactions on Wireless Communications 4(4) (July 2005)
3. Levendovszky, J., Bojarszky, A., Larloaci, B., Olah, A.: Energy balancing by combinatorial optimization for wireless sensor networks. WSEAS Transactions on Communication – SSN 7(2), 1109–2742 (2008)
4. Roggers, A., Corkill, D.D., Jennings, N.R.: Agent technologies for sensor networks. IEEE Intelligent Systems 24(2), 13–17 (2009)

5. Wang, Y., Lynch, J.P., Law, K.H.: A wireless structural health monitoring system with multithreaded sensing devices: design and validation. Struct. Infrastruct. Eng. 3(2), 103–120 (2007)
6. Treisman, M., Faulkner, A.: A model for the generation of random sequences by human subjects: Cognitive operation or psychophysical process? Journal of Experimental Psychology: General 116, 337–355 (1987) ISSN: 0096-3445
7. Moore, E.H.: On the reciprocal of the general algebraic matrix. Bulletin of the American Mathematical Society 26(9), 394–395 (1920)
8. Rahman, F., Shabana, N.: Wireless Sensor Network based Personal Health Monitoring System. WSEAS Transactions on Communications 5(V), 966–972 (2006)
9. Levendovszky, J., Jereb, L., Elek, Z., Vesztergombi, G.: Adaptive statistical algorithms in network reliability analysis. Performance Evaluation 948, 1–12 (2002)
10. Kiss, A., Levendovszky, J., Jereb, L.: Stratified sampling based network reliability analysis. In: Proceedings of Eight Internatioanl Conference on Telecommunications Systems, Nashville, TN, USA (March 2000)
11. Levendovszky, J., Vesztergombi, G., David, T., Jereb, L.: Improved importance sampling based reliability analysis for communication networks. In: Proceedings of the Eight IFIP ATM & IP 2000 Workshop, Ilkley, UK (2000) (in press)
12. Rushdi, A.M.: Performance indexes of a telecommunication network. IEEE Trans. Reliab. R 37, 57–64 (1998)

Efficient Core Selection for Multicast Routing in Mobile Ad Hoc Networks

Dai Tho Nguyen[1,2,*], Thanh Le Dinh[2], and Binh Minh Nguyen[2]

[1] UMI 209 UMMISCO IRD/UPMC
nguyendaitho@vnu.edu.vn
[2] University of Engineering and Technology
Vietnam National University, Hanoi
144 Xuan Thuy Street, Cau Giay District, Hanoi, Vietnam

Abstract. PUMA (Protocol for Unified Multicasting through Announcements) is a mesh-based multicast routing protocol for mobile ad hoc networks distinguishing from others in its class by the use of only multicast control packets for mesh establishment and maintenance, allowing it to achieve impressive performances in terms of packet delivery ratios and control overhead. However, one of the main drawbacks of the PUMA protocol is that the core of the mesh remains fixed during the whole execution process. In this paper, we present an improvement of PUMA by introducing to it an adaptive core selection mechanism. The improved protocol produces higher packet delivery ratios and lower delivery time while incurring only little additional control overhead.

Keywords: Mesh-Based Multicast, Core Selection, Mobile Ad Hoc Networks, Performance.

1 Introduction

Multicasting is a technique of communication that enables one or more network nodes to send data to many intended destinations simultaneously in an efficient manner by avoiding redundant data delivery. The IP-specific version of the general concept of multicasting is called IP multicast. Concerns regarding billing, bandwidth management and security have caused the operators of most Internet routers to disable IP multicast. Therefore, a world-wide backbone testbed for carrying IP multicast traffic on the Internet, known as MBONE, has been built. It allows researchers to deploy and test their experimental multicast protocols and applications.

In parallel to research activities, multicast communication has been implemented for a large number of applications, including video conferencing, whiteboard, desktop sharing, and content distribution. Initially, multicast applications could only be used in intranets or the MBONE multicast network. Recently, a number of Internet service providers began to offer support for multicast capabilities in their backbones.

---

* Corresponding author.

T.V. Do et al. (eds.), *Advanced Computational Methods for Knowledge Engineering*,
Advances in Intelligent Systems and Computing 282,
DOI: 10.1007/978-3-319-06569-4_31, © Springer International Publishing Switzerland 2014

In the early years, the research and development efforts were concerned with wired networks with the only dynamic effects coming from infrequent occurrence of node or link failures. As the emergence of wireless technology has bought a new dimension to the world of network services and applications, the focus of multicast routing has almost shifted to the area of wireless networks. Wireless networks provide a convenient way of communicating because they eliminate the barriers of distance, cost, and location.

Mobile ad hoc networks (MANETs) are one of the kinds of wireless networks that attract the most research attention. The nodes in such networks are free to move and organize themselves arbitrarily. This inherent characteristic, among others, makes designing communication protocols in MANETs very challenging.

Multicasting is an important communication primitive in MANETs. Many ad hoc network applications, such as searches and rescues, military battlefields, and mobile conferencing typically involve communication between one and many other people. There have been a number of multicast routing protocols proposed for MANETs. Traditionally, multicast protocols set up trees for the delivery of data packets. The use of meshes is a significant departure for multicasting over wireless networks.

Meshes have a high tolerance towards failures of nodes or links due to interference or mobility. Among most representative mesh-based multicast routing protocols, PUMA [10] has been shown to be the most stable and efficient one, compared to others such as MAODV [5] or ADMR [1]. The most noticeable aspect of PUMA is that it uses a very simple and effective method to establish and maintain the mesh, this results in a low control overhead. PUMA uses a single control message type, called multicast announcement, to maintain the mesh in a core-based approach. All transmissions are broadcast and no unicast protocol is needed. However, the method of core selection in PUMA has a serious drawback: The node with the highest identifier is selected as the core of the multicast group. Furthermore, core changes only happen when the mesh undergoes partition or the old core leaves the mesh.

Inspired by the idea of core selection from [7], in this paper we introduce improvements to PUMA in two aspects. First, we create a function to determine which nodes should be the core of the multicast group. The core node should be able to reach every receiving node as quickly as possible. Next, we propose a core migration procedure to cope with the mobility of the mobile ad hoc networks so that the core remains fit to the group. Our work has made PUMAs performances approximately 20 percent better in term of delivery time. The new implementation suffers from roughly 2 to 5 percent increase in total control overhead, which can be considered as the compensation for its superior delivery time and packet delivery ratio. We implement and evaluate our improved PUMA via simulations with ns-2 [13].

The rest of this paper is organized as follows. Multicast routing in MANETs is reviewed in Section 2 with an in-depth description of PUMA [10] and the core selection procedure in [7]. Section 3 describes our improvements to PUMA.

Simulation results are shown in Section 4. Section 5 presents our conclusion and future works.

2   Related Work

There are two main approaches to multicast routing in MANETs: *tree-based* and *mesh-based.*

In the tree-based approach, either multiple source-trees or a single shared-tree is constructed. This approach is bandwidth efficient since only a minimum number of copies per packet are sent along the paths of the tree. However, trees can easily break with the failure of a single wireless link due to interference or mobility. The reconstruction of the tree will take a certain a mount of time, during which multicasting is not possible. Examples of tree-based multicast routing protocols include AMRIS [11], MAODV [5], and TBM$_k$ [9].

On the other hand, the mesh-based approach may construct numerous paths between a pair of source and destination. The structure is thus more resilient to frequent link failures. In exchange, there may be multiple redundant copies of a same packet disseminated throughout the mesh. Examples of mesh-based multicast routing protocols include ODMRP [12], CAMP [6], PUMA [10], and LSMRM [8].

In ODMRP [12], advertising packets are periodically generated by source nodes and then flooded in the network. Each receiver uses backward learning to responds to these packets. The paths from the receivers to the sources form a mesh. ODMRP produces high packet delivery ratio and throughput even under high mobility. Its disadvantage is that the control overhead grows fast with large network size.

CAMP [6] creates a mesh for each multicast group and ensures that the shortest paths from receivers to sources are part of the mesh. However, it requires a naming service to obtain multicast group addresses, and assumes the availability of routing information from a unicast routing protocol.

In LSMRM [8], the mesh is constructed by using route request and route reply packets. Link stability is also concerned when building the mesh. Paths are found based on selection of nodes that have high stability of link connectivity.

PUMA [10] uses a receiver-initiated mesh-based approach. Each receiver joins a multicast group by using the address of a special core node without having to perform a network-wide flooding of control and data packets from every source of the group. Each receiver is connected to the elected core along all of the shortest paths between the receiver and the core. All nodes on shortest paths between any receiver and the core form the mesh collectively.

PUMA uses solely broadcasting with multicast announcement messages for communication among nodes. It does not require the use of any unicast protocol to work. Each multicast announcement contains a sequence number, the groups address (group ID), the cores address (core ID), the distance to the core, a mesh member flag which is set when the sending node is a member of the mesh and lastly, a parent preferred by the neighbor to reach the core. With information retrieved from multicast announcements, nodes elect the core for the mesh, select the best routes to reach the core from outside of a multicast group, maintain the mesh, and notify other nodes about the meshs state. When a receiver wants to join a multicast group, firstly, it determines whether it has received a multicast announcement for that group or not. If the node has, then it uses the core indicated in the message, specify the same core group, and start sending multicast announcement message. Otherwise, it considers itself as the core of the group, and start sending multicast announcement indicating itself as the core group and zero distance to core from itself. A multicast announcement whose core ID is higher is considered better. Nodes broadcast the best multicast announcement they have received from their neighbours. If a receiver joins the group before other receivers, then it becomes the group's core. If multiple receivers at the same time join the group then the node with maximal ID becomes the core.

The core stability is significant for the effectiveness in performance of PUMA. Frequent core changes lead to high communication overhead, and would lead to a dramatic rise in the number of packet drops. Nodes might detect a partition if they repeatedly fail to receive multicast announcements from the groups core.

Gupta et al. [7] presented an adaptive core selection and migration method for multicast routing in MANETs. They introduced the concept of *median* and *centroid* of a tree and make use of these to migrate the core. Their method ensures that the core is moved hop-by-hop to the multicast group node the nearest to the centroid node. However, it might lead to the increase in control overhead and packet loss because the mesh need to be reorganized during the core movement.

3   Our Proposition

3.1   Motivation

As we mentioned earlier, PUMA [10] has been shown to be stable and efficient. However, its core selection method has some disadvantages. First, it selects the node with the highest identifier to become the core of the multicast group. Second, core changes only happen when the mesh undergoes partition or the old core leaves the mesh. Third, the core is migrated hop-by-hop so that noticeable communication overhead is resulted. Inspired by the idea of core selection from [7], we propose an improvement to PUMA with an adaptive core selection mechanism.

3.2   Protocol Description

PUMA [10] uses only one type of control message, multicast announcement, to maintain the mesh. A multicast announcement is first originated from the core node and then is sent to its neighbours. Every time a node V receives a multicast announcement, it uses this message to update its connectivity list. The connectivity list helps each node to discern which is the shortest path to reach the core node. Node V then sends its own multicast announcement to its neighbours.

Table 1. Procedure for handling a multicast announcement message at each node

```
Input:
N: Number of nodes
ma: Multicast announcement message
this: This node
Begin
   if this.id = coreId then
      if this.maxChildWeight > N/2 then
         bestChild ← this.getBestChild()
         Send BECOME-CORE(this.id, bestChild)
      fi
   fi

   if ma.type = BECOME-CORE then
      if ma.destination = this.id then
         bestChild ← this.getBestChild()
         if confirmWait = bestChild then
            Send NEW-CORE(this.id)
            confirmWait ← null
         else if this.weight > N/2 then
            Send BECOME-CORE(this.id, bestChild)
            confirmWait ← bestChild
         else if this.isReceiver then
            Send NEW-CORE(this.id)
         else
            Send BECOME-CORE(this.id, ma.source)
         fi
      else
         this.updateWeightTable()
      fi
   fi
End

procedure getBestChild():
Begin
   for each child c of this
      if c.weight > CurrentBestChild.weight then
         CurrentBestChild ← c
      fi
   return CurrentBestChild
End

procedure updateWeightTable():
Input: id, weight
Begin
   for each child c of this
      if c.id = id then
         c.weight ← weight
      fi
End
```

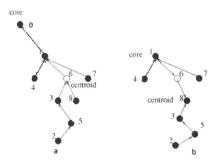

Fig. 1. Example of core migration

To reduce the communication overhead when implementing the core migration method, multicast announcements are used to help each node to calculate its weight. A weight field is added to each multicast announcement. When node V receives a multicast announcement from node U, V first checks if U is one of its children. If so, V uses the weight included in the message to update Us weight in Vs connectivity list. When V wants to send a multicast announcement, it sets the weight of the message to the sum of those of all its children.

The first step of our core migration process is similar to [7]. However, in order to make the core move faster to the new node, improvements are made to the remaining steps as follows. Every time node V receives a BECOME-CORE message from node U, V determines whether the centroid is in one of its sub-trees. If so, it forwards the message to the root node of that sub-tree. Otherwise, V determines if it is centroid node. If yes, it considers itself as the centroid node and then sends a NEW-CORE message to the network claiming itself to be the core of the group. If V is a forwarding node, i.e. the centroid is U, then V sends a RESPONSE message back to U. Upon receiving RESPONSE from V, U knows that it is the centroid and broadcasts the BECOME-CORE message accordingly.

This approaches still ensures that the core is moved to the nearest multicast group node. However, core change happens only once in order to avoid packet loss and additional communication overhead.

The pseudo code of our improved PUMA protocol is given in Table 1.

An example of the core movement is given in Figure 1. In this example, each node is labelled with its identifier. In Figure 1.a, the core node 0 determines that the centroid belongs to the sub-tree rooted at node 1. Therefore it sends a BECOME-CORE message whose destination is node 1, and sets the CONFIRM-WAIT for node 1 to true to imply that it has sent a message to node 1. Node 1 receives the message and then repeats the process to send the message to node 6. When node 6 receives the message, it determines that it is suitable to be the core. However, node 6 is a forwarding node, so it sends the message back to node 1. When node 1 receives the message, by checking the CONFIRM-WAIT, it knows that the core cannot move any further, so it becomes the core of the group. In Figure 1.b, node 8 is the centroid, and at the same time is a receiver, so it becomes the core of the group.

4   Simulation

P2MAN [2] is chosen to test and analyze the performances of the different versions of PUMA. P2MAN leverages on the PUMA protocol. Its function consists of delivering contents at the application layer. Sidney Doria developed both P2MAN and PUMA. Both source codes are publicly available [3], [4].

We compare the performances of two variants of our improved protocol with the original PUMA protocol. In the first variant (PUMA v1), we keep the core migration process as the same as in [7], while the second variant (PUMA v2) corresponds to the final version of our proposed protocol as described in the previous section.

4.1   Simulation Settings

Simulation settings for evaluation of communications overhead and packet delivery ratio is given in Table 2 and those for evaluation of delivery time is given in Table 3.

Table 2. Simulation Settings for Evaluation of Communications Overhead and Packet Delivery Ratio

Parameter	Value
Number of rounds	10
Total nodes	30
Simulation duration	400 time units
Simulation area	500m x 500m
Node placement	Fixed
Radio range	250m
Pause time	0 s
Bandwidth	2 Mbps
Data packet size	512 bytes
MAC Protocol	802.11 DCF
The number of senders	Varies
Traffic load	Varies

Table 3. Simulation Settings for Evaluation of Delivery Time

Parameter	Value
Number of rounds	10
Total nodes	100
Simulation duration	600 time units
Simulation area	1000m x 1000m
Node placement	Random
Radio range	250m
Pause time	0 s
Bandwidth	2 Mbps
Data packet size	512 bytes
MAC Protocol	802.11 DCF
Content size	Varies
Mobility	Varies

4.2   Communication Overhead

Figure 2 shows the communication overhead with a 30-member multicast group, the traffic load of 10 pkts/s, the velocity of 0 m/s, and a varying number of senders. Figure 3 shows the communication overhead with a 25-member multicast group, 5 senders, the velocity of 0 m/s, and a varying traffic load.

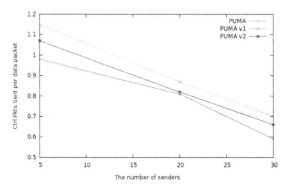

Fig. 2. Communication overhead with varying the number of senders

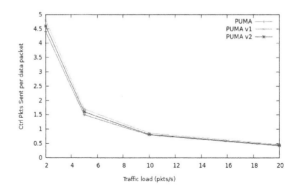

Fig. 3. Communication overhead with varying traffic load

Simulation results show that the communication overhead increases slightly when our improvements are applied. The reason is that we integrate the node weight calculation process in multicast announcements. Therefore, the control messages contain not only information for maintaining the mesh but also for calculating node weights.

4.3   Packet Delivery Ratio

Figure 4 shows the packet loss ratio in simulation settings with a 25-member multicast group, the traffic load of 10 pkts/s, the velocity of 0 m/s, and a varying number of senders. Figure 5 shows the packet loss ratio with a 25-member multicast group, 5 senders, the velocity of 0 m/s, and a varying traffic load.

Author Index

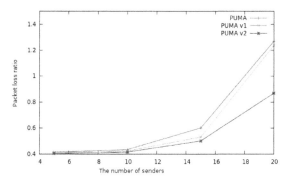

Fig. 4. Packet loss ratio with varying the number of senders

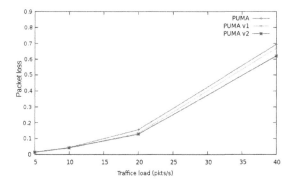

Fig. 5. Packet loss ratio with varying traffic load

In both series of simulations, PUMA v2 performs better with a lower packet loss ratio. PUMA v1 decreases the packet loss ratio a bit, however, it still cannot compensate its control overhead increment. Furthermore, as shown in Figure 8, the packet loss ratios of PUMA v1 and PUMA become closer and closer to each other as the number of senders grows.

4.4   Delivery Time

Figure 6 shows delivery time in simulation settings with 1 sender, 20 receivers, the velocity of 0-1 m/s, and a varying content size. With the content sizes of 50 kb and 100 kb, the delivery times are slightly different. However, when the content size is larger, the delivery time becomes longer, the difference in delivery times between the three versions of PUMA become more and more vivid, as shown in the figure. PUMA v2 outperforms PUMA v1 due to its better core migration method.

Figure 7 shows the delivery time in simulation settings with 1 sender, the velocity of 0-1 m/s, the content size of 100 KB, and a varying number of receivers. Simulation results show that the differences in delivery time between PUMA,

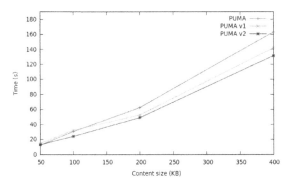

Fig. 6. Delivery time with varying content sizes

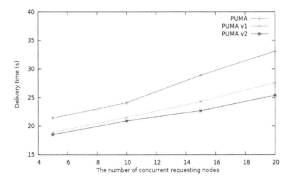

Fig. 7. Delivery time with varying the number of concurrent requesting nodes

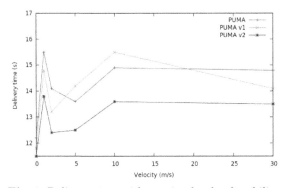

Fig. 8. Delivery time with varying levels of mobility

PUMA v1, and PUMA v2 grow larger when the number of receivers increase. PUMA v2 and PUMA v1 produce delivery times close to each other.

Figure 8 shows the delivery time in simulation settings with 1 sender, 20 receivers, the content size of 100 KB, and a varying velocity. One significant feature of P2MAN is that the delivery time does not depend on the mobility of the network. To put it simpler, no matter how fast each node move, it does not affect the content delivery time. We have run tests to compare the average delivery time between three versions of PUMA. As illustrated, even though delivery time does not depend on the mobility of nodes, PUMA v2 still proves itself more effective in content distribution throughout the network.

5   Conclusion and Future Work

In this paper, we have proposed a simple method to improve the PUMA [10] multicast routing protocol. The new protocol has a slight increase in total control overhead, but better performances in terms of packet delivery rate and content delivery time. The underlying idea is to choose the most appropriate node to become the core of the group, by using a simple yet effective function to calculate the weight of each node, and then move the core the closest possible to the true core node. The core is not moved hop by hop, but directly to the new position. This latter technique allows to reduce the control overhead yet further improve the packet delivery rate.

A research perspective to explore is to separate the weight calculation function from multicast announcement messages. This might lead to some increase in control overhead. However, the core node would be selected faster and the mesh would be better organized in the fastest possible time.

Acknowledgments. This work was partially supported by the VNU Scientist links.

References

1. Dewan, T.A.: Multicasting in Ad-hoc Networks. CSI5140F, University of Ottawa (2005)
2. Doria, S., Spohn, M.A.: A multicast approach for peer-to-peer content distribution in mobile ad hoc networks. In: Proc. of the IEEE Conference on Wireless Comm. and Networking, pp. 2920–2925 (2009)
3. Doria, S., Spohn, M.A.: Puma multicast routing protocol source code for ns-2, http://sourceforge.net/projects/puma-adhoc
4. Doria, S., Spohn, M.A.: Peer-to-manet source code for ns-2, http://sourceforge.net/projects/p2man
5. Royer, E.M., Perkins, C.E.: Multicast Ad hoc On-Demand Distance Vector (MAODV) Routing, http://tools.ietf.org/html/draft-ietf-manet-maodv-00.txt
6. Garcia-Luna-Aceves, J.J., Madruga, E.L.: The core-assisted mesh protocol. IEEE Journal on Selected Areas in Communications 17(8), 1380–1394 (1999)
7. Gupta, S.K.S., Srimani, P.K.: Adaptive Core Selection and Migration Method for Multicast Routing in Mobile Ad Hoc Networks. IEEE Trans. on Parallel and Dist. Syst. 14(1) (2003)

8. Rajashekhar, B., Sunilkumar, M., Mylara, R.: Mesh Based Multicast Routing in MANET: Stable Link Based Approach. Int. Journal of Comp. and Elect. Eng. 2(2), 371–380 (2010)
9. Sangman, M., Sang, J.L., Chansu, Y.: Adaptive multicast on mobile ad hoc networks using tree-based meshes with variable density of redundant paths. Wireless Networks 15, 1029–1041 (2009)
10. Vaishampayan, R., Garcia-Luna-Aceves, J.J.: Efficient and robust multicast routing in mobile ad hoc networks. In: Proceedings of the IEEE Int. Conf. on Mobile Ad-Hoc and Sensor Systems, pp. 304–313 (2004)
11. Wu, C.W., Tay, Y.C., Toh, C.-K.: Ad hoc Multicast Routing protocol utilizing Increasing id-numberS (AMRIS), http://tools.ietf.org/id/draft-ietf-manet-amris-spec-00.txt
12. Yi, Y., Lee, S.J., Su, W., Gerla, M.: On-Demand Multicast Routing Protocol (ODMRP) for Ad Hoc Networks, http://nrlweb.cs.ucla.edu/publication/download/693/draft-ietf-manet-odmrp-04.txt
13. The Network Simulator ns-2, http://www.isi.edu/nsnam